LASERS

VOLUME 3

LASERS

A Series of Advances

Edited by ALBERT K. LEVINE

DIVISION OF SCIENCE AND ENGINEERING
RICHMOND COLLEGE OF THE CITY OF NEW YORK
STATEN ISLAND, NEW YORK

and

ANTHONY J. DeMARIA

QUANTUM PHYSICS LABORATORY
UNITED AIRCRAFT RESEARCH LABORATORIES
UNITED AIRCRAFT CORPORATION
EAST HARTFORD, CONNECTICUT

VOLUME 3

1971

MARCEL DEKKER, Inc.,   New York

COPYRIGHT © 1971 by MARCEL DEKKER, Inc.

ALL RIGHTS RESERVED

MARCEL DEKKER, INC.
95 Madison Avenue, New York, New York 10016

LIBRARY OF CONGRESS CATALOG CARD NUMBER: 66-11288
ISBN No.: 0-8247-1413-X

PRINTED IN THE UNITED STATES OF AMERICA

PREFACE

Since the earliest days of man's utilization of the electromagnetic spectrum, there has been a steady drive toward the production and use of coherent electromagnetic energy of higher frequencies. This tendency resulted mainly from the realization that an increase in transmitted information, directivity, and efficiency was available with increasing carrier frequency and from the desire to relieve the crowding and interference among existing frequency bands. Another important incentive toward generating coherent radiation of higher frequency derived from the interest in utilizing such radiation for probing the atomic domains of solids, liquids, and gases by employing techniques such as nuclear magnetic resonance, paramagnetic resonance, and cyclotron resonance. In response to these needs, active devices such as vacuum tubes, transistors, magnetrons, klystrons, traveling wave tubes, parametric amplifiers, and tunnel diodes were devised for the coherent generation of higher frequencies. These devices made available coherent radiation to approximately 100 GHz; with the use of harmonic generators this figure was extended by approximately an order of magnitude. Almost without exception, as soon as higher frequency devices became available, researchers rushed to utilize them in experiments probing the interaction of coherent radiation with matter.

The physical dimensions of the resonators used to select the oscillating frequency of conventional oscillators in the higher frequency range are of the order of the wavelength of the radiation generated. As a result, the construction of resonators to the small dimension required at submillimeter wavelengths becomes extremely difficult. In the late 1940's and early 1950's it became apparent to workers in the field that the old method of scaling down existing devices for higher frequency generation was becoming impossible to apply. In the search for alternate methods, researchers came to realize that the large supply of natural resonators in the form of atomic and molecular systems could be utilized to amplify and even generate coherent electromagnetic energy.

The operation of NH_3 beam maser by Gordon and associates in 1954 ushered in a new category of active devices falling within a rapidly expanding field called quantum electronics. In 1958 Schawlow and Townes published a classical paper suggesting the use of the maser principle (with appropriate modification) for the generation of coherent infrared, visible, and ultraviolet radiation. The realization of the ruby laser by Maiman in the latter part of 1960 made available for the first time a light beam having characteristics usually associated with radio and microwave radiation. Laser action has since been extended to solids (crystalline and noncrystalline insulators and semiconductors), organic liquids, gases, and vapors, and thousands of discrete wavelengths varying from the vacuum ultraviolet to the submillimeter portion of the infrared have been attained. The quest for new methods of generating coherent radiation, at ever higher frequencies, is still in effect, with emphasis at present on the generation of coherent radiation in the far ultraviolet and soft x-ray regions.

Few developments in science have excited the imagination of scientists as has the laser. In the ten years since its first realization in the form of pulsed coherent emission from single-crystalline ruby, the field has grown at a rate rarely experienced in science. The availability of these intense, coherent optical radiation sources made possible for the first time the experimental investigation of optically generated plasmas, optical harmonic generation, stimulated scattering effects, photon echoes, self-induced transparency, optical adiabatic inversion, picosecond optical pulses, holography, optical shocks, self-trapping of optical shocks, optical parametric amplification, optical ranging to the moon, and extremely high resolution spectroscopy. In addition to the inherent attraction of exploring, characterizing, extending, and exploiting a new physical phenomenon, research in lasers was stimulated by the early experimental verification that coherent radiation could be generated in crystalline systems different from ruby and in other optically transparent media such as gases, glasses, and liquids; thus, a large research effort in materials science was joined to the extensive phenomenological investigations.

The laser device field today encompasses numerous disciplines: solid state, molecular, and atomic physics; spectroscopy; optics; acoustics; electronics; semiconductor technology; plasma physics; vacuum technology; organic and inorganic chemistry; molecular and atomic kinetics; thin film technology; glass working technology; crystallography; and more recently fluid dynamics, aerodynamics, and combustion physics. In sum, even without considering applications, the field has grown so fast and proliferated so broadly that the tendency for scientists to specialize within it is virtually complete. As a result, today probably no individual would profess to be authoritative over the whole field of laser devices, or even knowledgeable of most of the significant literature.

Lasers is a series of critical reviews that evaluate the progress made in the field of lasers. The contributing authors are scientists who are intimately involved in expanding the research frontiers in their specialities and, therefore, write with the authority that comes from personal contribution.

The series provides the background, principles, and working information that is needed by physical and biological scientists who seek to use lasers as a tool in their research and by engineers who wish to develop the laser phenomenon for commercial and military applications. Moreover, these critical reviews are sufficiently intensive that they can be used by a specialist in one portion of the laser field to bring himself up to date authoritatively in other areas of the field.

April, 1971   ALBERT K. LEVINE
                                                    ANTHONY J. DeMARIA

CONTRIBUTORS TO VOLUME 3

M. BASS, *Research Division, Raytheon Company, Waltham, Massachusetts*

P. K. CHEO,* *Bell Telephone Laboratories, Inc., Whippany, New Jersey*

T. F. DEUTSCH, *Research Division, Raytheon Company, Waltham, Massachusetts*

HENRY KRESSEL, *RCA Laboratories, Princeton, New Jersey*

M. J. WEBER, *Research Division, Raytheon Company, Waltham, Massachusetts*

Present address: United Aircraft Research Laboratories, East Hartford, Connecticut.

CONTENTS OF VOLUME 3

Chapter 3. **Dye Lasers**   **269**
M. Bass, T. F. Deutsch, and M. J. Weber

CONTENTS OF OTHER VOLUMES

Volume 1

Volume 2

LASERS

VOLUME 3

Chapter 1 · SEMICONDUCTOR LASERS

HENRY KRESSEL

RCA Laboratories
Princeton, New Jersey

1

I. Introduction

The *p–n* junction laser has followed an historical evolution familiar in other semiconductor devices. The discovery in 1962(*1–3*) was followed by studies of operating principles and comparisons between first-order theory and experiment. Finally, successive technological innovations moved the devices from the laboratory to the stage where serious systems applications were possible. This review will concentrate on the three-year period since 1967, during which important technological innovations have occurred. The fundamental aspects of laser theory as well as the earlier work on diode lasers has been extensively reviewed in the literature up to March 1967(*4–10*). In addition, the theoretical background useful in understanding laser operation may be found in various textbooks(*11*).

While diode lasers, which are of the utmost practical interest at this time, receive most of the attention in this chapter, a review of the state of the art in electron beam and optically pumped semiconductor lasers is also given. However, no attempt is made to cover these subjects exhaustively.

II. General Laser Concepts

In this section we review a few of the key concepts concerning laser action in semiconductors, without any attempt to consider the literature in detail. Extensive theoretical treatments of this subject can be found elsewhere and the objective here is to summarize the concepts needed for the subsequent description of devices.

A. DIRECT AND INDIRECT BANDGAP SEMICONDUCTORS

In direct bandgap semiconductors (the only type in which stimulated emission has been observed), both photon emission and absorption can occur without the need for a third quasi particle (a phonon) to conserve momentum. This is because the lowest conduction band minimum and highest valence band maximum are at the same wave vector ($\bar{k}$) in the Brillouin zone. Figure 1 shows the schematic diagram of electron energy vs. $\bar{k}$ in a semiconductor, such as GaAs, where the smallest bandgap energy $E_g = E_c - E_v$ is at $\bar{k} = [000]$.

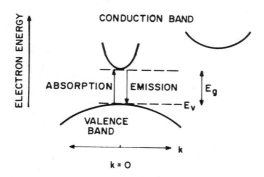

Fig. 1. Photon absorption and emission in a direct bandgap semiconductor.

In indirect bandgap semiconductors, the conduction band minimum and valence band maximum are not at the same $\bar{k}$ value. Hence, photon emission and absorption require the participation of phonons to conserve momentum. A schematic diagram of an indirect bandgap semiconductor such as GaP or AlAs is shown in Fig. 2. In these semiconductors the lowest-lying conduction band minima are along $\bar{k} = \langle 100 \rangle$, of which there are 6 equivalent valleys, while the highest valence band maximum is at $\bar{k} = [000]$.

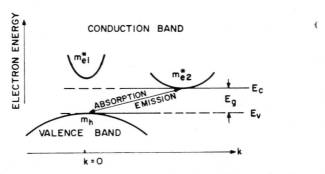

Fig. 2. Photon emission and absorption in an indirect bandgap semiconductor. The effective electron masses $m_{e_1}^*$ and $m_{e_2}^*$ refer to the two conduction band minima shown. In general, $m_{e_2}^* > m_{e_1}^*$. In the case of GaP and AlAs, the conduction band minima occur along $\bar{k} = \langle 100 \rangle$. The m_h is the density-of-states effective hole mass.

By mixing different bandgap compounds, a continuous variation of the band structure is possible. For example, in $GaAs_{1-x}P_x$, the change from direct to indirect bandgap transition occurs at about $x \cong 0.45$, with $E_g \cong 1.96$ eV (12). Another alloy system of great technological interest is $Al_xGa_{1-x}As$ in which the crossover occurs at $x \cong 0.34$ $(13,14)$. Figure 3 shows the variation of the bandgap energy of $Al_x Ga_{1-x}As$ as a function of x. The composition at which the $\langle 100 \rangle$ and $[000]$ conduction band minima are at the same energy marks the crossover from a direct to an indirect semiconductor ($E_g \sim 1.92$ eV).

As illustrated in Fig. 2, the effective masses in the different nonequivalent conduction band minima are not equal and generally $m_{e_2}^* > m_{e_1}^*$. Hence, the electron mobility is lower in the indirect than in the direct composition range as was shown in Ga(AsP) (15).

Lasing in indirect bandgap semiconductors is improbable because the lowest-energy band-to-band transition probabilities are much smaller than in direct semiconductors. Because of the relatively long lifetime of electrons in the indirect minima, there is time for nonradiative processes to occur. Furthermore, the stimulated recombination rate is a function of the band-to-band absorption coefficient $(11,16)$. Since this coefficient is lower for indirect than direct transitions, the potential laser gain is correspondingly reduced (17). Theoretical discussion of the possibilities of lasing in indirect bandgap semiconductors will be found also in Refs. $(18,19)$.

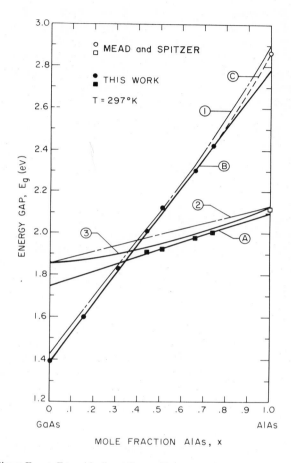

Fig. 3. The direct $\Gamma_{15} \to \Gamma_1$ and indirect $\Gamma_{15} \to X_1$ energy gaps for $Al_xGa_{1-x}As$. Line A is the straight-line fit to the photoresponse indirect energy gap values designated by the squares. Line B is the straight-line fit to the photoresponse direct energy gap values given by the circles. Line C is the direct energy gap calculated from the empirical relation [A. G. Thompson and J. C. Woolley, *Can. J. Phys.*, **45**, 255 (1967)]:

$$E_{gd}(x) = E_{gd}(GaAs) + Bx + 0.3x^2/\{\tfrac{1}{2}[E_{gd}(GaAs) + E_{gd}(AlAs)]\}^{1/2}$$

with B evaluated at $x = 1$, $E_{gd}(GaAs) = 1.395$ eV, and $E_{gd}(AlAs) = 2.86$ eV. Line 1 is the direct energy gap calculated from the quadratic relation with $E_{gd}(GaAs) = 1.435$ eV and $E_{gd}(AlAs) = 2.90$ eV. Line 2 is a straight-line approximation to the indirect energy gap with $E_{gi}(AlAs) = 2.13$ eV and the difference in Γ_1 and X_1 as 0.43 eV. Line 3 is given by the quadratic relation with $E_{gi}(GaAs) = E_{gd}(GaAs) + 0.42 = 1.855$ eV and $E_{gi}(AlAs) = 2.13$ eV. From (*14*).

B. POPULATION INVERSION AND LASING CRITERIA

Optical gain is possible when population inversion has been achieved; i.e., the probability of photon emission with energy $h\nu$ is greater than the inverse process of absorption at the same photon energy. Analysis of laser action in semiconductors is greatly complicated by the fact that the density-of-states distribution affects the theoretically predicted behavior. These matters will be considered below. We begin with the basic requirements for laser action.

1. In thermodynamic equilibrium, the probability of an electron occupying a state with energy E is given by the Fermi–Dirac distribution function

$$f = \left(1 + \exp\frac{E-F}{kT}\right)^{-1} \tag{1}$$

where F is the Fermi level and T the temperature.

When minority carriers are injected into a semiconductor (for example, electrons into p-type material), the condition of thermodynamic equilibrium no longer holds. However, a steady state distribution of carriers in the conduction and valence bands, independently, can be assumed to occur. Hence, a quasi-Fermi level for electrons, F_c, and for holes, F_v, is defined, where

$$f_c = \left(1 + \exp\frac{E-F_c}{kT}\right)^{-1} \tag{2}$$

Similarly, for the valence band, the probability for a state at a given energy to be empty (containing a hole) is

$$1 - f_v = 1 - \left(1 + \exp\frac{E-F_v}{kT}\right)^{-1} \tag{3}$$

It was shown by Bernard and Duraffourg(20) that the condition for net gain at a photon energy $h\nu$ is

$$F_c - F_v > h\nu \tag{4}$$

The carrier distribution must be degenerate for this condition to be satisfied. In order to obtain lasing in degenerate p-type material, for example, it is therefore necessary to inject a sufficient density of electrons to fill states in the conduction band until the quasi-Fermi level F_c has been raised (and F_v lowered) sufficiently for condition (4) to be satisfied. If the density-of-states variation dN/dE in the conduction band valley is large, this may require a very high injection level (high threshold current density). As will be discussed later (Section

II.C), the addition of a high density of donors to the p-type material (i.e., compensation) adds a *tail* of states below the conduction band which lowers dN/dE. The density-of-state variation in the tail is generally approximated by

$$\rho \propto \exp \frac{E}{E_0} \tag{5}$$

instead of the parabolic distribution $\rho \propto E^{1/2}$ of the unperturbed conduction band. If the constant E_0 is sufficiently large, the density-of-states variation with energy is much smaller in the band tail than in the conduction band. Hence, degeneracy, particularly at high temperatures, is achieved with a smaller density of injected electrons.

The lasing criterion (4) applies, of course, to both homogeneous samples (electron-beam or optically pumped) and p–n junctions. In a p–n junction, the density-of-states distribution as a function of distance is shown in Fig. 4. We assume that both the n and p sides of the junction are degenerate, with the applied voltage V sufficient to nearly

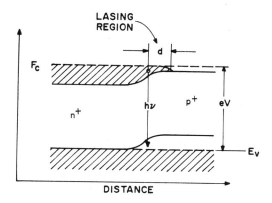

Fig. 4. Energy bands in a p–n junction under very high forward bias voltage V. Both the n and p sides of the junction were initially degenerate. As shown, a high density of electrons is injected into the p side, with negligible hole injection into the n side. Both the electron and hole populations are substantially degenerate over the active region d where the condition $F_c - F_v > h\nu$ is satisfied.

eliminate the barrier at the p–n junction. The injected electrons are distributed over a distance d (a few microns) in which, if the injection level is high enough, the population is inverted. Hole injection into the n side is small in such diodes, partly because the bandgap in the p side is effectively reduced by the high acceptor concentration; hence, the injection efficiency of electrons is nearly unity(21).

2. The second laser condition is that the optical gain match the losses over one transit of the beam through the cavity. With a Fabry–Perot cavity of length L, with reflectivity R at the lasing wavelength (0.32 for cleaved GaAs facets), this condition is

$$R \exp[(g-\alpha)] = 1 \qquad (6a)$$

or

$$g = \alpha + \frac{1}{L} \ln\left(\frac{1}{R}\right) \qquad (6b)$$

where g is the optical gain coefficient (which is a function of the current density), and α is the optical loss in the cavity which includes free carrier absorption and absorption in the surrounding noninverted regions.

3. Finally, the last key laser equation is the one relating the external quantum efficiency with the internal loss. A generally useful equation experimentally confirmed(295), is(22)

$$\eta_{ext} = \eta_{int} \frac{\ln(1/R)}{\alpha L + \ln(1/R)} \qquad (7)$$

where η_{int} is the internal quantum efficiency. For a comprehensive theoretical discussion of the laser efficiency we refer to Pilkuhn(23). Also of interest is the internal photon flux distribution in a Fabry–Perot cavity, which has been experimentally studied(296).

C. LASER GAIN

The relationship between the subthreshold gain and the current density at room temperature in the diode (i.e., injected electron density in the active region) has been the subject of extensive discussion and review in the literature. We review here some of the major conclusions.

1. For a modified simple two-level laser, the gain is a linear function of the current density J and the following relationship holds(24):

$$g = \beta J^b \qquad (8a)$$

where $b = 1$ and the gain constant β is given by

$$\frac{1}{\beta} = \frac{8\pi e \, n_2{}^2 \Delta\nu}{\eta_{int}\lambda^2\gamma} d \qquad (8b)$$

In this expression, n_2 is the refractive index in the lasing region of width d, λ is the wavelength, $\Delta\nu$ is the recombination line width, and γ

is a temperature-dependent factor which takes into account the distribution of carriers in the bands ($\gamma = 1$ at $T = 0$).

It is important to note that the gain constant β depends on the overlap between the regions of optical mode confinement and of inverted population. If the modal distribution extends over a region wider than the inverted population region, then the gain constant is reduced since only part of the optical wave propagating along the waveguide in the junction vicinity will be amplified.

2. The calculation of the gain in a real semiconductor must consider the density-of-states distribution. Depending on whether a parabolic, Gaussian, or exponential density of electron states is assumed, varying relationships between the current density and the gain are calculated. Furthermore, best agreement with experiment is obtained by assuming that the transition probability is independent of the initial and final states.

Assuming a parabolic density-of-states distribution, the calculated gain is *not* a linear function of J ($b > 1$) except at low temperatures (Fig. 5)(25). But with a Gaussian distribution of the form

$$\rho \propto \exp\left(\frac{E}{E'}\right)^2 \tag{9}$$

Stern(25) found that even at higher temperatures there is a less steep dependence (Fig. 5) and the degree of linearity increases with increasing doping of the active region (see Table 1). The case of $N_A = 6 \times 10^{18} \, \text{cm}^{-3}$ and $N_D = 3 \times 10^{18} \, \text{cm}^{-3}$ is closest to the experimental conditions in typical homojunction and single heterojunction lasers.

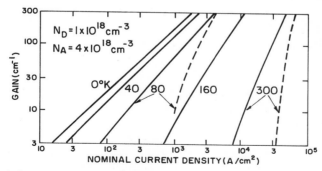

Fig. 5. Variation of the gain with nominal current density (1-μ-wide active region) for recombination in a region with 1×10^{18} donor and 4×10^{18} acceptor atoms per cm³. The dashed curves are results calculated with a parabolic band approximation [G. Lasher and F. Stern, *Phys. Rev.*, **133**, A553 (1964)]. The solid curves are calculated using the Gaussian band-tail approximation. From(25).

TABLE 1

Theoretical Gain Constant β and Exponent b
($g = \beta J^b$) at 300°K for Various Doping
Levels(25)

Doping in active region (cm^{-3})				
N_D	N_A	$N_A - N_D$	β (cm/A)^a	b
3×10^{18}	6×10^{18}	3×10^{18}	1.1×10^{-4}	1.86
1×10^{19}	1.3×10^{19}	3×10^{18}	2.1×10^{-3}	1.34
3×10^{19}	3.3×10^{19}	3×10^{18}	13×10^{-3}	1.04

^{a}The following assumptions are made:
  (a) perfect electrical and optical confinement in active region 2 μ wide;
  (b) gain required = 30–100 cm^{-1};
  (c) Gaussian density-of-states distribution;
  (d) $\eta_{int} = 1$.

A refined band-tail density of states of distribution was made by Halperin and Lax(26) which includes the Gaussian and the exponential distribution as limiting cases. A calculation of the gain dependence on the current density has been made(27) using the Halperin and Lax theory and the results are qualitatively similar to those obtained by Stern. The major difference is that the band tails are less pronounced with the Halperin and Lax model than with the Gaussian approxima- tion. Thus, for a given doping level in the active region, the value of b in Eq. (8a) will be higher at room temperature, leading to reduced linearity between the gain and the current density.

There has been no direct and conclusive experimental determination of the value of b. Most reported conclusions concerning b were based on plots of the threshold current density vs. the reciprocal of the cavity length in which, if a linear dependence is obtained, the implication that $b = 1$ is assumed. However, it is doubtful whether the experimental data are sufficiently precise to determine departures from linearity up to a value of perhaps $b = 2$. As discussed by Pilkuhn(10), and in Section X.G of this chapter, the assumption of $b = 1$ at room tempera- ture appears to be a useful working approximation at this time (and no more) when comparing the quality of lasers of roughly comparable cavity length since the approximate gain and loss parameters can then be simply deduced from experimental data. Additional research con- cerning the dependence of the gain on the current density is clearly desirable. To avoid metallurgical problems which lead to random variations in laser quality of devices having varying lengths, it would be desirable to concentrate on the study of only a few devices in which the facet reflectivity is changed by known amounts. Another approach would be to study amplification in diodes used as amplifiers pumped by a similar diode placed close to one facet.

3. To minimize the threshold current density J_{th}, it is obviously necessary to obtain a value as large as possible of β. Good optical confinement is one factor. While theoretically derived numerical estimates of β are difficult to make because of the approximations, some clearly established trends exist. The most important factors affecting the temperature dependence of the gain for a given current density are the doping level and the degree of compensation in the active region (constant width recombination region). Using rather intuitive arguments, Dousmanis et al.(29) showed that an increase in the value of E_0 in Eq. (5) (which results from an increase in the doping level in the active region) reduces the temperature dependence of J_{th}, and leads to increasingly lower J_{th} values at 300°K (assuming that the laser loss remains constant). The basic reason is that the ease of obtaining a degenerate carrier population at high temperatures increases as dN/dE decreases.

Stern(25) calculated the current density required to obtain a given gain value at various temperatures assuming different levels of doping and compensation (Gaussian band-tail approximation). Figure 6 shows his curves for the current density J required to obtain $g = 100$ cm^{-1} assuming a 1-μ wide recombination region, with $\eta_{int} = 1$ and perfect optical confinement to the region of inverted population. The trend is

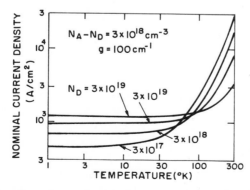

Fig. 6. Temperature dependence of nominal current density required to reach a gain of 100 cm^{-1} for four compositions with $N_A - N_D = 3 \times 10^{18}$ cm^{-3} (active region 1 μ wide). From(25).

clear: with increasing doping the current density at 300° K required to obtain this gain is reduced. This is well illustrated in Table 1 which shows the theoretically calculated gain constant β for $g = 100 \text{ cm}^{-1}$ at 300° K, with an active region width $d = 2\ \mu$ for various values of the doping. The condition $N_A = 6 \times 10^{18} \text{ cm}^{-3}$ and $N_D = 3 \times 10^{18} \text{ cm}^{-3}$ corresponds approximately to typical laser doping levels.

Experimentally, Winogradoff and Kessler[30] first demonstrated the advantage of compensation in reducing the 300° K J_{th} value.

Other recent theoretical laser studies will be found in Refs. [31–34]. Experimental E_0 values are given in Ref. [35].

4. Because the narrow inverted region in the vicinity of the $p–n$ junction is bound by highly absorbing noninverted regions, some wave-guiding is required to permit optical propagation with losses low enough for lasing to occur at reasonable current densities. This subject will be considered in detail in Section V.

5. It is possible to increase substantially the gain for a given J by reducing the width of the active region d to a value smaller than the diffusion length. This can be accomplished by placing a potential barrier on the p side of the junction close to the $p–n$ interface, thus blocking the flow of electrons away from the junction region. Such *electron confinement* can be obtained by use of a $p^+–p$ heterojunction with a higher bandgap energy on the p^+ side[36,37]. The $p^+–p$ barrier is present in homojunction epitaxial lasers, while the heterojunction is introduced in the *close-confinement* single and double heterojunction types of lasers (see Section III for a description of these structures). Assuming good optical confinement, the increase in β clearly will depend upon the electron diffusion length. Values of $1–2\ \mu$ have been reported in highly doped GaAs similar to that in the junction region[38]. However, higher values also have been reported in some laser structures[39].

III. Practical Laser Configurations

This section describes the basic laser configurations, leaving a technological discussion to Section IV. The laser diodes of interest are: homojunction lasers; single heterojunction ("close-confinement," or CC, structure) lasers; double heterojunction lasers; multiple lasing structures (homojunction and heterojunction); large optical cavity structure (LOC).

A. HOMOJUNCTION LASERS

This is the standard laser which consists of a p^+pn sandwich in which the p region is rather closely compensated (Fig. 7a). This device can be made solely by diffusion, but epitaxial growth methods combined with diffusion give better-defined doping discontinuities. Typically, the hole concentration in region 1 is about 2×10^{19} cm^{-3}, while

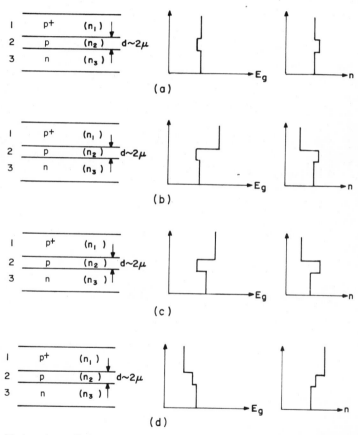

Fig. 7. Various laser diode structures. (a) Homojunction diode with well-defined recombination region, such as is formed by epitaxial growth methods. (b) Single heterojunction diode of the "close-confinement" type. (c) Double heterojunction diode. (d) Heterojunction diode where the bandgap energy of the p^+ region is *lower* than that of the p region ("inverse" close-confinement type). The variation of the bandgap energy and the refractive index n_1, n_2, and n_3 in the three laser regions is shown.

the electron concentration in region 3 is $2-4 \times 10^{18}$ cm^{-3}. In Fig. 7a
the effective bandgap energies in the various laser regions are shown
as slightly different because of bandedge changes owing to impurities
in highly doped GaAs. The compensated active region 2 has the lowest
effective bandgap energy.

B. SINGLE HETEROJUNCTION (CLOSE-CONFINEMENT STRUCTURE) LASERS

In this device, first developed by Kressel and Nelson(36) and
independently by Panish et al.(40), the bandgap energy in region 1 is
made higher than in regions 2 and 3 by the use of suitable composition
(AlGa)As alloys grown by liquid phase epitaxy (LPE). The band
diagram vs. distance is shown in Fig. 7b. The p^+p heterojunction near
the $p-n$ junction accomplishes three important functions to varying
degrees better than the previous p^+p interface: (a) improvement of the
optical confinement in the active region, (b) reduction of the optical
absorption in the p^+ region 1 at the lasing wavelength, and (c) improvement of the confinement of the injected electrons to the active region
2 because of the increased potential barrier at the p^+p heterojunction.
The theory of the device will be discussed in Section V and some
technological aspects in Section IV. We will denote this device by
"SH-CC" to differentiate it from the devices described in Section
III.D below.

C. DOUBLE HETEROJUNCTION LASERS(DH)

Alferov and others(41,128b,278,279) have described a laser structure using LPE (AlGa)As in which both the p^+ and the n regions have
a higher bandgap energy than the active p region that they adjoin
(Fig. 7c). This structure, originally proposed by Kroemer(42), was
intended to inject electrons and holes into the active region.

These devices were made by the sequential LPE growth of three
epitaxial layers. This structure differs from the SH-CC type in that a
double high barrier heterojunction is provided and that it does not rely
on the diffusion of Zn to form the active region, the active region being
grown epitaxially to the desired thickness of a few microns or less. The
double heterojunction lasers have even better optical confinement than
the single heterojunction ones because of the increased refractive index
discontinuity at the $p-n$ interface.

D. HETEROJUNCTION LASERS WITH STIMULATED EMISSION AT TWO WAVELENGTHS SIMULTANEOUSLY

By using a heterojunction structure of the type shown in Fig. 7d, it is possible to obtain lasing in both the p^+ and the p regions simultaneously (43). In effect, this structure is the inverse of the CC device because the electrons, not being confined to the higher bandgap energy region 2 at high J values, spill over into the lower bandgap region 1, because no barrier exists. Furthermore, optical pumping may occur. For example, simultaneous lasing was obtained at 8450 and 7280 Å. This effect has been observed only at low temperatures, probably because the optical losses are too high at elevated temperatures. By adjustment of the bandgap of regions 1 and 2, a whole range of simultaneous wavelengths can be obtained.

Earlier simultaneous lasing at two wavelengths was reported in special GaAs lasers(44). In these devices the separation between the lasing peaks was about 400 Å, with the effect being observed at low temperatures only. Here also, the barrier to electron flow from region 1 into region 2 no longer existed, in contrast to the conventional LPE lasers, making possible lasing in region 1 if the active region width is under 5 μ. This requirement is based, of course, on the need to have a significant density of electrons diffusing into region 1.

E. LARGE OPTICAL CAVITY (LOC) HETEROJUNCTION LASER

In this device, shown in Fig. 8, the electron-hole pair recombination region is made deliberately narrower than the waveguide region(276). The objective is to distribute the optical flux over a relatively large

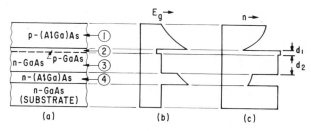

Fig. 8. Schematic of the large optical cavity (LOC) laser structure: (a) cross section of the diode; (b) bandgap energy, and (c) index of refraction. The electron-hole recombination occurs in the narrower region 2, while the waveguide region consists of regions 2 + 3. The n-type GaAs region 3 consists of relatively lightly doped, low absorption coefficient material [from Ref.(276)].

region in order to improve the resistance of the device to catastrophic degradation (Section VIII). The internal laser losses are low because the absorption coefficient in the n-type material adjoining the p-type recombination region is very low at the lasing photon energy. Thus high values of η_{ext} are possible at room temperature (Section X).

IV.   Some Aspects of Diode Technology

A. DIFFUSION TECHNIQUES

A single-step Zn diffusion (850°C for 1 hr, for example, using a Zn_3As_2 source) into n-type GaAs (2×10^{18} cm^{-3}) yields lasers suitable for 77°K operation with a minimum of processing difficulty (assuming the quality of the substrate is satisfactory, see Section IX). However, such devices will typically have J_{th} values in excess of 10^5 A/cm^2 at 300°K. The reason for the poor performance at high temperature is believed to be inadequate electron and optical confinement in the active region which results in the laser gain coefficient being low compared to that of the epitaxial devices (Section X)(45). By using a second diffusion step, Carlson(46) succeeded in improving the room temperature performance of his lasers. This improvement may result from the fact that the impurity distribution of these lasers approaches the p^+pn configuration which is desirable for high temperature operation.

B. VAPOR PHASE EPITAXY (VPE)

In this technique, both the n and the p regions are sequentially grown onto melt-grown substrates by vapor transport.

A technique for the VPE growth of GaAs(47) which uses arsine as the As source was used for the lasers reported by Tietjen et al.(48). The n region was first grown, followed by a narrow compensated p-type region formed by simultaneous doping with Se and Zn. Finally a p^+ region was grown. These devices have, therefore, the p^+pn structure desirable for room temperature operation (Section V). Their J_{th} values of $\sim 40,000$ A/cm^2 are comparable to those of conventional LPE lasers.

It is of interest to note that VPE lasers were the first used to demonstrate the requirement of compensation of the active region to improve room temperature laser performance(30).

C. LIQUID PHASE EPITAXY (LPE)

In this method, from Nelson(49,50), GaAs is grown from a Ga solution containing GaAs and the desired dopant in the apparatus shown in Fig. 9. The melt and the substrate wafer are brought up to the desired growth temperature (~ 900–$1000°$C); then the furnace is tipped and the melt rolls onto the substrate. Other variants of this furnace arrangement obviously are possible, in which the wafer is brought into contact with the melt at the desired moment. For example, a vertical furnace has been described in which the wafer is mounted on a rod which is inserted into the melt (51) or a sliding boat apparatus (52). The crystalline properties of the GaAs grown is controlled by factors such as the cooling rate and the substrate quality.

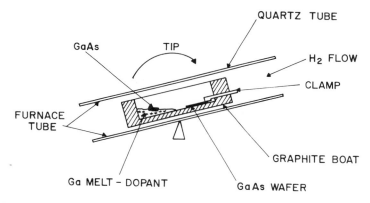

Fig. 9. Schematic showing the basic liquid phase epitaxial growth technique using a tipping furnace as developed by Nelson (49).

Lasers have been made by two methods: growth of an n-type layer on a p^+ substrate or growth of a p^+ layer onto an n-type substrate ($45,49,50$). In both cases, heat treatment is required either during or after the growth to move the p–n junction away from the interface into the initially n-type region by about $2\ \mu$. This heat treatment serves two purposes: (1) The interface is generally strained and contains imperfections. Thus, it is desirable to have the active region away from it. (2) The waveguide configuration p^+pn is formed in the heat treatment process. In fact, without some heat treatment, laser action is frequently not observed ($54b$).

The preferred dopant concentration for $300°$K operation are $\sim 3 \times 10^{18}$ cm^{-3} for the n-type region and about $\sim 2 \times 10^{19}$ cm^{-3} in the p^+-type region. The effect of variations in the acceptor concentrations on LPE lasers were studied by Susaki et al. (53).

A third method of laser fabrication consists of growing both the n-type and the p-type regions by LPE. This is the method used for (AlGa)As lasers (see Section XII).

Because of varying growth rates, the crystal orientation affects the laser properties to some extent as described by Beneking and Vits (*54a*) who investigated (111) and (100) growth planes.

The basic process for fabricating single heterojunction CC lasers is similar to that of homojunction LPE lasers, the only difference being that the p^+ region consists of (AlGa)As instead of GaAs(*36,40*). Following the deposition of the p^+ layer, the wafers are heat treated to displace the Zn into the substrate, a distance of about 2 μ (for optimum laser performance), thus forming the compensated waveguide region in which lasing occurs. The effect of variations in the width of the active region, bandgap energy discontinuity, and doping on the laser properties is discussed in Section V in conjunction with device theory. A cross section of the completed device revealed by etching is shown in Fig. 10.

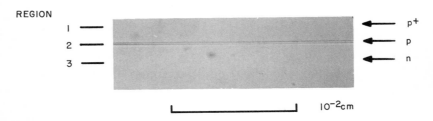

Fig. 10. Photomicrograph of sectioned laser diode showing regions 1, 2, and 3 as designated in Fig. 7. The material was etched to delineate the p^+–p and p–n interfaces (*82*).

Multilayer heterojunction lasers are generally grown in an apparatus in which several bins containing different dopants are present. Either the wafer can be kept stationary and the various melts adjusted to come in contact with the wafer to grow the required layers as the furnace temperature is reduced, or the wafer can be successively inserted into the bins. The idea is to grow the layers without the need to cool down the furnace between layers (*52*).

Because of the ease of growing (AlGa)As by LPE, this appears to be the preferred method of fabricating CC lasers. However, there is no reason, in principle, why vapor phase epitaxy cannot be used for the same purpose.

It should be noted that because of the excellent lattice-constant match between GaAs and AlAs, varying composition alloys can be deposited epitaxially to form the heterojunction. With Ga(AsP) on GaAs the lattice constant mismatch is considerably more severe; hence strains are introduced at the heterojunction and dislocations are formed. These degrade the laser performance (Section IX), and the advantages of the CC structure are probably lost.

D. DOPANTS

Lasers have been made using various donors (Te, Se, Si, and Sn). The main requirement is that the donor solubility be sufficiently high to yield an electron concentration of about $2-3 \times 10^{18}$ cm^{-3}. Differences in laser performance observed with different donors are generally owing to metallurgical factors (Section IX). While Zn has been the most widely used acceptor, Mg and Be which introduce 0.03-eV acceptor levels in GaAs have been used to prepare lasers[55], and also Si and Ge[56] which introduce a shallow acceptor in GaAs[57,58] have been used.

With respect to deep centers in GaAs (acceptors with ionization energies greater than 0.03–0.04 eV), lasing transitions in diodes recently have been observed involving the deeper (~ 0.1-eV) acceptor level due to Si in GaAs[56]. Previously only by optical excitation of p-type Si-doped GaAs samples[59] has it been possible to obtain stimulated emission via this level. It was originally identified in spontaneous luminescence studies by Kressel et al.[60] who showed that this level exists in LPE GaAs in addition to the ~ 0.03-eV acceptor.

Finally, we call attention to the Proceedings of the 1966 and 1968 GaAs Conferences[61,62] which contain numerous papers concerning GaAs device and materials technology.

The laser facet orientation is technologically important. The effect of changing the junction plane orientation with respect to the cleaved facets forming the Fabry–Perot cavity has been studied. The threshold current density is increased with increasing angular deviation and the experimental values have been theoretically explained by Deutsch[63].

E. LASER DIODE CONSTRUCTION

High power levels (consistent with reliability, Section VIII) can be obtained either by arrays of individual lasers in series or by the fabrication of single wide diodes. Both approaches have been used.

1. High Power Single Lasers

With increasing laser width, the possibility of transverse super-radiance becomes increasingly important, despite the fact that the laser sides are roughened by use of a saw, because the reflectivity is not reduced to zero. Hence, with increased drive a current density will be reached equal to the lasing threshold transverse to the desired lasing direction. Such cross lasing "robs" the desired direction of much of its power. To prevent cross lasing, it is possible to groove the junction, in effect forming a number of parallel junctions. An alternative scheme consists of utilizing the "wing structure" in which a strongly absorbing region is in parallel with the lasing region (Fig. 11).

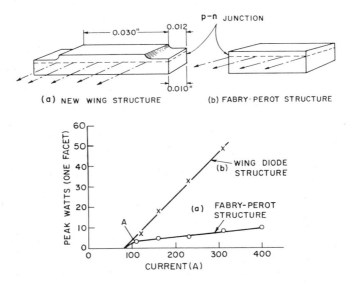

Fig. 11. Wide-diode structure showing a comparison between (b) the conventional Fabry–Perot structure and (a) the wing structure. The power output of the Fabry–Perot structure tends to saturate because of cross lasing. This does not occur in the wing structure because the additional losses in the transverse direction are too high for stimulated emission to occur (*64*).

Thus, the threshold current density for cross lasing is effectively infinite. Pulsed peak power outputs to $\sim 100\,\text{W}$ (at 350 A) have been obtained with such a structure, using diodes about 60 mils wide (*64*) at $300°\,\text{K}$.

2. *Laser Diode Arrays*

The use of large area diodes is limited by the need for high current pulsers and material uniformity. A convenient method of designing high power sources is to series-connect smaller lasers (6–10 mils wide, for example) cut from single bars of material. The basic technique consists of soldering a metallized cleaved sliver of the material on a metallized block ceramic (see Fig. 12a). The diodes are then defined by sawing through the sliver and partly through the ceramic substrate, thus electrically isolating the diodes from each other. The lasers are then series-connected by running wires from the top of one diode to the ceramic base of the adjoining diode. Thousands of watts of peak power can be conveniently obtained with such arrays. To obtain high average power, refrigeration is required. Either mechanical refrigerators (capable of operation to 77° K) or thermoelectric refrigerators (where the lower temperature limit for economic operation is about 240° K) can be used (277).

Stacked diode arrays designed for high-power operation (low-duty cycle) have also been reported (131) (Fig. 12b).

3. Stripe Geometry Lasers

In some experiments, as well as for continuous wave operation, it is desirable to obtain very narrow lasers (a few microns wide) in which good lateral confinement is present.

A structure which simplifies the contact and heat sink problems has been described by Dyment and D'Asaro (65) and is shown in Fig. 13. The junction is a shallow (2-μ) one with oxide everywhere on the surface of the wafer except where the lasing region is to be. Following junction definition, the wafer surface is metallized. Contact to the junction is made only in those regions without the oxide, thus defining the laser. The diffused region depth must be shallow and the Zn concentration relatively low in order to prevent lateral spreading of the current. Extensive experiments dealing with CW operation as well as Q switching and spectral studies, have been made with such structures, as will be described in subsequent sections. A planar stripe geometry laser in which the junction region is defined by Zn diffusion through an oxide mask also has been described by Becke (66).

4. Multilayer Lasers

GaAs lasers consisting of a $p-n-p-n-p-n$ structure grown by vapor phase epitaxy, in which the spacing between the active regions is of the order of microns, have been described. Experimental far-field emission data suggest optical coupling of the active regions, resulting in a very narrow (~ 0.5 deg) beam divergence (i.e., thick emitting region) (67).

Stacked lasers are reported by Muss et al. (68).

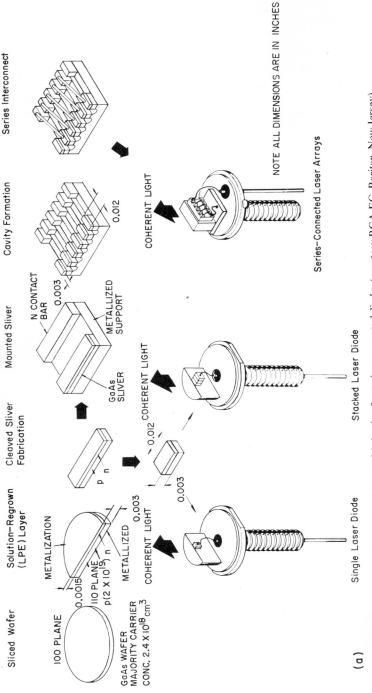

Fig. 12. (a) Laser preparation and array fabrication for series-connected diodes (courtesy RCA-EC, Raritan, New Jersey).

22

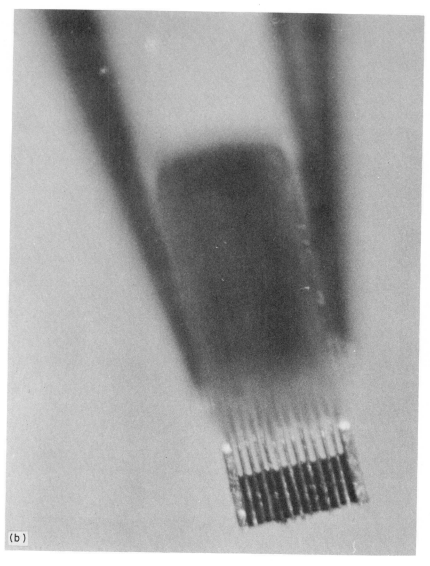

(b)

Fig. 12. (b) 12-stack laser array. Laser chips are 20×20 mils. The laser junctions are on 3-mil centers with a 1-mil string saw cut through the junction from the p-side. Radiating area is 40×20 mils (*131*).

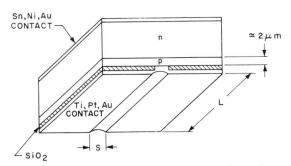

Fig. 13. Stripe geometry laser, showing diffused p region and the silicon dioxide which defines the stripe metallic contact of width $S = 12.5 \mu$ and length $L = 380 \mu (65)$.

V. Optical Waveguiding and Laser Loss

A. THEORY

1. *Concepts*

The concept of optical waveguiding in the junction region was considered by several authors(69–75). Stern(74) suggested that the index of refraction varies with position in the direction perpendicular to the junction plane because of the changes in the bandedge with doping. The exact treatment of this problem is difficult because of gradients in the dopant distribution in the active region. A simplification of the problem consists of treating the diode structures shown in Fig. 7 in which three distinct regions are defined. The following analysis is particularly applicable to heterojunction or homojunction epitaxial lasers. A waveguide analysis more appropriate to diffused lasers, in which the index of refraction variation is more gradual, also has been made(76a,b).

We will concentrate on an analysis of the devices in Fig. 7, a and b, where region 1 is p^+ (uncompensated) with bandgap energy E_{g_1}. The active region 2 is compensated p type with bandgap energy E_{g_2} and region 3 is n type with bandgap energy $E_{g_3} \cong E_{g_2}$. Region 2 is an optical waveguide of thickness d because its index of refraction n_2 is higher than either n_1 or n_3, as explained below. The index of refraction is known to depend on the bandgap energy of the material, its carrier concentration, and compensation(74). To enhance the optical confinement to the lasing region, it is advantageous to reduce both n_1 and n_3 relative to $n_2(77)$.

The simplest method of decreasing n_1 at the lasing wavelength is to increase the bandgap energy of region 1 relative to that of region 2. The structures shown in Fig. 7 are a simplification of the real device. While we will assume a uniform n_2 value, it is in fact not uniform because the acceptor concentration varies with distance away from the p–n junction. Nevertheless, the main features of the laser performance variation with diode structure can be deduced from the theoretical arguments, to be presented below, which are based on the computer solutions of Anderson(78) for TE wave propagation along the waveguide structure of Fig. 7b. (TM propagation is quite similar.) His solutions considered only the differences in the real part of the index of refraction; the neglect of differences in the imaginary component simplifies the analysis but does not significantly affect the numerical results. Furthermore, the numerical calculations developed by Anderson consider only the fundamental transverse waveguide mode. Since the index of refraction of the p^+ region of the heterojunction lasers at the lasing wavelength is much lower than the index of refraction of the p^+ region used in the homojunction lasers, there is the possibility of higher-order modes' existing in the cavity(79). However, the present calculations are based on the assumption that the first-order (fundamental) mode is the predominant lasing mode. From Anderson's analysis it follows that if n_1, n_2, and n_3 are nearly equal, then the laser threshold condition is given by

$$(-\alpha_2 - \alpha_{fc}) \cong \frac{1}{F_1} (\alpha_3 + \alpha_1 F_2) \qquad (10)$$

where α_3 and α_1 are the absorption coefficients at the lasing wavelength in regions 3 and 1, respectively; α_{fc} is the free carrier absorption coefficient in the active (inverted) region 2; $(\alpha_2 + \alpha_{fc})$ is the total absorption coefficient in region 2, which must be negative for lasing to occur; F_1 and F_2 are functions of the waveguide geometry. The expression (10) is sufficiently general to include asymmetry in the index of refraction between the two sides of region 2. Values of F_1 and F_2 were plotted by Anderson for varying values of the normalized cavity width D ($D = 2\pi d/\lambda_0$, where d is the actual cavity width and λ_0 is the wavelength in vacuum) for specific values of the asymmetry parameter η. These plots are shown in Figs. 14 and 15.

The waveguide asymmetry is defined by η (for small differences between n_1, n_2, and n_3)

$$\eta = \frac{n_2 - n_1}{n_2 - n_3} \qquad (n_2 > n_3 \geqslant n_1) \qquad (11)$$

When the negative absorption coefficient or gain $(-\alpha_2)$ is a function

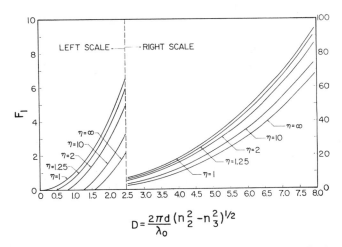

Fig. 14. Plots of the function F_1 in Eq. (10) as a function of the normalized waveguide width D for various values of η. The actual waveguide width is d and λ_0 is the wavelength in vacuum(78).

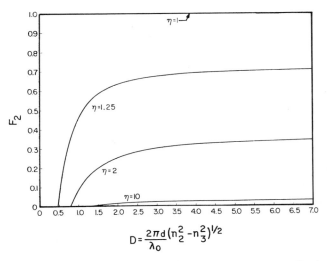

Fig. 15. Plots of the function F_2 in Eq. (10) as a function of the normalized waveguide width D for various values of η. The actual waveguide width is d and λ_0 is the wavelength in vacuum(78).

of the current density J^b then, in terms of the specific expression for the loss through cavity ends with reflection coefficient R within a cavity length L,

$$J^b_{th} \cong \frac{1}{\beta} \left\{ \frac{1}{F_1} (\alpha_3 + \alpha_1 F_2) + \alpha_{fc} + \frac{1}{L} \ln \frac{1}{R} \right\} \qquad (12)$$

where b is a constant for a given doping level in the active region. The *total* internal laser loss coefficient α is given by

$$\alpha = \frac{1}{F_1} (\alpha_3 + \alpha_1 F_2) + \alpha_{fc} \qquad (13)$$

The quantity calculated in this section is

$$\alpha' = \alpha - \alpha_{fc} = \frac{1}{F_1} (\alpha_3 + \alpha_1 F_2) \qquad (14)$$

which is the loss through the leaky *walls* of the waveguide. The value of α' can be calculated for various cavity lengths (and corresponding lasing photon energy) by choosing the appropriate index values n_1 and n_3 and absorption coefficients α_1 and α_3.

Equation (12) overstates the contribution of α_{fc} if the optical confinement is poor, since only a fraction of the modes are propagating through the active region. In the single and double heterojunction devices, however, the confinement is sufficiently high that the consequent error is small.

2. *Choice of Laser Parameters*

Reasonable values of α_3, α_1, n_1, n_2, and n_3 for homojunction and SH-CC LPE lasers are tabulated in Tables 2 and 3 (at 300° K and 77° K). These values were obtained as follows. Since no absorption data are available for p-type (AlGa)As, the absorption coefficient α_1 was estimated from GaAs data by assuming a shift by 0.07 eV of the band-edge energy (i.e., increase in 300° K bandgap energy from 1.43 to 1.50 eV). The absorption coefficient of 10 cm^{-1} for the n-type material is probably somewhat too large at 77° K for a concentration of 2×10^{18} cm^{-3}.

The value of n_1 at 1.375 eV was estimated from the calculated plots of Stern(74) (Fig. 16) which were based on experimental absorption data(80). Likewise, n_3 may be estimated with reasonable accuracy from experimental absorption data. The estimate of n_2, on the other hand, is more uncertain since region 2 is compensated but not uniformly Zn-doped. In view of the displacement of the band-edge to

TABLE 2

p^+ Material Parameters in Laser at Lasing
Photon Energy at 300° K and 77° K for Two
Values of the Bandgap Energy (82)

T (°K)	$h\nu_L$ (eV)	E_{g_1} (eV)	α_1 (cm^{-1})^a	$n_1{}^b$
300	1.375	1.43^c	350	3.59
300	1.375	1.50	100	3.55
77	1.460	1.515^c	30	3.59
77	1.460	1.585	30	3.55

$^a p^+ = 1.5 \times 10^{19}$ cm^{-3}.
^{b}See Fig. 16.
^{c}No Al in p^+ region.

TABLE 3

300° K Laser Parameters for Two Values of the
Bandgap Energy (82)

Parameter	$E_{g_1} = 1.43$ eV^a	$E_{g_1} = 1.50$ eV
p^+	1.5×10^{19} cm^{-3}	1.5×10^{19} cm^{-3}
α_1	350 cm^{-1}	100 cm^{-1}
α_3	10 cm^{-1}	10 cm^{-1}
$n_2 - n_3$	0.01	0.01
$n_2 - n_1$	0.01	0.05
n_1	3.59	3.55
n_2	3.60	3.60
n_3	3.59	3.59
η	1.0	5.0

^{a}No Al in p^+ region.

lower energies as a result of compensation, it is to be expected that
n_2 will be higher than either n_1 or n_2 at the lasing wavelength. In
estimating $n_2 - n_3$, we use $n_2 = 3.60$, a value which gave satisfactory
agreement with experiment in a study of the near-field and far-field
emission patterns of SH-CC lasers (79).

3. Dependence of Loss on Laser Parameters

In view of the numerical uncertainties, the calculated loss values are
of semiquantitative interest only. Nevertheless, they do suggest the
laser behavior to be expected experimentally.

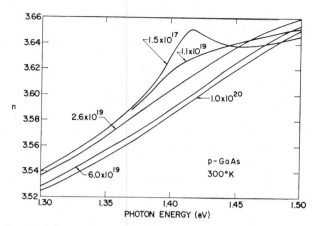

Fig. 16. Refractive index of p-type GaAs as a function of photon energy for various doping levels, as calculated by Stern(74).

Consider the dependence of α' on η and Δn. Figure 17 shows plots of α' as a function of Δn ranging from 0.005 to 0.02 (with $d = 2\,\mu$). Values of $\alpha_1 = 350\,\mathrm{cm^{-1}}$ and $150\,\mathrm{cm^{-1}}$ were chosen, which correspond to typical practical values. Notice that with $\alpha_1 \sim 350\,\mathrm{cm^{-1}}$ the loss α' decreases by a factor of 2.7 ($\Delta n = 0.01$), simply by increasing η (i.e., $n_2 - n_1$) by a factor of 3.2; Δn can be increased by use of the double heterojunction structure (Fig. 7c), thus practically eliminating optical spread in region 3.

We next illustrate the dependence of α' on η. Figure 18 shows that the most significant reductions in α' occur when η is increased from 1 to 2. Further increase in η has a relatively smaller effect on α'. The effect on α' of changing α_1 is shown in Fig. 19. The variation is essentially linear.

Finally we consider the effect of changing the active region thickness d. Assuming that its carrier population is fully inverted, it can easily be shown that α' should decrease with increasing d. However, the simple theory used here is inadequate to predict its variation because only the fundamental mode is considered. With increasing d, higher order modes become preferable, and their propagation loss must be calculated(79).

It is important to note(78,280) that the value of d cannot be reduced to arbitrarily small values if $\eta > 1$ because the optical confinement is lost below a certain value which depends on the waveguide properties. Thus, the laser loss can increase because of the spread of light into the surrounding regions, and the gain constant will decrease.

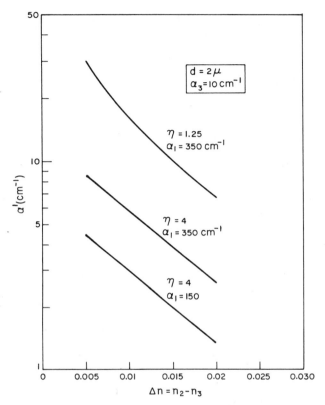

Fig. 17. Theoretical laser absorption coefficient α' as a function $n_2 - n_3$ for two values of η and α_1 (82).

The preceding analysis was based on a constant lasing photon energy. However, with increasing cavity length, the required laser gain is reduced and hence the threshold Fermi level separation ΔF is lowered. The result is that the lasing photon energy generally decreases in a given material with increasing cavity length, as experimentally shown in Ref. (81). As a result, the refractive index difference between the various laser regions will depend to some extent on the cavity length. Furthermore, the absorption coefficient in regions 1 and 3 will also change with changing photon energy and hence cavity length. Thus, the value of α' will depend to some extent on L. Because of the decrease in α_1 with decreasing photon energy, it is expected that α' will also decrease with increasing L. Since the free carrier absorption in the active region α_{fc} is independent of the photon energy in the

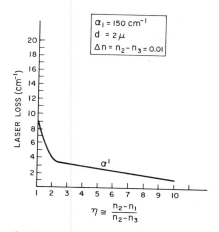

Fig. 18. Theoretical laser absorption coefficient α' as a function of η (82).

narrow range of interest in typical laser operation, the over-all laser loss may or may not depend on L in a significant way, depending on the relative contributions of losses through the cavity wall and in the active region itself. In lasers with good optical confinement, therefore, the laser loss should not vary significantly with the cavity length.

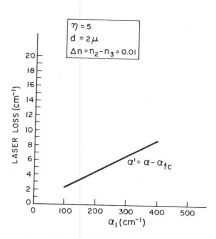

Fig. 19. Theoretical laser absorption coefficient α' as a function of the absorption coefficient at the lasing wavelength in the p^+ region, α_1 (82).

4. Summary of Theoretical Predictions

The main features of the expected laser behavior based on the preceding theoretical arguments are as follows:

1. Increasing the bandgap energy of region 1 in single hetero-junction lasers, which increases $n_2 - n_1$ and hence η, decreases α_1 and α'. Higher E_{g_1} also increases the exterior *spontaneous* efficiency because of reduced internal absorption. However, increasing the bandgap energy difference $\Delta E_g = E_{g_1} - E_{g_2}$ beyond a certain point will have a marginal effect on α' for two reasons, (1) because the relative decrease in refractive index is smaller as we move away from the bandedge (Fig. 16), and (2) because an increase of η beyond ~ 2 has a minor effect on the laser loss (Fig. 18).

2. Increasing the p^+ concentration increases α_1 and hence the waveguide loss (Fig. 19).

3. At $77°$ K, the α' difference between structures with and without the heterojunction will be small because α_1 is hardly affected by the ΔE_g increase (Table 2).

4. There may be a decrease of α with increasing d, in the range of d values small enough to maintain population inversion. Counteracting this, however, the gain term β should also decrease because of the greater recombination volume.

5. With n_1 fixed, an increase in $\Delta n = n_2 - n_3$ will decrease α'. The refractive index n_2 may be increased by increasing the doping and compensation in region 2, thus shifting the absorption edge to lower energy. An increase in compensation may be achieved by increasing the substrate donor level. With a higher donor doping we also expect that the gain for a given current density will increase at $300°$ K because of the deepening of the bandtails (Section II).

6. There exists a minimum value for d below which lasing will not be observed because of the loss of optical confinement in SH-CC lasers.

B. COMPARISON OF SIMPLE WAVEGUIDE THEORY AND EXPERIMENT IN SINGLE HETEROJUNCTION (CC) LASERS

1. Laser Loss Coefficient

A detailed study of the loss coefficient in SH-CC lasers as a function of key device parameters has been made by Kressel et al.(82). We summarize below some of the main results.

The experimental results show that α increases rather than decreases with increasing d, in disagreement with the theoretically expected behavior for a uniformly doped and inverted active region. The discrepancy is attributed to two factors: (1) when d is greater than the electron diffusion length, the condition of uniform gain in region 2 is not satisfied, and (2) the neglect of higher order modes. Furthermore, because of a gradient in Zn concentration, region 2 is not uniformly doped. For d values of 2 μ or less, comparable to the diffusion length, we expect that the simple waveguide model will be a reasonable approximation. For simplicity, all comparisons of theory and experiment for α were made for $d = 2 \mu$.

The other observations of Ref. (82) are qualitatively consistent with the behavior predicted in Section V.B:

(a) The laser loss was essentially independent of cavity length in the narrow range studied (lengths between 10 and 30 mils). Also since plots of J_{th} vs. $1/L$ were linear within the experimental scatter, it was concluded that a linear dependence of the gain on the current density was a reasonable approximation for the devices studied (i.e., $b \cong 1$) which simplified the analysis.

(b) As shown in Fig. 20, α is sharply reduced from 89 to 35 cm^{-1} when E_{g_1} is increased by 0.1 eV. A further increase in E_g to 0.24 eV has a much smaller relative effect, since α is reduced to only 28 cm^{-1}. Also, the spontaneous exterior quantum efficiency is increased by a factor roughly proportional to the decrease in the absorption coefficient in the p^+ region (Table 2).

(c) Increasing the p^+ concentration for a given E_{g_1} increases α.

(d) At 77° K, the single heterojunction does not significantly reduce the laser loss. Figure 21 shows a reduction from 11 to 8 cm^{-1} due to the heterojunction.

(e) A higher donor and/or acceptor concentration increases the gain constant β. However, for a given donor concentration at $d = 2 \mu$, the difference in β between the diode with and the one without the heterojunction is within a factor of 2 (Figs. 20 and 21).

Since the calculated laser loss values depend on the choice of refractive index values, quantitative comparisons are hindered by uncertainty in the values of the index of refraction in the various laser regions. As noted in Section V.B, the absorption coefficients in regions 1 and 3 can be estimated with reasonable accuracy on the basis of published values. However, the refractive index value is difficult to estimate in the active region 2. Table 3 presents estimates of the various index values. From these, the following η values are calculated: when region 1 was of GaAs ($E_{g_1} = 1.43$ eV), $\eta \cong 1.0$ while with $E_{g_1} \cong 1.5$ eV, $\eta \cong 5$.

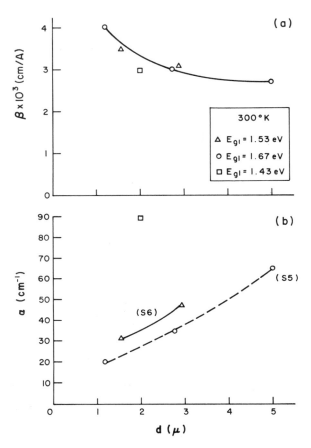

Fig. 20. (a) Laser gain constant β (in units of 10^{-3}) and (b) absorption coefficient α for three values of bandgap energy in the p^+ region: $E_{g_1} = 1.43$ eV (GaAs), 1.53 eV and 1.67 eV. The doping level in the p^+ region is 1.5×10^{19} cm^{-3} in all cases (82).

In lasers where the optical confinement is very high, the active-region free-carrier absorption α_{fc} becomes the major contribution to the observed laser loss. However, a quantitative estimate of this value is difficult because the doping in the active region is not uniform. From the absorption data of Hill(83) in homogeneous GaAs, the absorption owing to free holes (i.e., the "plateau" region below the absorption edge) is $\sim 0.6 \times 10^{-17}$ cm^{-1}/hole (300° K) and $\sim 0.17 \times 10^{-17}$ cm^{-1}/hole (77° K), for hole concentrations of 1.7×10^{19} and 7×10^{18} cm^{-3}. Even lower values are expected for electron absorption. The average hole concentration in the active region is in the mid-10^{18} cm^{-3} range.

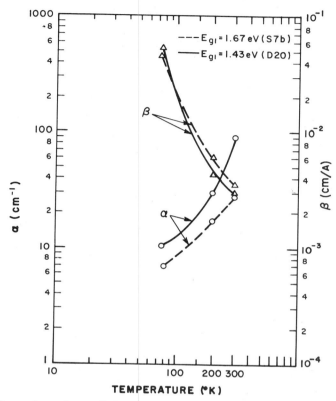

Fig. 21. Laser absorption coefficient α and gain constant β as a function of temperature for a SH-CC and a homojunction LPE laser(82).

Assuming a value of 5×10^{18} cm^{-3}, $\alpha_{fc} \cong 30$ cm^{-1} at 300° K and ~ 8.5 cm^{-1} at 77° K. Note that the injected electrons in the active region make but a small contribution. For example, assuming $d = 2\,\mu$, $J = \sim 10^4$ A/cm^2, and an electron diffusion coefficient of ~ 50 cm^2/sec, the injected electron density is only about 3×10^{17} cm^{-3}. If the Zn concentration in the active region is increased, we would expect an increase in α_{fc} (assuming a constant donor concentration). The theoretical laser loss values in Tables 4 and 5 are calculated on the basis of the above estimates of α_{fc} in the case of lasers with a p^+ concentration of 1.5×10^{19} cm^{-3}. It is clear, however, that the margin of error in this estimate could be substantial because of the nonuniform Zn distribution in the active region, and in fact, α_{fc} could be smaller than the estimate.

TABLE 4

Experimental and Calculated Laser Absorption Coefficient (82)

$p^+ = 1.5 \times 10^{19} \text{ cm}^{-3}, d = 2 \mu, n = 2.7 \times 10^{18} \text{ cm}^{-3}, \alpha_3 = 10 \text{ cm}^{-1}, \Delta n = 0.01$

E_{g_1} (eV)	α_1 (cm^{-1})	η	$\alpha' = \alpha - \alpha_{fc}$ (cm^{-1})	α (cm^{-1})[a]	α (cm^{-1})	T (°K)
			Theory		Experiment	
1.43[b]	350	1.0	20	50	89	300
1.53	100	5.0	2.3	32	35	300
1.515[b]	30	1.0	2	11	11	77
1.62	30	5.0	1	10	8	77

[a] $\alpha_{fc} = 30$ cm^{-1} at 300° K and 9 cm^{-1} at 77° K (see text).
[b] No Al in p^+ region.

The theoretically predicted values of α were compared with experimental results. Consider the dependence of α on E_{g_1} at 300° K and 77° K. For $E_{g_1} = 1.43$ eV (homojunction laser), Table 4 shows that with $\eta = 1$ and $\Delta n = 0.01$, the calculated $\alpha' \approx 20$ cm^{-1}. Assuming $\alpha_{fc} \approx 30$ cm^{-1} at 300° K, $\alpha = \alpha' + \alpha_{fc} \cong 50$ cm^{-1}. This compares to the experimental value of 89 cm^{-1}. With $E_{g_1} = 1.53$ eV, a value of $\alpha \cong 32$ cm^{-1} is calculated, compared to the experimental value of 35 cm^{-1}. At 77° K, the agreement between theory and experiment is reasonable.

With regard to dependence of α on the hole concentration in region 1, Table 5 compares the predicted and observed values of α at 300° K

TABLE 5

Effect of p^+ Concentration on α, Theory and Experiment (300° K and 77° K) [from Ref. (82)]

$(d = 2 \mu, n = 2.7 \times 10^{18} \text{ cm}^{-3}, E_{g_1} = 1.67, \Delta n = 0.01, \eta = 5, \alpha_3 = 10 \text{ cm}^{-1})$

T (°K)	p^+ (cm^{-1})	α_1 (cm^{-1})	$\alpha' = \alpha - \alpha_{fc}$ (cm^{-1})	α (theory) (cm^{-1})	α (exper.) (cm^{-1})
300	1.5×10^{19}	100	2.3	32[a]	27
	2.3×10^{19}	150	3.1	48[b]	48
77	1.5×10^{19}	30	~1	10[a]	8
	2.3×10^{19}	45	~1	15[b]	10

[a] $\alpha_{fc} = 30$ cm^{-1} at 300° K and 9 cm^{-1} at 77° K.
[b] $\alpha_{fc} = 45$ cm^{-1} at 300° K and 14 cm^{-1} at 77° K.

and 77°K. The increase in the p^+ concentration from 1.5×10^{19} to 2.3×10^{19} cm^{-3} increases α_i and α_{fc} proportionately. The predicted values of α are 32 and 48 cm^{-1} respectively, which compare to experimental values of 27 and 48 cm^{-1}, respectively.

It may be concluded that semiquantitative agreement with simple theory for the laser loss values is obtained using reasonable values of the various material constants and free-carrier loss coefficients in the active region. The qualitative trends are correct with respect to variations in the bandgap energy and Zn doping. Furthermore, the model correctly predicts the reduction in the laser loss with decreasing temperature. The main source of quantitative uncertainty is attributed to the estimates of the free-carrier absorption coefficient in the active region.

Hayashi and Panish(39) have reported loss values in their SH-CC lasers ($E_{g_1} = 2$ eV, $d \sim 2 \mu$) of 30–40 cm^{-1} at 300°K and 10 cm^{-1} at 77°K, which are consistent with those in Ref. (82). These authors also reported that their plots of J_{th} vs. $1/L$ were linear. However, Goodwin and Selway(84) found a marked deviation from linearity in their data. Thus, it appears that experimentally a variety of J_{th} dependencies on cavity length may be observed. These differences may be partly due to the fact that the gain and the current density are not always linear (and the degree of optical confinement varies with $h\nu$), with the exact dependence changing with doping in the active region, which in turn depends on the laser fabrication procedure. Finally, much experimental scatter, particularly for long lasers, is due to randomly distributed imperfections in the material. This makes any precise determination of J_{th} vs. L very difficult. Ideally, as previously mentioned, the best procedure consists of measuring J_{th} before and after applying an ideal reflective film to one facet of the cavity which changes the effective cavity length L without introducing additional materials complications.

The fact that a minimum d value exists for single heterojunction lasers has been established experimentally by several investigators (40,288,289). Typically, very high threshold current densities are observed if d is less than about 1 μ, which is consistent with the waveguide model discussed above. As expected in view of the increased $n_2 - n_3$, much smaller d values have been successfully achieved with double heterojunction lasers with a resultant decrease in the threshold current density (see Section X).

2. Near- and Far-Field Laser Emission Patterns

The concept of laser waveguiding has been experimentally demonstrated rather early in the history of diffused injection lasers and diodes [see, for example, Yariv(11)].

Byer and Butler(79) have studied the dependence of the near-field and far-field emission pattern in SH-CC lasers on the active region width d. They found that the near-field emission pattern shows only the fundamental transverse mode for $d = 2.5\,\mu$, but that a third-order mode is evident for $d = 5\,\mu$ (Fig. 22). These authors calculated the laser loss α' using a more sophisticated model than Anderson's because it included the imaginary component of the refractive index. The values of the refractive indices and absorption coefficients in the regions 1, 2, and 3 used in the calculations were close to those shown in Table 3. They found good agreement between the experimental and theoretical far-field (Fig. 23) and near-field (Fig. 22) emission patterns for $d = 5\,\mu$ (300° K). In the homojunction lasers, the third-order mode is not observed because its propagation loss coefficient is very high — 190 cm⁻¹, as compared to only ~4 cm⁻¹ in the SH-CC laser. Figure 24 shows the theoretical and experimental far-field pattern for a CC laser with $d = 2.5\,\mu$. The agreement with theory is good, except for the fine structure which requires further theoretical refinements to explain. Note that the angular spread at the half intensity points is about 20°. This value agrees with $\lambda/d \sim 0.9/2.5 \sim 20°$.

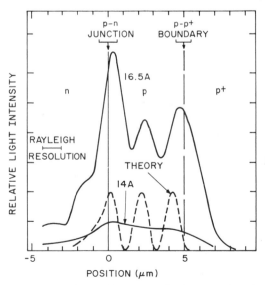

Fig. 22. Intensity distribution along a line perpendicular to the junction at the emitting facet of a SH-CC laser for two current densities. $I_{th} = 14\,A, d = 5\,\mu (79)$.

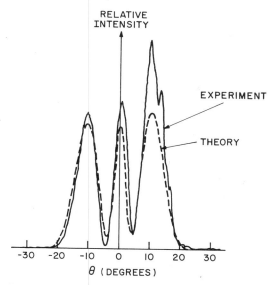

Fig. 23. Far-field radiation pattern for a SH-CC laser operating in the third-order mode at $I = 16$ A. The θ is in the plane perpendicular to the laser plane and $\theta = +90°$ corresponds to the positive x axis, $d = 5\,\mu$ (79).

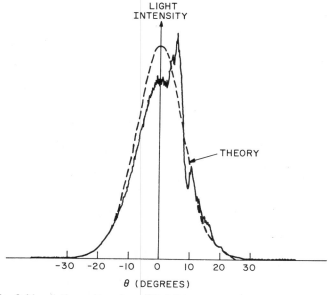

Fig. 24. Far-field radiation pattern for a SH-CC laser operating in the fundamental mode $(d = 2.5\,\mu)$(79).

A further point of interest is that Byer and Butler found that the laser emission is polarized (mostly TM).

The simple three-layer waveguide model is deficient in that it predicts that the fundamental transverse cavity mode will always have the lowest transmission loss and, hence, high-order modes should not be observed experimentally. As noted above, high-order modes are

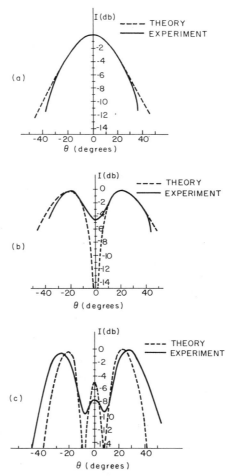

Fig. 25. Comparison of the experimental and theoretical far-field patterns for double heterojunction lasers with varying thickness of the waveguide region: 0.7 μ, fundamental transverse mode only; (b) 1.1 μ, 2nd order mode only; (c) 2.8 μ, 3rd order mode with small admixture of the 4th order mode. The theoretical curves are based on the model which divides the waveguide region into two regions of different gain(283). The emission is polarized (mostly TE).

indeed observed in thick cavities; and while the *shape* of the far-field pattern can be predicted, the fact that a high-order mode will dominate is not predictable with the simple model. A more refined model has, therefore, been proposed by Butler(*282*) based on the idea that the waveguide region is not uniformly inverted; hence, the waveguide region is divided into two parallel regions with different gain. This model does predict that high-order modes will dominate under certain conditions which depend on the total width of the waveguide region, the relative widths of the two gain regions and the refractive index discontinuities at the outside boundaries. The results of this theory have been experimentally tested in double heterojunction lasers(*283*) with satisfactory results. Figure 25 shows a comparison between the theoretically predicted mode and its far-field pattern and the experimentally observed pattern for three values of the total waveguide region thickness *d*. Note that the third-order mode dominates in a cavity 2.8-μ thick, while only the fundamental mode was observed in the SH-CC laser of comparable width (Fig. 24). The reason for the difference is the much larger refractive index discontinuity at the *p–n* interface in the double heterojunction laser as compared to the single heterojunction one. Note in Fig. 25 that the angular spread at the half-intensity point of the double heterojunction lasers is about 40°, or about twice the spread of the SH-CC device operating in the fundamental transverse cavity mode.

C. FAR-FIELD PATTERNS IN OTHER LASERS

Detailed experimental and theoretical studies dealing with wave-guiding and cavity modes in diffused lasers have been reported by several authors(*76,85–89*). In narrow-stripe geometry lasers, far-field patterns with theoretically predicted Gaussian patterns have been observed(*87–89*) and the cavity modes can be identified.

VI. Power Conversion Efficiency

The power conversion efficiency η_p of electrical energy to light is a key device parameter when high power values are required, especially in portable systems:

$$\eta_p = \frac{P_o}{I^2 R_s + E_g I/e} \tag{15}$$

where R_s = diode series resistance, I = diode current, and P_o = power output.

The relationship between the power efficiency and various laser parameters ($\alpha, \beta, \eta_{int}, L, R_s, A_d, R, W$) has been theoretically calculated by Sommers(90) for a cavity of length L and width W and the results presented in a compact graphical form. His calculation neglects internal heating and is based on the validity of Eq. (12) with $b = 1$.

The essential part of the treatment is that the parameters relating to the material and processing can be separated from the cavity parameters (cavity length and reflectivity), and that the best design of the cavity depends only on the material and the desired power output P_o/W.

The quality of the material and processing is characterized by a single parameter δ, which is independent of the cavity design. The experimental value of δ determines the optimum value of the reduced cavity parameter x_m, which defines the best value of both the cavity length and the reflectivity, for a specified P_o/W.

The material parameter is

$$\delta \equiv \frac{\beta E_g}{\alpha R_s A_d} \tag{16}$$

Figure 26 shows the optimum geometrical parameter x_m as a function of δ and Fig. 27 shows the maximum power efficiency for a given δ. The optimum conditions are then

(a) cavity length

$$L_m = \left(\frac{\beta}{\eta_{int}E_g}\right)\left(\frac{P_o}{W}\right)\left(\frac{1}{x_m^2}\right) \tag{17}$$

(b) reflectivity

$$\left[\ln\left(\frac{1}{R_1}\right) + \ln\left(\frac{1}{R_2}\right)\right]_m = \frac{2\beta P_0}{\eta_{int}E_g W}\frac{1}{x_m} \tag{18}$$

(R_1 and R_2 are the reflectivities of the ends of the cavity.)

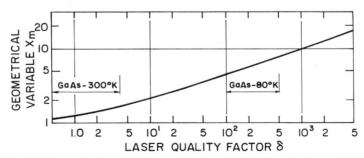

Fig. 26. Geometrical variable x_m as a function of the laser quality factor δ(90).

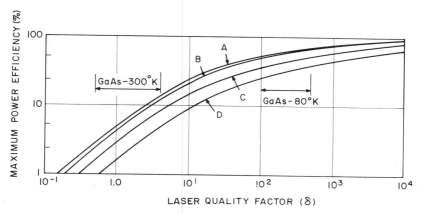

Fig. 27. Diode power conversion efficiency as a function of δ. Curve A is the maximum efficiency, which requires both the optimum cavity length and reflectivity. The other curves show the departure from maximum resulting from either the length or the reflectivity departing from its best value while the other cavity parameter is maintained at its extreme. For curve B, either the length or the reflection parameter ($\ln 1/R_1 + \ln 1/R_2$) is changed by the factors ($\frac{1}{2}$ or 2); in curves C and D, the factors are ($\frac{1}{5}$ or 5) and ($\frac{1}{10}$ or 10), respectively (90).

(c) current density

$$J_m = \left(\frac{\alpha}{\beta}\right)(1 + x_m)^2 \tag{19}$$

In practice, it is often desirable to use facets either cleaved and untreated, or one cleaved and the other with a reflective coating. In either case, the cavity length can be properly adjusted for the round trip reflection, and performance maintained at optimum. The loss to be expected for nonoptimum geometry can be calculated from the general treatment, as well as the adjustment of the cavity length for the changed reflection coefficients. It is found that the maximum power efficiency is little affected by improper reflectivities and L.

The measured values of the power efficiency are found to be in good agreement with the predicted ones based on the material parameters. Consider, for example, a GaAs laser with the following properties at 300° K:

$$R_s A_d = 4 \times 10^{-5} \ \Omega \ cm^2$$
$$\alpha = 20 \ cm^{-1}$$
$$\beta = 4 \times 10^{-3} \ cm/A$$
$$E_g = 1.43 \ eV$$
$$\eta_{int} = 0.6$$

From Eq. (16) we obtain $\delta = 7.2$. From Fig. 27, $x_m \cong 2$ and from Fig. 27 the maximum predicted power efficiency δ is approximately $0.2\eta_{int} = 12\%$. From Eq. (19) we estimate that the optimum operating current J_m is 45,000 A/cm² (at room temperature this is generally between 2.5 and 3 times the threshold current density). The optimum cavity length L_m [from Eq. (17)] is about 10^{-2} cm (with $P_o/W = 400$ W/cm).

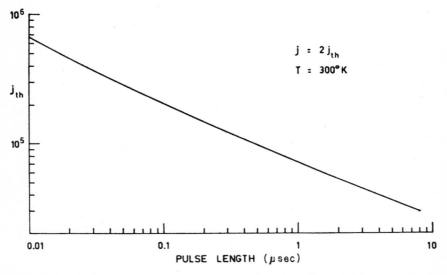

Fig. 28. Theoretical maximum pulse length as a function of threshold current for a rectangular-drive current pulse of amplitude $J = 2J_{th}$ (*101*).

Experimentally, we find for such a device that the power efficiency is about 10% with a cavity length of 2.5×10^{-2} cm operating at $J = 3 \times J_{th}$ (*82*). The agreement between theory and experiment is therefore good.

VII. Thermal Properties

A. THEORETICAL CONSIDERATIONS

The thermal properties of the devices determine to a large extent the maximum *average* power obtainable from a chip of given dimension (neglecting the degradation limits). Thermal problems are increasingly important as the ambient temperature is increased because of several factors.

1. Because of the increase in J_{th}, the operating current density increases and hence the $I^2 R_s$ loss increases.
2. The internal laser absorption increases, hence additional heat is generated in the vicinity of the active region.
3. The thermal conductivity of the semiconductor decreases.

In principle, if the device geometry and material parameters are known, it is possible to calculate the junction temperature and hence the maximum possible power. In practice, matters are complicated by the assumptions which have to be made about the effective heat source and the thermal transfer properties at material interfaces. Various calculations have been reported in the literature with emphasis on determining the conditions for CW operation.

A simple approximate criterion for CW operation has been derived by Gooch(91) which is based on measured laser parameters:

(a) At low T when $I \times V \gg I^2 R_s$

$$\frac{I_{th} V \theta (\eta_{int} - 1)}{T} < 0.15 \qquad (20)$$

(b) At high T when $I^2 R_s \gg I \times V$

$$\frac{I_{th}^2 R_s \theta}{T} < 0.067 \qquad (21)$$

In the above expressions, θ is the thermal resistance of the laser, T is the ambient temperature, R_s is the diode series resistance, and V is the junction voltage.

The value of θ can be conveniently determined from the increase in junction temperature as a function of the power dissipation in the diode (dc operated). The simplest method consists of measuring the shift in the spontaneous emission peak (which follows the thermal shift of the bandgap energy) while increasing the current in the device. Since only 1% of the optical power is radiated from the device in the spontaneous mode, one may assume that the power dissipated in the diode equals IV. Hence,

$$\theta = \frac{\Delta T}{IV} \qquad (22)$$

Other techniques for measuring θ have been described by Gooch(91). The value of R_s can be obtained from the slope of the IV curve at high current. A special curve tracer designed for determining R_s under pulsed operation has been described by Pankove and Berkeyheiser(92).

CW operation has been discussed by several authors(93–99). Dyment et al.(93) studied stripe geometry lasers (Section IV) and concluded that a stripe width of about 13 μ was optimum for the highest possible CW operating temperature and that CW operation at 300°K theoretically requires J_{th} < 15,000 A/cm².

Treatments of the maximum possible power available from a laser (neglecting catastrophic damage) have been presented by Pilkuhn and Guettler(100).

Pulsed rather than CW operation is of greatest general interest. Broom(101) has calculated the junction temperature rise during a pulse assuming that the power is dissipated in a region about 4 μ wide (i.e., the active region and its immediate surroundings where light is absorbed). In this analysis the key parameter is the thermal diffusion length L_t

$$L_t = \sqrt{\kappa t} \qquad (23)$$

where κ is thermal diffusivity and t is the time. The value of $\kappa = K/C\rho$ (K is the thermal conductivity = 0.5 W/cm deg for GaAs and ρ is the density = 5.37 g/cm² for GaAs). Hence, for GaAs, $\kappa = 0.29$ cm² sec at 300°K and $L_t = 1$ μ for $t = 40$ nsec. For short pulses, the heating at the junction is essentially adiabatic since heat flow is negligible and the junction temperature rise can be estimated from the power dissipated in a given volume.

For longer pulses and heat sinks close to the active region, thermal diffusion during the pulse limits the temperature rise. Broom considered both single and double heat sinks and obtained numerical solutions for either a single or double heat sink a distance of about 20 μ from a heated region 4 μ wide (300°K ambient). Assuming a dependence of J_{th} on temperature of the form

$$J_{th} = J_{th}(0)\left(1 + \frac{\Delta T}{T}\right)^3 \qquad (24)$$

where $J_{th}(0)$ is the initial J_{th} value, the maximum pulse length over which lasing will occur when the laser is operated at $J = 2J_{th}(0)$ is shown in Fig. 28.

The above calculation assumes that the duty cycle is low enough so that the junction temperature at the beginning of each pulse is the ambient value. In practice this condition will be satisfied if (a) the distance l between the active region and the heat sink is substantially greater than the thermal diffusion length L_t, and (b) the semiconductor is mounted on a heat sink maintained at the ambient temperature. For example, assuming $l = 20$ μ and $L_t = 2l$, we find from Eq. (23) that the laser "off time" between pulses needs to be about 800 μsec. The pulse repetition rate should not, therefore, exceed about 1 kHz if the junction is to be at the ambient temperature at the beginning of each pulse.

B. LASER PACKAGING

The simplest laser package is the TO-46 transistor header fitted with a gold-plated copper block on which the laser is mounted facing the direction of the leads. However, if the back facet of the laser does not have a reflective film, the power measured includes some of the light reflected from the header and appropriate corrections must be made if the total power emitted by the laser is to be accurately determined. This package can be covered with a glass-topped cap, which is useful when the laser is operated immersed in various cryogenic liquids. Freons, in particular, discolor the facets because of films formed on the semiconductor. Because TO-46 headers are not easily heat sunk, operation is not advisable at rates in excess of a few kilohertz at 300° K.

A laser package, shown in Fig. 29, with improved thermal properties has been introduced by the RCA-Solid State Division. The screw base permits ready insertion into heat sinks.

Fig. 29. Commercial laser package used for mounting individual units (RCA TA7606 type).

For very high duty cycle operation, when a heat sink to both sides of the chip is required, the package shown in Fig. 30 has been used by Nelson and Kressel(*64*). The laser is held in place by the pressure generated by the epoxy in drying. The inside top and bottom of the package are tin plated to improve the contact between the metal and the semiconductor.

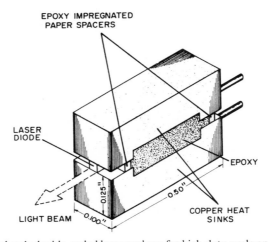

Fig. 30. Nelson's double-ended laser package for high-duty cycle operation(*64*).

Laser mounts for CW operation were described by Konnerth and Marinace(*103*). Both the laser and the mount are plated with indium and the contact is made by cold welding. A flared end prevents obstruction.

The use of high thermal-conductivity diamond heat sinks in conjunction with the stripe geometry lasers designed for CW operation was described by Dyment and D'Asaro(*65*). Such heat sinks are of general utility in all devices where thermal dissipation is a problem.

VIII.   Laser Reliability

Two distinct types of failure mechanisms have been identified: (1) "catastrophic" decrease in the power output that is due to mechanical damage of the facets — the onset of damage is a function of the *optical flux density* in the active region, and may occur even with a single pulse of the appropriate energy and length; (2) gradual decrease in power due to increasing threshold current density and decreasing differential quantum efficiency—the degradation rate here is a function of the *current density* of operation.

A. Catastrophic Failure

The failure is due to the formation of tiny cracks or grooves in the junction vicinity where the optical flux density is the highest. Figure 31 shows a typical example of the facet appearance as observed with an optical microscope after catastrophic damage. The same type of facet damage observed with a higher-magnification scanning electron microscope is shown in Fig. 32. Operating at 77°K with long pulses, Dobson and Keeble(*104*) observed free gallium in some portions of the junction, suggesting that dissociation of the GaAs had occurred. These authors suggested that the damage occurs by the optical absorption of the light near the diode surface in a region where the absorption coefficient is above average. In their view, the temperature in this region may rise by nearly 1000°C, thus leading to dissociation. An alternative failure mechanism has been proposed(*102*) in which some of the optical

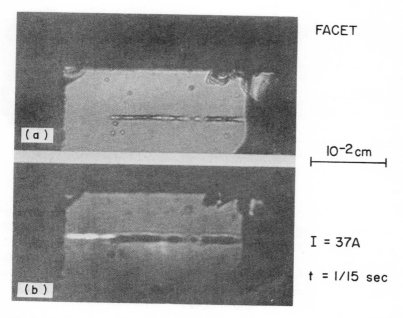

Fig. 31. Appearance of laser facet following catastrophic damage. (a) No current applied to the laser; (b) current applied, showing lasing in the undamaged part of the junction (*102*).

Fig. 32. Scanning electron microscope observation of damaged homojunction LPE GaAs laser facet, following catastrophic failure at 300° K. The sample surface is at an angle of 45° with respect to the electron beam to show the accumulation of material on the facet.

energy is transferred, via stimulated Brillouin scattering, into acoustic phonons. The flux density of 10^6–10^7 W/cm² in the active region is high enough for this to occur. A shock wave may be formed which would mechanically damage the surface. Alternatively, the phonon train may be absorbed at the surface, also leading to a substantial localized temperature rise. While there is no definite way of differentiating between the above-proposed mechanisms, recent experiments are inconsistent with the first model(298). A third explanation, based on critical thermoelastic stresses, was proposed by Kruzhilin et al.(105).

The power level where catastrophic failure occurs (measured W/cm of facet) depends on the following:

1. *Temperature of operation.* Figure 33 shows the peak power at failure for homojunction LPE lasers between 300°K and 77°K. The failure point decreases with decreasing temperature. Thus, the *peak* power attainable from a device is less at 77°K than at 300°K. This behavior does not hold in heterojunction lasers where the waveguide region is better defined.

2. *Pulse length.* The longer the pulse, the lower the peak power attainable before catastrophic damage. For example, Table 6 shows the critical flux level for three values of the pulse width at 300°K and 195°K. Note that while for a 25-nsec pulse the failure point is at 700 W/cm at 300°K, it decreases substantially for a pulse length of 100 nsec (*106*).

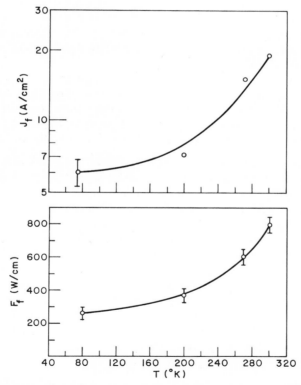

Fig. 33. Temperature dependence of the current density (J_f) and of the power output F_f (in W/cm of facet) for the onset of catastrophic damage in homojunction LPE lasers (*102*).

TABLE 6

Threshold for Catastrophic Failure in SH-CC Lasers for Various Pulse Lengths (*106*)[a]

T (°K)	Pulse width (nsec)	$J_f{}^b$ (10^3 A/cm²)	Critical optical flux F_f (W/cm)[c]
300	25	190	700
	100[d]	160	525
195	100[d]	93	440
	1200	40	195

[a] Laser dimensions: 3.3 × 11 mils, selected from a single wafer.
[b] Current density at failure.
[c] Power output per unit junction width (± 5%).
[d] The pulse is Gaussian with 100 nsec half-width.

3. *Width of the active region.* In view of the dependence of the damage on the flux density, it is expected that, if the optical emission is more closely confined in the active region, the damage threshold (as measured in W/cm) will decrease. This is in fact the case with SH-CC lasers in which the mode confinement is better than in homojunction LPE lasers. Note that Fig. 33 shows that the failure point for 100-nsec pulse operation of a homojunction laser is at about 800 W/cm at 300° K as compared to ~ 500 W/cm for the SH-CC laser (Table 6) (*82,106*).

To increase the threshold power for damage, the power density at the facets should be reduced. One method for doing this involves using a curved junction, as described by Dobson and Keeble (*104*), which is applicable to diffused lasers, but such a structure decreases the laser efficiency. Another possibility is the LOC structure (Section III).

The simplest method of increasing the attainable peak power level consists of the application of antireflective films to one laser facet while the other facet is covered with a reflective film. This decreases the reflectivity and hence the ratio of the power inside and outside the device. Table 7 shows the improvement in the peak power level at failure obtained in a group of diodes from a single lot of SH-CC lasers (*106*). The average peak power at failure is increased from 560 to 986 W/cm with the antireflective film. Such films also increase J_{th} and η_{ext} because of a decrease in reflectivity (*298*).

B GRADUAL DEGRADATION

The causes for the gradual degradation in power output observed at high current densities are exceedingly complex, judging by the results

TABLE 7

Threshold for Catastrophic Failure in SH-CC Lasers with and without Antireflective Films on the Facets. The Lasers Are 13 mils Long(*106*).

Group	Coating	Diode No.	J_f (10^5 A/cm²)	F_f (W/cm)	Av F_f (W/cm)[a]
A	None	16	1.47	762	
		18	1.26	518	
		19	0.63	308	560
		20	1.21	653	
B	Reflective	21	1.00	739	
	coat on	22	1.15	560	
	one facet	24	1.25	942	727
		25	0.95	666	
C	Reflective	26	1.00	1088	
	coat and	28	1.25	682	
	antireflective	29	1.75	1042	986
	coat	30	1.85	1132	

[a]Pulse 100 nsec wide.

reported to date. The study of homojunction LPE lasers led to the following conclusions(*107*):

1. The degradation process is a bulk phenomenon not amenable to control by simple surface coatings such as SiO. A characteristic feature is that uniformity in the near-field emission patterns decreases.

2. Gradual degradation may occur with no evidence of mechanical changes on the laser facets. The possibility that microcracks are forming in some cases in the interior of the device cannot, however, be excluded. Such cracks may sometimes be formed by the repeated thermal cycling in pulsed operation.

3. The formation of nonradiative recombination centers in the junction vicinity appears to be one of the most important factors responsible for the reduction in the *internal* (and hence external) *differential quantum efficiency*, as well as for increase in the threshold current density. Evidence of the formation of nonradiative centers similar in activation energy to those formed in electron bombardment (vacancies) with high energy electrons (1 MeV) has more recently been obtained by Nuese and Schade(*108*).

[However, the defects formed by irradiation are not identical to those formed in the course of diode operation. In particular, the irradiation damage anneals out much more readily than the defects responsible for diode degradation.]

Evidence was also found in some lasers of an increase in the internal optical loss with operating time.

4. The laser degradation rate was shown to be a superlinear function of the current density, and the *optical* flux density does not significantly affect the degradation rate (*109*).

5. There is a great deal of random variation in laser life. Some homojunction LPE laser diodes provided with reflective coatings have supplied 400 W/cm of power from the emitting facet at 8000 pulses/sec for hundreds of hours at 300° K. For a typical solution-grown laser such as that described by Nelson (*50*), this corresponds to operation between two and three times the threshold current, i.e., current density of about 10^5 A/cm². Other similarly made lasers last only about 100 hr. In CW operation of diffused lasers at 77° K, noncatastrophic degradation due to deposits forming on the laser facets has been investigated. Operation for periods of 1000 hr has been obtained under optimum conditions (*103*).

At 77° K, where the current density of operation is considerably reduced (because of the reduction in threshold current density), safe operation at high duty cycles is easily obtained. No degradation in power was observed for diodes operated for over 300 hr with a duty cycle of 8% at about 6500 A/cm² ($6 \times J_{th}$) (*107*).

6. The degree of uniformity of the lasing emission has an important bearing on the degradation rate for a given current density of operation. Since the lasing uniformity is believed to be inversely related to material imperfections in the laser such as dislocations and precipitates, the extension of laser life may depend on the ability to fabricate devices relatively free of flaws. The fact that a great deal of variation in laser life has been observed further suggests that imperfections randomly distributed in lasers control their life.

The effect that certain imperfections have on increasing the degradation rate was directly demonstrated in experiments where diodes were tested which contained known densities of dislocations or precipitates. It was shown (*110*) that a high density of dislocations ($\geqslant 10^5$ cm⁻²) leads to accelerated degradation, although the initial laser properties are relatively unaffected. These observations are important because dislocations are frequently introduced in the process of *p–n* junction formation itself.

The effect of dislocation was shown as follows. An *n*-type, Si-doped wafer with $n \approx 2 \times 10^{18}$ cm⁻³ (Bridgman grown) was divided into two parts. One was hot pressed in vacuum while the other was kept as a control. X-ray topographs showed that the dislocation density increased from the 10^3 cm^{-2} range to values in excess of 10^5 cm^{-2} in portions of the deformed wafer. Lasers were fabricated after deformation, using the homojunction (non-CC) LPE process. Table 8 shows the characteristics of the deformed and control lasers (each group consisting of five devices) before and after operation at 1.5×10^5 A/cm^2 for 140 hr. Although the initial differences in threshold current density and efficiency are small, the control lasers hardly changed during operation while the threshold current density of the deformed devices doubled and their efficiency decreased by about 30%. Figure 34 shows the time dependence of the output of the plastically deformed and control lasers.

Tests of the same types of diodes operated in the *spontaneous* mode at the same current density indicated a reduction of the spontaneous efficiency of $24 \pm 5\%$ in the first 100 hr for the deformed diodes vs. a change of about 6% for the control devices. These changes are consistent with the laser efficiency results. Thus imperfections undetected by prior measurements may be present and result in rapid degradation. Moreover faulty processing may lead to lasers with a poor lifetime because diffusion-induced imperfections may form even in initially perfect substrates(*111*).

Controlled experiments(*110*) have shown also that a high density of Ga_2Te_3 precipitates (10^{16} cm^{-3}) in the material leads to accelerated degradation.

There is also a substantial difference in the degradation between diodes doped with different acceptors. For example, Be doped diodes degrade faster than either Mg or Zn doped ones(*110*).

Since, in lasers, changes in the quantum efficiency are measured at very high current densities ($> 10^4$ A/cm^2), the increasing nonradiative

TABLE 8

Characteristics of Homojunction GaAs LPE Lasers before
and after Life Test(*106*)

	J_{th} (A/cm^2)		η_{st} (%)	
	Before	After[a]	Before	After[a]
Not deformed (control)	7 ± 0.9	7.1 ± 0.7	12.4 ± 1.4	12.6 ± 1.4
Plastically deformed[b]	8 ± 1.1	17 ± 5	9.4 ± 2	6.3 ± 1.2

[a] 140 hr operation at 150,000 A/cm^2, 100 nsec pulse width, 4 kHz. Each group consisted of 5 diodes simultaneously tested.
[b] Dislocation density $\geqslant 10^5$ cm^{-2}.

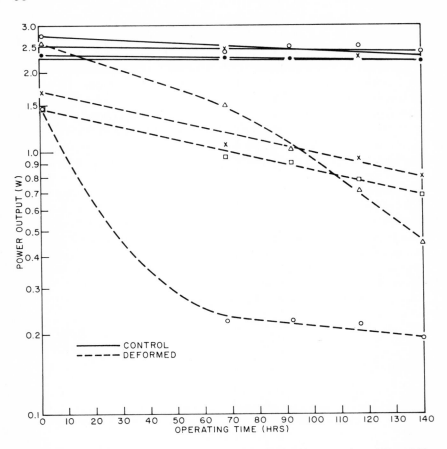

Fig. 34. Power output as a function of operating time (100-nsec pulse width, 4-kHz repetition rate) of lasers made from as-grown and plastically deformed substrates(*106*).

tunneling or generation-recombination currents are negligible compared to the diffusion current. Thus, changes observed in the exterior quantum efficiency reflect either an internal decrease in the quantum efficiency or increasing optical absorption, or both.

The simplest hypothesis which fits the observations is that of acceptor displacement [Weisberg-Gold hypothesis(*112,113*)], the probability of which is enhanced in the vicinity of strain field associated with certain imperfections. The fact that the degradation rate of diodes with Be, Mg, or Zn acceptors increases with decreasing covalent radius is consistent with this model, since a smaller atom should be more easily displaced.

Once displaced, leaving behind nonradiative centers(114), the atoms may reach a low energy configuration in the imperfection vicinity, resulting in clusters which constitute microplasma sites and increase the internal absorption in diodes operating in the stimulated mode, in agreement with previous experimental observations of laser degradation. The lack of capacitance change is explained by the localized nature of the atomic displacement. The details of how the displacement occurs (i.e., how energy is transmitted to the atom) are unknown.

An hypothesis has been presented(112) based on the idea that the energy released by nonradiative electron-hole recombination is capable of displacing impurity atoms (at least with finite probability) into interstitial sites.

It should be added that precipitates other than Ga_2Te_3 can exist in junctions. Their effect on degradation may depend on the degree of lattice strain associated with them, as discussed in Ref. (110).

IX. Crystal Imperfections

Imperfections, either present in the original material used for the device fabrication or introduced in the course of junction formation, may seriously affect the laser efficiency. Table 9 lists the imperfections affecting particular basic laser parameters. We will be concerned with the following imperfections: dislocations; precipitates; and point defects such as traps, vacancies and interstitials introduced, for example, by high energy electron irradiation, and contaminants

TABLE 9

Imperfections Affecting Specific Laser Parameters

Laser Parameter	Imperfection
Internal quantum efficiency η_{int}	1. Small precipitates 2. Nonradiative centers (vacancies)
Absorption coefficient α	1. Precipitates 2. High free-carrier density 3. Nonplanar junctions
Nonuniform emission	1. Nonuniform dopant density 2. Precipitates 3. Nonradiative centers 4. Nonplanar junctions
Lasing delays	1. Traps (nature unknown)

A. DISLOCATIONS

The correlation between dislocations and laser performance has been the subject of detailed study(*115*). Dislocation clusters in the substrate may lead to nonplanar junctions because of enhanced Zn diffusion along the dislocation cores. It was shown that the average laser efficiency is not reduced as long as the dislocation density in the substrate is below 10^4–10^5 cm^{-2}. It is desirable, however, to use substrates with dislocation densities as low as possible because of the effect that they have on laser reliability (Section VIII). Furthermore, the possibility of getting more nearly uniform near-field emission patterns is increased (*116*).

The question of dislocations is particularly important in alloys because when layers are epitaxially deposited on substrates with significantly different lattice constants, a high density of misfit dislocations may be present in the epitaxial films(*117*).

Dislocations may be introduced even in initially perfect substrates in the process of Zn diffusion(*118*). Dislocation formation is the result of strains introduced by the diffusant with a different atomic radius from that of the host atoms. This subject has been extensively studied in Si following the work of Prussin(*119*). Much less is known about diffusion-induced imperfections in GaAs, but estimates may be made using the theoretical expressions derived by Black and Lublin(*120*). The concentration of such imperfections may be expected to depend on the method of junction formation, in particular the diffusion temperature, the diffusion gradient, and annealing treatments following diffusion(*111*).

B. PRECIPITATES

Direct evidence (based on electron transmission microscopy) has been obtained for the presence of Ga_2Se_3(*121*) and Ga_2Te_3(*122*) precipitates in some samples of highly Se- and Te-doped ($> 10^{18}$ cm^3) GaAs, respectively. When present in significant density in the junction region ($\sim 10^{16}$ cm^{-3}), these precipitates seriously increase the internal optical loss. It was also shown that enhanced degradation occurs in diodes containing Ga_2Te_3 precipitates (Section VIII).

Extensive experiments have indicated that Si-doped GaAs prepared by the horizontal Bridgman growth technique yields the most consistent high quality laser diodes(*115*). This has been attributed to the fact that these crystals are in general free of precipitates. However, in the tail end of such ingots one occasionally finds a high density of small inclusions(*115*), with a lattice constant close to that of SiO_2(*123*).

Lasers made from materials containing these inclusions are very inefficient.

C. Point Defects

1. *Traps*

Traps of an unknown origin, present in some diodes, are believed responsible for the time delays between the laser current flow and the onset of stimulated light emission (see Section XI).

2. *Electron Irradiation*

Nonradiative recombination centers have been shown to be introduced into GaAs as a result of irradiation with 1-MeV electrons(*114*). These centers are presumably vacancies and interstitials formed by the displacement of Ga and/or As atoms. The laser efficiency is substantially reduced by an integrated flux density of the order of 10^{16} cm^{-2} (*107*).

The degree of damage for a given integrated flux density depends on the compound. Nuese et al.(*124*) have studied the effect of electron irradiation on spontaneous Ga(AsP) diodes and found that the annealing temperature depends on the P composition: the greater the P content the lower the annealing temperature required to remove the damage. Figure 35 shows the annealing temperature required as a function of the phosphorus content in Ga(AsP)(*124*).

The effect of Co^{60} irradiation on LPE laser characteristics has also been studied(*284*). The results obtained indicate that the relatively simple nonradiative centers formed anneal out quite easily, in contrast to the centers formed in the course of high current density laser operation. This suggests that the latter defects are probably different in nature.

Several methods have been described which are useful in determining the relative density and ionization energy of traps in the junction vicinity. A capacitive method for determining the density and the ionization energy of traps in the space charge region was described by Itoh and Nishizawa(*125*) in which light of varying photon energy is incident on a back biased junction at low temperature and the change in the diode capacitance is measured. This change is indicative of both the density and the ionization of the centers in the space charge region. A method in which the traps are filled at low temperatures and the diode is gradually raised in temperature has been described by Weisberg and Schade(*126*).

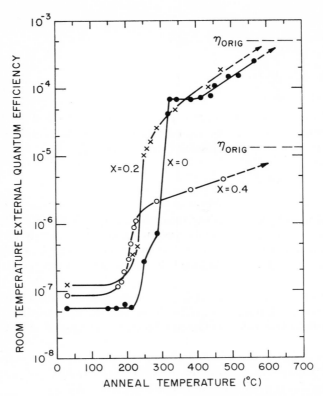

Fig. 35. Effects of isochronal annealing on the external quantum efficiencies of vapor-grown $GaAs_{1-x}P_x$ electroluminescent diodes. Prior to annealing, the diodes were irradiated with a 1-MeV electron beam with a flux of 10^{17} electrons/cm². The decrease in the annealing threshold temperature with increasing $GaAs_{1-x}P_x$ alloy composition, x, is thought to be owing to the relative ease of replacing interstitial phosphorus atoms, rather than arsenic atoms, back into their usual crystal lattice sites (124).

3. Metallic Contaminants

A common contaminant is Cu which, if present in sufficient concentration, will inhibit lasing. The same is probably true of Fe and Ni.

Aside from possible nonradiative processes associated with individual Cu centers, clusters of the metal also are likely to form at dislocation sites which increase the internal laser optical loss.

In general, devices grown by LPE are expected to contain relatively low Cu concentrations because the Ga melt is an excellent sink for the metal. Copper may be introduced, however, in subsequent heat treatments in evacuated ampules if precautions are not taken to prevent contamination.

D. Techniques for Substrate Selection

Table 10 lists the techniques useful for flaw detection. The rapid, nondestructive selection of n-type substrate material suitable for the fabrication of good quality lasers is possible using a combination of photoluminescence measurements, X-ray topography or dislocation etching, and infrared transmission with an ordinary metallurgical microscope fitted with an infrared viewer.

The quality test involves essentially three steps (115):

1. Determination by infrared transmission measurements that the crystal is free of gross impurity inhomogeneities.
2. Determination of the dislocation density. A value of $10^3 \, \text{cm}^{-2}$ is a safe value if the dislocations are distributed fairly uniformly (dislocation clusters are most detrimental), but the lower the better.
3. Measurement of the relative photoluminescence (PL) efficiency at 77° K. This technique is useful in quickly eliminating crystals containing a high density of small precipitates.

The PL measurement consists of illuminating the GaAs surface with light having an energy in excess of the bandgap energy, thus generating hole–electron pairs within $\sim 1 \, \mu$ of the surface. These pairs recombine, emitting radiation characteristic of the recombination centers in the material. The radiative recombination processes (in the energy range

TABLE 10
Techniques Used for Laser Substrate Flaw Detection

Imperfection	Detection technique
Junction nonplanarity	Etching
Small precipitates	Transmission electron microscopy
Large precipitates	IR transmission
Nonuniform dopant distribution	IR transmission
	X-ray topography
Dislocations	Etching
	X-ray topography
Nonradiative centers	Photoluminescence
Crystal strains	X-ray topography
	Etching

where efficient detection is possible using a photomultiplier ($h\nu > 1$ eV) compete with either nonradiative recombination processes or processes involving very deep centers (such that $h\nu < 1$ eV). A measure of the substrate quality is given by the relative photoluminescence efficiency, which is the ratio of the light output to the incident light.

For example, the emission spectra of typical Si-doped, melt-grown GaAs is shown at 77° K in Fig. 36. Two peaks are observed — the near-band edge band A_1 (~ 8200 Å), and a much broader band A_2 between 10,000 and 10,500 Å (not corrected for detector response). Band A_1 involves electron transitions to either free holes or holes trapped on *shallow* Si acceptors (0.03 eV), while A_2 involves electron transitions to deep centers (ionization energy of 0.2–0.3 eV).

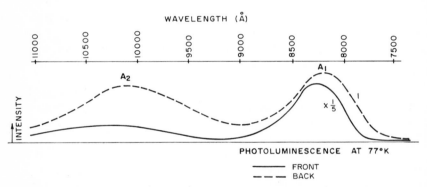

Fig. 36. Photoluminescence spectra at 77° K of front end (seed end) and back part of Si-doped Bridgman GaAs ingot ($N \sim 2 \times 10^{18}$ cm^{-3}). The intensity of the emission from back part is about $\frac{1}{5}$ that of the front. Also, the ratio A_2/A_1 is higher in back than in front. Data are not corrected for S-1 photomultiplier surface response (*115*).

The relative PL intensity correlates with laser performance as shown in Table 11 for homojunction LPE lasers made from several Si-doped ingots. We see that the most significant part of the PL measurement is the relative intensity of band A_1; a low value of band A_1 correlates with poor laser performance. This is also shown in Fig. 36. The low value of PL efficiency was shown(*115*) to be due to small clusters present in such high density that the average distance between them may be less than the minority carrier diffusion length. For example, for a cluster density of 10^{16} cm^{-3}, the average distance between clusters is under 0.1 μ, which is smaller than the minority carrier diffusion length of ~ 1 μ. Thus, assuming that the recombination at the cluster is nonradiative, a hole has an excellent chance of recombining nonradiatively. On this basis, the gradual decrease in the PL intensity with increasing distance

TABLE 11

Photoluminescent Efficiency of Substrate Correlated with Homojunction LPE Laser Performance (*115*).
(Si–doped substrate $\sim 2 \times 10^{18}$ cm^{-3})

Crystal	Part of ingot	PL band amplitudes (relative)		Average laser threshold (A/cm^2)	
		A_1	$A_2{}^a$	300° K	77° K
2336	Front	5.3	2.7	64,000	1,600
	Back	0.28	1.4	> 100,000	30,000
2346	Front	5.0	2.3	48,000	1,200
	Back	0.28	2.0	> 100,000	27,000
2341	Center	0.28	2.0	> 100,000	40,000
	Back	0.67	1.7	> 100,000	8,000
2350	Front	4.6	2.5	65,000	1,400
	Back	0.94	1.9	> 100,000	7,000

^{a}Not corrected for photomultiplier response.

from the seed end of the ingot, shown in Fig. 37, is explained by an increasing cluster density. It is noteworthy that these precipitates are too small to be detected by infrared transmission.

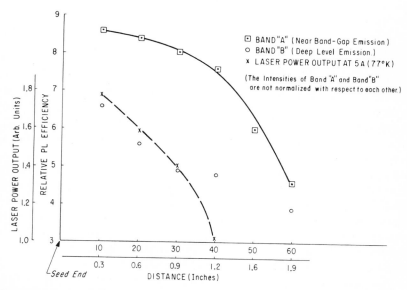

Fig. 37. Photoluminescence efficiency and laser power output as a function of substrate distance from the seed end of Bridgman-grown, Si-doped crystal (*115*).

These clusters will degrade also laser performance because they constitute infrared absorbing centers which increase the optical loss in the junction region. Furthermore, the current–voltage characteristics will be affected because such clusters in the space charge region will give rise to microplasma sites.

As mentioned previously, precipitates not detectable by infrared transmission are commonly found in Te-doped GaAs when examined by electron microscopy. These will similarly reduce the PL efficiency if present in high density (122).

Studies of the correlation between substrate PL measurements and diffused laser performance made from Te-doped substrates also have been reported (127); it was found that an annealing treatment prior to PL measurement improves the correlation with laser performance.

Finally we call attention to Ref. (128a) for discussion of the effect of imperfections on electroluminescent diodes.

X.   Properties of State-of-the-Art GaAs Lasers

In this section we survey important properties of various state-of-the-art GaAs lasers. Threshold current density, efficiency, gain and loss parameters, and peak power (CW and pulsed) are included. Time-dependent properties are considered in Section XI.

A. Threshold Current Density

In homojunction LPE lasers, the lowest J_{th} value reported at 300° K is 26,000 A/cm² [cavity length of 15 mils (45)]; 40,000 A/cm² is a more typical value for these devices. Figure 38 shows a typical curve of power output vs. diode current. Diffused lasers have typically J_{th} values of 40,000–100,000 A/cm², depending on the details of the fabrication process.

The lowest J_{th} value reported in single heterojunction CC lasers (36) is 8000 A/cm² (12-mil cavity length); 10,000 A/cm² is a more typical value which can be reproduced under manufacturing conditions. These devices, when properly doped, are capable of laser operation up to very high temperatures. Operation at 380° K can be routinely obtained if the active region is doped sufficiently so as to obtain a reasonable refractive index discontinuity at the p–n junction interface.

The addition of a second (AlGa)As–GaAs heterojunction permits the fabrication of very narrow lasing regions without loss of mode guiding (Section V). Furthermore since the lasing region is epitaxially

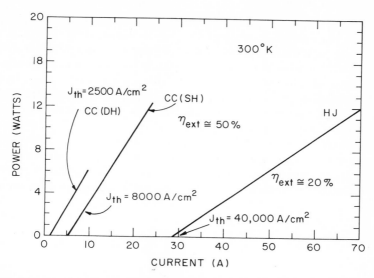

Fig. 38. The pulsed power output of a state-of-the-art SH-CC laser compared to that of a homojunction LPE laser made from a similar high-quality substrate. A low threshold double heterojunction laser is shown for comparison. The devices have equal dimensions.

grown, it can be highly doped and compensated, which helps increase the gain coefficient for a given diode current density. The room temperature threshold current densities depend on the width of the active p-region, at least between 2 and 0.5 μ. While narrower regions are possible, their control is rather difficult at this time. Table 12 shows that with a 0.5-μ wide GaAs p-region, the J_{th} is 1400 A/cm² at room temperature in a Fabry–Perot cavity device(281). In devices where all four sides of the device are cleaved (total internal reflection with no

TABLE 12

Representative Values of J_{th} and F_f for Heterojunction
Lasers [Ref. (281)]

Laser type	$d(\mu)$	J_{th} (A/cm²)	F_f (W/cm)[b]
Single heterojunction	~2	8000	500–700
Double heterojunction[a]	~2	5800	~400
Double heterojunction[a]	0.8	2400⎫	100–200
Double heterojunction[a]	0.5	1400⎭	

[a] $L = 15$–20 mils, similarly doped lasing regions.
[b] $\Delta t \approx 100$ nsec.

power being emitted), J_{th} values of about 1000 A/cm² can be obtained (285).

The J_{th} temperature dependence of the heterojunction lasers is considerably smaller than in homojunction lasers, as shown in Fig. 39 which compares a homojunction laser with a SH-CC laser made from a similarly doped substrate. The SH-CC laser shows an exponential temperature dependence of J_{th}, which is sometimes observed in homojunction LPE lasers as well(129).

Large optical cavity (LOC) lasers have been made with threshold current densities as low as 1700 A/cm² for thin ($\sim 1\,\mu$) waveguide regions(286). An unusual feature of these devices is that lasing has been observed at room temperature with waveguide regions as wide as $\sim 30\,\mu$.

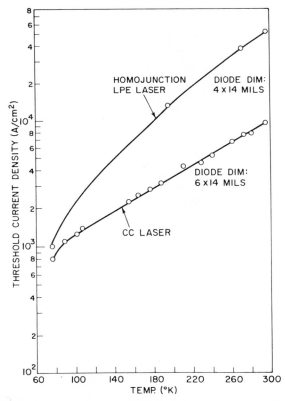

Fig. 39. Threshold current density vs. temperature of a SH-CC laser and a homojunction LPE laser having equal Fabry–Perot cavity lengths.

Figure 40 shows the temperature dependence of J_{th} for a low threshold double heterojunction lasers and the corresponding variation in the external differential quantum efficiency. Note that despite the very low J_{th} value at room temperature, the ratio of the threshold current density at 300 and 77°K is still approximately 8–10, or similar to that of the SH-CC device. The reason is that the major factor influencing the temperature dependence of J_{th} is the decrease in the gain constant β with increasing temperature in devices with good optical confinement.

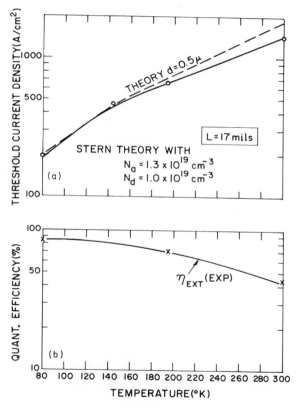

Fig. 40. (a) Temperature dependence of the threshold current density of a double heterojunction laser with an active region thickness of 0.5 μ; (b) temperature dependence of the external differential quantum efficiency η_{ext} for the same device. The theoretical curve is based on the calculations of Stern(290) using the Halperin and Lax bandtail approximation(26) for an active region doped with a donor concentration of 1.0×10^{19} cm^{-3} and an acceptor concentration of 1.3×10^{19} cm^{-3}(281).

B. Laser Efficiency

The efficiency values of interest are the differential quantum efficiency η_{ext}, the power conversion efficiency η_p and the over-all quantum efficiency. The highest observed values of differential quantum efficiency are ~ 52% at 300° K (*130*) and 80% at 77° K (*36*) in SH-CC lasers about 12 mils long. Double heterojunction lasers can have similar values, with a high of 60% in LOC lasers (*286*). In homojunction lasers, the room temperature efficiency is 15–20% for a similar length laser (*45*). At 77° K, however, the difference between the homojunction and CC type is relatively small. Thus, the efficiency is much more temperature dependent in the homojunction than in the heterojunction lasers (Fig. 41). The *over-all* quantum efficiency depends on J_{th} and how far above it the laser is operated. A value of 40% has been reported at room temperature (*130,286*).

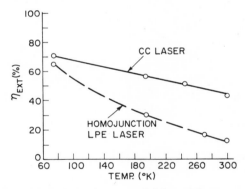

Fig. 41. Temperature dependence of the external differential quantum efficiency η_{ext} of a SH-CC laser and a homojunction LPE laser made from similar GaAs substrates.

The power conversion efficiency also depends on the drive current (Section VI). Figure 42 shows the variation of the power conversion efficiency of a good SH-CC laser between 77° and 300° K. Because of the $I^2 R_s$ losses, the efficiency of the better units is in the vicinity of 10% at room temperature, but increases to 50% at 77° K where the operating currents are greatly reduced. Power conversion efficiencies of ~ 20% have been obtained with LOC lasers at 300° K (*286*).

C. Peak Power Output

As mentioned in Section VIII, the peak power attainable from a given diode is limited by catastrophic damage. It is possible to obtain

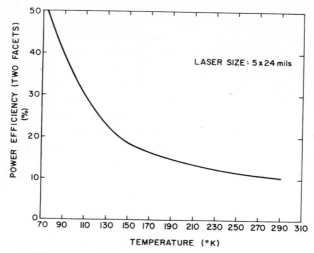

Fig. 42. Power efficiency as a function of temperature of a GaAs laser with the SH-CC structure.

higher peak power with short pulses than with long pulses. Peak power values of 105 W have been obtained with a pulse width of 25 nsec using diodes about 60 mils wide (*64*). Laser arrays have been reported (*131*) which emit about 1000 W at 300° K when operated with 25-nsec-wide pulses.

D. AVERAGE POWER

At 77° K, CW laser operation is possible with the emission of relatively high power values from single diodes, 1 W being a representative value. Values as high as 2–3 W also have been obtained (*132*) but long term operation is questionable under such conditions. The highest temperature for CW operation reported to date is in excess of 300° K with double heterojunction lasers (*285*). Outputs of ~ 120 mW were obtained with a power conversion efficiency of ~ 7% at 300° K (*299*).

Average power outputs in the milliwatt range can be obtained by pulsed laser operation at room temperature. A duty cycle of 10^{-3} is reasonable at $3 \times J_{th}$ for a well heat-sunk CC diode. At this current density a diode provided with a reflective film will emit about 300–400 W/cm of facet width. Thus, the average power will be 0.3–0.4 W/cm of facet width. With a typical laser 10 mils wide, the average power emitted will be between 7 and 10 mW.

The maximum duty cycle possible at room temperature depends on the packaging and the current density of operation. Figure 43 shows the peak power output of a SH-CC laser mounted on diamond heat sink with a radial cooling fin added to the package cap. The threshold current of that laser is 10 A (14,000 A/cm²). At $1.5I_{th}$ there is no falloff in output up to 19 KHz. With increasing drive, the heating effects become noticeable at decreasing duty cycle values. The average power output of the laser is about 11 mW at 35 A and 10 KHz(*133*). With lower threshold double heterojunction or LOC lasers, duty cycles of 1–10% are conveniently possible using single ended heat sinks if the operating current density is only a few thousand amperes per centimeter square.

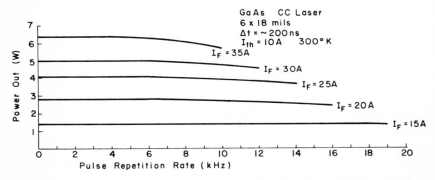

Fig. 43. Peak power output from a SH-CC laser as a function of repetition rate at various drive currents. The current pulse is Gaussian with a half-width of 200 nsec. The laser is mounted on a diamond heat sink (*33*).

E. RELIABILITY OF HETEROJUNCTION LASERS

Because of their lower operating current density at 300°K, heterojunction lasers have a much longer life than previous homojunction lasers (as long as the peak power is kept below the catastrophic damage point, Section VIII). Lasers are designed for operation at ~ 50,000 A/cm² and about 400 W/cm, thus below the catastrophic damage point for diodes without antireflective films (see Table 6). At a duty cycle of 4×10^{-4} (a typical useful value), the reported laser half-life is above 2000 hr (Fig. 44). However, as is the case with all electroluminescent diodes, the life of the CC lasers depends on the metallurgical quality of the junction region. In particular, the p^+p heterojunction may introduce strains and imperfections in the junction region. With regard to CW operation at 300°K, preliminary indications are that the total lasing life is presently < 100 hr at 5000 A/cm².

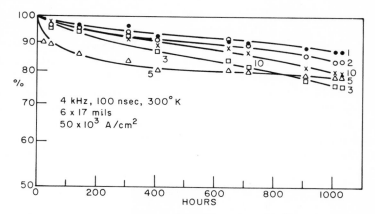

Fig. 44. Relative variation of power output with operating time of five lasers selected from a single wafer. The duty cycle is 0.04% and the current density of operation is 50,000 A/cm². The power output of such diodes, provided with a reflective film, is typically 400 W/cm of facet (1 W/mil). From Ref.(*106*).

The problem of catastrophic degradation can be troublesome in narrow active region double heterojunction lasers. Because of the near-perfect optical confinement, the optical flux density for a given value of emitted power (W/cm of emitting facet) increases nearly linearly with decreasing value of d. Table 12 shows that, for lasers similarly operated, decreasing d also decreases the failure limit F_f. Thus, the very low threshold lasers are not suited for high peak power pulsed operation and must be utilized with great care to prevent failure.

The advantage of the LOC lasers is that very high peak powers can be obtained without catastrophic failure. For example, with a cavity 7-μ thick, exterior differential quantum efficiency values of about 50% can be obtained at room temperature, and no failure was observed even with 2000 W/cm of emitting facet. This value is considerably higher than obtainable with other laser structures(*276*).

F. ANOMALIES IN HETEROJUNCTION LASERS

A number of effects have been observed in SH-CC lasers (rarely seen in homojunction LPE lasers) which may seriously affect their performance at room temperature or above, particularly if the laser cavity is short and the active region narrow ($\sim 1\ \mu$). In some CC lasers, J_{th} increases drastically at a given temperature despite the fact that the J_{th} vs. T dependence is normal below that temperature(*134*).

Similar effects have been reported in some diffused lasers [see, e.g., Ref. (*163*)]. This transition temperature may be either below or above room temperature. Generally, the narrower the active region, the lower the transition temperature is. Furthermore, the shorter the laser cavity in a given material, the lower the transition temperature. The *Q*-switching effects (Section XI) have been found to accompany this particular phenomenon(*134*).

A second effect probably related to the above is laser "quenching" at a critical *current density during* a current pulse, i.e., the laser turns on in the initial portion of the pulse and shuts itself off at a critical

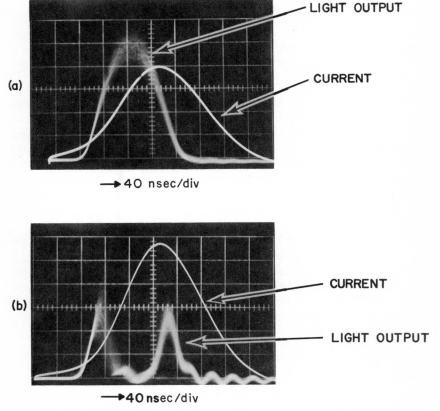

Fig. 45. Typical instabilities in optical output in certain SH-CC lasers above a critical current density as observed with F4000 phototube at 300° K. Optical pulse (a) below and (b) above the critical current density (43,000 A/cm² in this material). The curves for the light and optical pulses are displaced on the oscilloscope screen.

current density, then turns on again at the tail of the pulse when the current density drops below the critical value. This occurs near the critical temperatures and may be owing to heating during the pulse. This effect is illustrated in Fig. 45. If the laser cavity is sufficiently short, J_{th} will drastically increase as it approaches the critical current density (Fig. 46). Here again some diffused lasers exhibit similar effects. This effect is temperature dependent. A current density which quenches lasing at 300°K, for example, will not affect the laser a few degrees below room temperature. As is the case with the first effect discussed above, this phenomenon tends to occur in lasers with a narrow active region ($d \sim 1 \mu$).

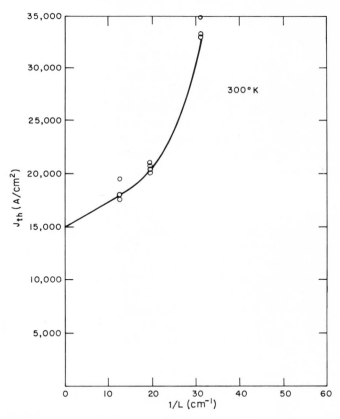

Fig. 46. The nonlinear variation in threshold current density as a function of the reciprocal of the laser length for a SH-CC laser exhibiting the instability effect. The instability threshold is at $\sim 34{,}000$ A/cm² for lasers grown on this wafer.

It is noteworthy that the above effects are not generally observed in double heterojunction lasers, which suggests that they may be related to optical mode guiding effects resulting from the fact that the refractive index discontinuity at the p–n junction interface is relatively small in the SH-CC lasers. As noted earlier in Section V, waveguiding in asymmetrical structures is a sensitive function of d. If the laser structure is such that mode guiding is barely possible at a particular laser wavelength, then a change in the Fabry–Perot cavity length, which changes the lasing wavelength, will also change the refractive index discontinuity. Furthermore, it is possible that slight changes occur in this discontinuity with changing temperature. However, an alternative explanation for some of the anomalies has been proposed which involves the injection of holes from the narrow p-region into the n-region(134).

G. LASER GAIN (β) AND LOSS (α) VALUES

As discussed in Section II, it is *theoretically* expected(25) that at high temperatures the gain is a true linear function of the current density only for highly doped junction regions. Furthermore, as mentioned in Section V, the loss coefficient α may depend to some extent on the precise value of the lasing photon energy and hence, for a given material, on the cavity length L. Because of these complications, the simple laser equation

$$J_{\text{th}}^{b} = \frac{1}{\beta}\left(\alpha + \frac{1}{L}\ln\frac{1}{R}\right) \tag{12}$$

with $b \cong 1$ and α independent of L is an approximation which is useful mainly in comparing lasers over a rather narrow range of L values.

In good quality heterojunction lasers, the differential quantum efficiency typically decreases with increasing L. Because of the anomalies discussed in the previous section in certain lasers, care must be taken to ensure that the laser behavior for all L values is normal. Otherwise, of course, no conclusions can be drawn concerning the dependence of the laser parameters from plots of either J_{th} or η_{ext} as a function of L.

Experimentally, Eq. (12) was found to be approximately obeyed with $b \cong 1$ (within the experimental scatter) in diffused lasers(45,135), solution-grown lasers [homojunction(45) as well as SH-CC(37,82,136, 288,289)], and double heterojunction lasers(128b) between 77° and 300°K. The results of Goodwin and Selway(84) in SH-CC lasers, however, showed marked deviation. Thus, in many experimental situations, α and β are useful quantities in comparing lasers made by various processes of roughly comparable L. Table 13 lists the reported values for these quantities in various lasers. The data of Susaki et al. (137) indicate an exceptionally wide range of α values depending on the laser cavity length. Such a wide spread apparently has not been found by others.

A noteworthy feature is that the gain constant β is found to increase with the Zn concentration (53), and that β is frequently lower for simple diffused than for LPE lasers of both types. The difference is attributed to poorer electron and optical confinement to the active region in the diffused lasers when a p^+p barrier is not present (Section II)(45). There does not appear to be a large difference in β between the homojunction and SH-CC lasers. However, the value of α is substantially lower in the heterojunction lasers for the reasons discussed in Section V.

Because of the possibility of some light propagation in the n-type region, depending on the details of the fabrication process, it is possible that β may vary from lot to lot in SH-CC lasers, despite the fact that the absorption coefficient remains moderate. This will occur, for example, if the n-type region is lightly doped.

TABLE 13

Gain (β) and Loss (α) Values in GaAs Lasers

Temp. (°K)	Laser process	α (cm^{-1})	β (cm/A)	References
300° K	Diffused	20	0.57×10^{-3}	Pilkuhn and Rupprecht (45)
		~ 60	~ 2.80×10^{-3}	Susaki et al. (137)
	Homojunction LPE	90	3.80×10^{-3}	Pilkuhn and Rupprecht (45)
		60–200	1.5–6×10^{-3}	Susaki et al. (53,137)
		100	3.00×10^{-3}	Kressel and Nelson (36)
	Double hetero. LPE	15	~ 3.00×10^{-3}	Alferov et al. (128b)
	CC LPE	~ 14	~ 3.00×10^{-3}	Quinn and Morton (136)
		20	3.00×10^{-3}	Kressel and Nelson (36)
		21–48	3–4.4×10^{-3}	Kressel et al. (82)
		30–40	4–6.0×10^{-3}	Hayashi and Panish (39)
77° K	Diffused	15	2.5×10^{-2}	Pilkuhn and Rupprecht (45)
		30–100	0.14	Susaki et al. (137)
	Homojunction LPE	14	4×10^{-2}	Pilkuhn and Rupprecht (45)
		8–100	10^{-2}–0.2	Susaki et al. (137)
		10	5×10^{-2}	Kressel et al. (82)
	Double hetero. LPE	5	2×10^{-2}	Alferov et al. (128b)
	CC LPE	8–12	3.2–5×10^{-2}	Kressel et al. (82)
		10	3.0–5×10^{-2}	Hayashi and Panish (39)

The threshold values obtained with double heterojunction lasers [substantially lower than those reported in Ref.(*128b*)], are due to very high values of β resulting from near-perfect optical and electron confinement to the very thin, highly doped active regions(*279*). Based on the observed values of differential quantum efficiency (which are comparable to those in SH-CC lasers), there does not appear to be a large difference in the internal absorption coefficient between the two types of devices, except for variations in free carrier absorption in the active region which can vary depending on the fabrication process.

A comparison has been made(*279*) of the observed temperature dependence of J_{th} of double heterojunction lasers with the theoretical calculations of Stern(*290*) using the Halperin and Lax(*26*) bandtail approximation rather than the simpler Gaussian model used in Ref.(*25*).

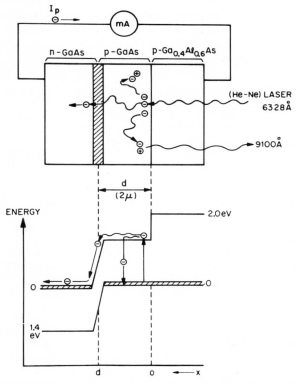

Fig. 47. Method for measuring the diffusion length of electrons in the active region of a SH-CC laser. The measured values must be corrected for the drift field owing to the Zn gradient(*39*).

Figure 40(a) shows a comparison of the theoretical and experimental J_{th} variation between 80 and 300° K assuming perfect electron and optical confinement and unity internal quantum efficiency at all temperatures. The agreement is quite good if it is assumed that the active region is very highly doped to a level in the 10^{19} cm^{-3} range. Such high doping is reasonable since this region is simultaneously doped during growth with Si and Zn, and the spontaneous emission spectra are consistent with high doping(279). Stern's revised theory predicts that the exponent b in Eq. (12) should be about 2. This has not been experimentally tested in the set of devices discussed above.

An important parameter in determining the role of electron confinement in a given active region width d is the electron diffusion length. A technique used for measuring this diffusion length in a SH-CC laser is described in Fig. 47. The light is incident on the p^+p interface and carriers which diffuse across the active region then recombine at the p–n junction giving rise to a current which is measured. In this particular measurement scheme, care must be taken to correct for the built-in electron drift field due to the Zn gradient. Diffusion length values of 3–6 μ were estimated by Hayashi and Panish(39). This value will depend on the doping level, and other investigators(38), using different measurement techniques, found that the diffusion length in the junction region of their devices is between 1 and 2 μ.

XI. Transient Phenomena

A. Delays

Delays (up to 100 nsec) between the beginning of current flow and of stimulated light emission have been observed by numerous investigators in diffused GaAs lasers. Delays observed in conventional LPE lasers have been found to be inconsequential (a few nsec at the most). The origin of the delays has been the subject of rather extensive research. The magnitude of the delays depends on the temperature of operation, the details of the fabrication process, and the drive conditions. Connected with these transient phenomena is the discovery of Q-switching effects in certain diffused laser structures, which will be described later.

Present theories of the delays assume that deep traps due to unknown imperfections are present in the active region. The models differ, however, in significant details.

1. Traps (density of about 10^{19} cm^{-3}) are postulated by Konnerth (138) which must be filled by injected carriers before population

inversion is achieved. Dobson et al.(*139*) have associated the traps with donor states tied to the ⟨100⟩ conduction band minima in GaAs.

2. Fenner(*140*) postulates that the traps initially *absorb* the light internally emitted, thus preventing lasing. The traps may be filled *either* by injected carriers *or* optically by photon absorption. When the traps are filled, their absorption coefficient drops and lasing occurs.

3. Starting from the saturable absorber model of Fenner, Dyment and Ripper(*141*) postulated a double acceptor model to account for the temperature dependence on the delay process. In this hypothesis, the trap can exist in three different states (Fig. 48). The first trap level is assumed to be located close to the valence band. It captures its first electron and goes into state Tr_2. The probability of the trap being in state Tr_2 depends on the position of the Fermi level and, therefore, is temperature dependent. This feature is required to explain the temperature dependence of the delay phenomena. By capturing a second electron, the trap transfers to state Tr_3. In state Tr_2, the trap is the saturable absorber postulated by Fenner since it is capable of absorbing photons of very nearly the lasing energy. When a sufficient number of traps are in the nonabsorbing state Tr_3, normal lasing occurs. A detailed mathematical treatment based on these traps has been presented by the above authors which accounts qualitatively for the observed phenomena in their diffused GaAs lasers, in particular the fact that the delays are longer at higher than at lower temperatures(*142,143*).

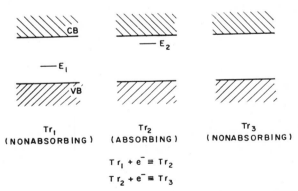

Fig. 48. Representation of the energy levels available for the capture of electrons in the states Tr_1, Tr_2, and Tr_3 of a single hypothetical trapping center(*141*).

It is noteworthy that the basic concept of saturable absorbers (models 2 and 3) has been experimentally demonstrated in Ga(AsP) lasers where long delays have been observed by Pankove(*144*). In these experiments, it was shown that the lasing delays could be eliminated by optically pumping the Ga(AsP) laser ($\lambda = 6440$ Å) with a lower photon energy (8500 Å) GaAs laser source, indicating that the traps could indeed be optically filled. It was not, however, possible to determine the ionization energy of the traps. A density of greater than 10^{17} cm^{-3} was indicated.

Bistable CW laser operation of diffused devices which exhibit delays was discussed in Ref.(*154*).

B. Q-SWITCHING

Lasing delays are clearly undesirable. However, some devices exhibiting delays have also exhibited interesting Q-switching effects(*163*). Thus, GaAs lasers may be sources of very short pulses, a potentially valuable feature. In these diffused devices the light burst is observed at the termination of the current pulse and the width is under 200 psec (*143*). The Q switching is observed only in a narrow operating range of both temperature and current, and has been explained by the double acceptor model outlined above(*143*). Briefly, Q-switching will occur under conditions such that lasing does not occur during the current pulse because the optical loss is too large. At the end of the current pulse, the traps transfer from the absorbing state Tr_2 to the nonabsorbing state Tr_1. Thus the internal optical loss rapidly decreases, and the remaining electrons in the conduction band recombine with holes in the valence band with the result that a narrow burst of stimulated light is emitted. Some SH-CC lasers also have exhibited Q-switching effects (*134,287*); whether the origin is the same as in diffused lasers is still uncertain.

C. OSCILLATION IN LASER OUTPUT

High frequency oscillation of the emitted light has been observed in some lasers while the current is constant. These pulses are of subnanosecond duration with repetition rates reaching several GHz. These phenomena are, in some lasers, owing to nonuniform injection across the $p-n$ junction and a detailed mathematical treatment has been presented by Basov et al.(*149a*). These authors experimentally demonstrated that the pulse repetition rate could be controlled in a structure consisting of two lasers isolated electrically but optically coupled by being placed in a common Fabry–Perot cavity. A similar structure was studied also by Lee and Roldan(*149b*).

D'Asaro et al.(*150*) reported microwave oscillations (0.5–3 GHz) in both the current and the light output in CW-operated stripe geometry lasers. The frequency of oscillation increased with drive current. The laser oscillations have been explained by multimode interactions in the laser cavity(151). The control of these oscillations is significant because it is possible to obtain 100% self-modulation of the light output of CW lasers by suitable control of the device operating conditions (151). Using an exterior laser cavity, Broom(152) obtained laser modulation in the frequency range of 0.6–2.2 GHz with a line width of 200 kHz.

By using the experimental configuration shown in Fig. 49, Paoli and Ripper(153) were able to lock into a desired oscillation frequency. Figure 49B shows the stabilized pulsations of the laser using a feedback loop provided by coupling the laser current leads to a microwave amplifier through a circulator. In this case locking was obtained on the second harmonic of the pulse rate with an amplifier gain of 7.5 dB. The line width of the pulse rate was reduced from its self-induced value of 600 kHz to less than 30 kHz and the pulse width was decreased from approximately 380 psec to less than the resolution limit of the detection system (180 psec). The frequency pulling evident in the figure occurs because the feedback loop provides a signal in phase with the self-induced modulation of the population inversion only at discrete frequencies separated by the inverse of the loop delay.

Other work dealing with laser oscillations will be found in Refs. (155–158). Work dealing with the control of excess noise in lasers was reported by Guekos and Strutt(297).

The theory of these oscillation phenomena involves consideration of the laser "rate equations" which describe the dynamic behavior of the carriers and photons in the active region(159). Theoretical studies of noise in semiconductor lasers based on these equations were made by Haug(160,161) and Haug and Haken(162).

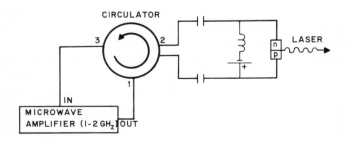

Fig. 49A. Experimentally observed frequency spectrum of the fundamental rate of pulsations in the output intensity from a stripe-geometry GaAs junction laser operating continuously at a current level 1.09 times threshold at 93°K. Schematic.

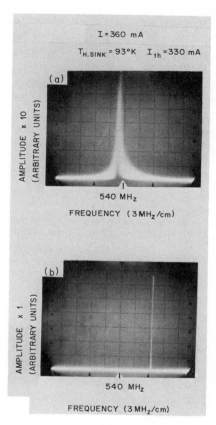

Fig. 49B. In (a), the pulsations were self-induced at a repetition rate of 539 MHz with a spectral width of 600 kHz. In (b), the pulsations were stabilized by the regenerative feedback of the injection current oscillations which occur simultaneously with the self-induced optical pulses. From Ref. (*153*).

It is noteworthy that spiking of the optical output is sometimes observed in homojunction LPE lasers the origin of which has not been explained. In particular, it was reported (*107*) that pulsation is seen in some lasers following noncatastrophic degradation. It is possible that these phenomena are associated with nonuniform injection due to changes which have occurred in the device during operation.

D. LASER MODULATION

It is possible to modulate the laser output at very high frequencies because the lifetime of the carriers is so short — 2 nsec or less (*145,146*).

Modulation up to 47 GHz was obtained by current modulation by Takamiya and Nishizawa(147). The estimate of Nishizawa(146) was that the lifetime decreases with injection, reaching values of the order of 10^{-11} sec at very high injection levels.

Modulation schemes, involving the use of ultrasonics, have been described which permit modulation to 4–5 GHz (the theoretical limit with this technique)(148).

A review of laser modulation techniques and results obtained will be found in Ref. (164a). This subject is, of course, of greatest relevance to the use of lasers in communication systems.

XII. Visible and Near-IR Lasers

Mixed crystals of Ga(AsP) and (AlGa)As yield diode lasers covering the wavelength range from 9000 to 6280 Å. The transition from a direct

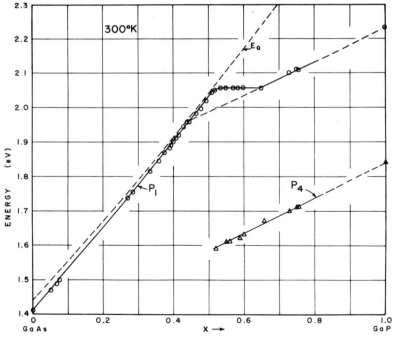

Fig. 50. Energy of the $GaAs_{1-x}P_x$ diode emission maxima at 300°K as a function of composition. In the composition range $x < 0.45$ and $x \geqslant 0.65$, the high-energy peak, P_1, is located within ~ 20 meV of the bandedge (dashed curves). In the composition range $0.45 < x < 0.65$, P_1 moves above the indirect bandedge. The low-energy emission peak, P_4, which becomes appreciable in intensity only for $x > 0.45$, moves parallel to the indirect bandedge(12).

to an indirect bandgap occurs at ~ 1.92 eV in (AlGa)As (Fig. 3) and 1.96 eV for Ga(AsP) at 300°K (Fig. 50). A major technologically important difference between the alloy systems is the fact that the lattice constants of AlAs and GaAs are nearly the same, $a_0 = 5.6605$ Å vs. $a_0 = 5.6533$ Å, while in GaP the $a_0 = 5.4506$ Å. Since the alloys are grown epitaxially on GaAs substrates, the lattice strain in a Ga(AsP) layer is greater than in an equal bandgap energy (AlGa)As layer. Because a significant lattice constant mismatch introduces misfit dislocations into the epitaxial layer(117,164b), the CC structure, which incorporates a heterojunction in close proximity to the lasing region, is not practical for Ga(AsP) lasers. As a result, it is experimentally found that (AlGa)As lasers are superior to those of Ga(AsP), especially at 300°K.

The diode lasing photon energy as a function of P composition is shown in Fig. 51 for Ga(AsP). A similar plot for (AlGa)As diodes is shown in Fig. 52 as a function of Al content. In both alloy systems, the lasing photon energy is approximately 0.05 eV below the bandgap energy, as is the case in GaAs.

A. GaAs$_{1-x}$P$_x$ Lasers

Laser action was first reported in Ga(AsP) by Holonyak and Bevacqua(165a). The best Ga(AsP) lasers were prepared by two-step vapor-phase epitaxy on GaAs substrates(48,165b). The p^+, compensated p, and n regions were sequentially deposited. As shown in Fig. 53, lasing was obtained at 6750 Å at 300°K but with an extremely high J_{th} of 900,000 A/cm^2 (40.5% P). At 77°K lasing was obtained at 6350 Å. These lasers are superior to previous Ga(AsP) lasers prepared by diffusion(166–169), particularly since these epitaxial lasers were the only ones to lase at 300°K over a wide compositional range. This is probably due to improved waveguiding in the epitaxial structure (see Section V) and better junction metallurgy. Peak power outputs to 10 W were obtained by Pankove et al.(165b) at 300°K ($\lambda \cong 7200$ Å).

Ga(AsP) has been prepared also by solution growth(170) but no lasers have been reported with such material.

B. Al$_x$Ga$_{1-x}$As Lasers

The best (AlGa)As lasers have been made by LPE and incorporate one heterojunction(13,171) or two heterojunctions(279). Figure 53 shows the dependence of J_{th} on the lasing wavelength at 300°K and 77°K of (AlGa)As lasers prepared by LPE with(13,171) and without

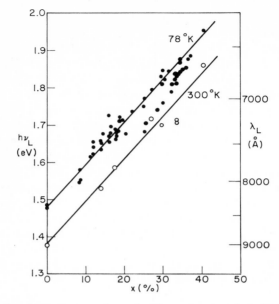

Fig. 51. Lasing photon energy and wavelength as a function of x in $GaAs_{1-x}P_x$ (48).

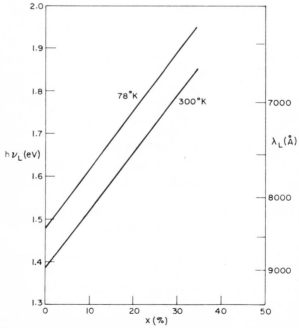

Fig. 52. Lasing photon energy and wavelength in $Al_xGa_{1-x}As$ (171).

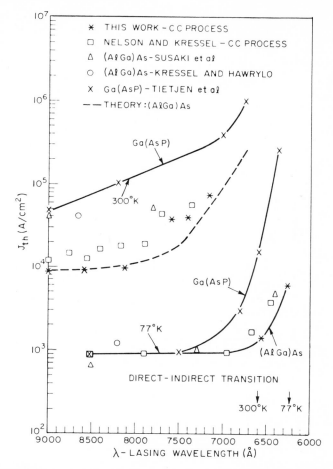

Fig. 53. Threshold current density J_{th} as a function of the lasing wavelength λ_L at 300° and 77° K. All of the diodes have roughly comparable cavity lengths (about 10–15 mils). Present laser data are compared to previous data by Kressel and Hawrylo(43), Susaki et al.(172), Tietjen et al.(48), and Nelson and Kressel(13). The theoretical curve was calculated as described in the text. Not shown on the figure are the data points of Rupprecht et al.(173) for a laser emitting at ~ 8000 Å with $J_{th} > 10^5$ A/cm² ($T = 273°$ K). The plot is from Ref.(171).

(43) the single heterojunction and by diffusion into an LPE-*n*-type region(172). Lasing only up to 273° K was reported by Rupprecht et al.(173) in LPE lasers grown by a technique in which the junction is formed during the growth by the addition of Zn to a Te-doped melt. The performance of these devices probably was limited by the doping profile obtained by this fabrication technique.

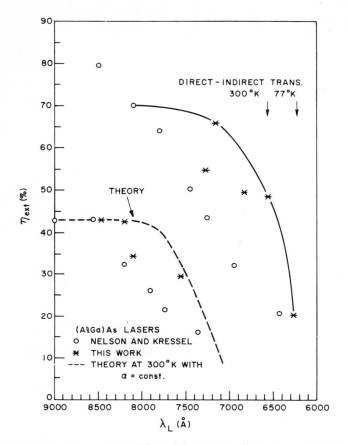

Fig. 54. Variation of the external differential quantum efficiency η_{ext} with lasing wavelength λ_L at 300° and 77° K of (AlGa)As lasers with the close-confinement structure. The theoretical curve was calculated as described in the text, based on the η_{int} variation(*171*).

The exterior differential quantum efficiency as a function of λ_L in SH-CC lasers is shown in Fig. 54 at 300° and 77° K.

The observed laser performance was compared by Kressel et al. (*171*) with the theoretically predicted one assuming that the limiting process is the thermal depopulation of the direct [000] conduction band minimum by electron transfer to the 6⟨100⟩ minima (i.e., the decrease in the internal quantum efficiency). Figure 55 shows the variation of the internal quantum efficiency as estimated from exterior efficiency measurements of spontaneous diodes. Using the computer solution of Maruska and Pankove(*174*), and assuming that the direct-to-indirect

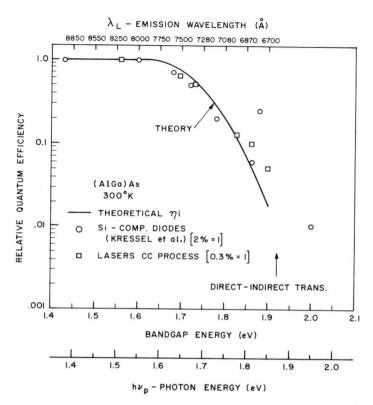

Fig. 55. Variation of the exterior quantum efficiency (spontaneous emission) as a function of E_g, λ_L, and $h\nu$. The solid line is the theoretical variation of the internal quantum efficiency (see text). The Si-compensated diode data are from H. Kressel, F. Z. Hawrylo and N. Almeleh, *J. Appl. Phys.*, **40**, 2248 (1969). The plot is from Ref.(*171*).

transition occurs at 1.92 eV with $m_{e_2} \cong 0.35 m_0$, the experimental efficiency data can be fitted quite well, as shown in Fig. 55.

Considering only the change of λ and η_{int}, and neglecting any changes of α with increasing Al content, the 300°K theoretical J_{th} curve is shown in Fig. 53, where the curve is fitted to the lowest experimental J_{th} value. The experimental J_{th} variation follows the predicted behavior quite well, with the deviations partly due to metallurgical factors. Figure 54 shows the theoretically predicted variation in η_{ext} at 300°K based on Fig. 55 with constant L and α but varying η_{int}. The good experimental agreement which is observed with the simple theory may be partly fortuitous because above the lasing threshold the lifetime for carriers in the $\bar{k} = 0$ minimum decreases with increasing stimulation, thus favoring radiative recombination. Hence, the "effective" internal quantum efficiency above threshold may in fact be higher than the value observed below threshold.

Figure 56 shows the J_{th} temperature dependence of lasers with increasing Al content. At the Al composition very near the direct-to-indirect bandgap transition, lasing is not possible above 100°K at 6600 Å (*171*).

Although (AlGa)As has been also grown by vapor-phase epitaxy (*175*), lasers so prepared have not yet been reported.

A key factor affecting the laser power efficiency is the diode series resistance (Section VI). With increasing Al content, the difficulties of achieving low series resistance contacts may be expected to increase. The data of Ref. (*171*) show that the diode resistance (in the form $R_s A_d$) at 300°K and 77°K of λ_L is essentially constant to a 300°K

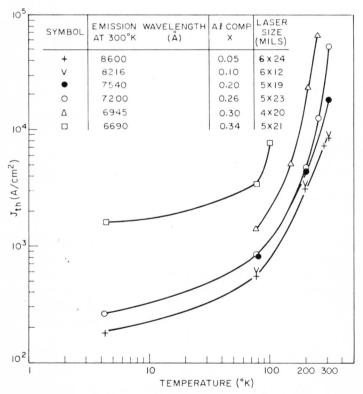

SYMBOL	EMISSION WAVELENGTH AT 300°K (Å)	Al COMP X	LASER SIZE (MILS)
+	8600	0.05	6 x 24
v	8216	0.10	6 x 12
●	7540	0.20	5 x 19
○	7200	0.26	5 x 23
△	6945	0.30	4 x 20
□	6690	0.34	5 x 21

Fig. 56. Temperature dependence of J_{th} in SH-CC lasers with different Al composition in $Al_x Ga_{1-x} As$ (*171*).

emission wavelength of $\sim 8000 \, \text{Å}$ ($R_s A_d \approx 5 \times 10^{-5} \, \text{ohm cm}^2$) with the greatest scatter near the direct–indirect crossover. Improved technology should reduce this resistance in the future.

Double heterojunction (AlGa)As lasers have been described(279) in which the thin ($< 1 \, \mu$) active region was doped with Zn and Si. By varying the (AlGa)As composition in this active region, the lasing wavelength can be shifted. Except for metallurgical difficulties, the relative performance of such lasers should vary with wavelength as shown in Figs. 53 and 54, but with lower threshold current densities if the d values is sufficiently small. For example, at $\lambda_L \cong 7400 \, \text{Å}$, J_{th} values at $300° \text{K}$ of 3000–4000 A/cm^2 have been obtained(281) instead of about 30,000 A/cm^2 with the SH-CC structure (Fig. 53).

XIII. Infrared Lasers

A. III–V COMPOUNDS

Lasers were reported shortly after GaAs in a series of narrower bandgap III–V compounds: InAs(176,177,291), GaSb(178a,179), InSb(180), InP(181), (GaIn)As(182), In(AsP)(183,292). Table 14 lists all materials in which diode laser action has been reported.

The technology of these IR lasers has received relatively little attention and the devices so far fabricated do not, therefore, necessarily represent the ultimate performance. As was evident in the progress of GaAs lasers, the materials technology required to obtain high performance laser diodes is extremely sophisticated. A review of III–V compound preparation was presented by Bailey(184).

A comprehensive study of diffused InAs lasers ($h\nu \approx 0.39 \, \text{eV}$) including the effect of magnetic fields (which reduce J_{th}) was made by Melngailis and Rediker(185). The maximum temperature at which lasing was observed was $150° \text{K}$, but the limitation may be technological. Indications are that the internal quantum efficiency at $77° \text{K}$ is lower than in GaAs at this time. Maslennikova et al.(186) estimated that $\eta_{int} = 3\%$ with $\alpha = 17 \, \text{cm}^{-1}$ and $\beta = 10^{-2} \, \text{cm/A}$. These devices were made by diffusing Cd. Self-modulation effects similar to those in GaAs lasers also were observed(187). For other work on InAs lasers, see Refs. (188) and (189).

A detailed study of GaSb lasers was published by Chipaux and Eymard(190) who obtained CW operation at $5° \text{K}$. Lasing was observed at 0.8, 0.775, and 0.79 eV at $77° \text{K}$. Other studies of GaSb lasers were made more recently(191,192,178b).

TABLE 14

Compounds in Which Laser Diodes Have Been Reported
(See text for references)

Compound	Approximate wavelength (μ)
(AlGa)As	0.628–0.90[a]
(GaAs)P	0.64–0.90[a]
GaAs	0.85–0.90[a]
InP	0.9[b]
Ga(AsSb)	0.9–1.5
GaSb	1.5
(InGa)As; (InAs)P[b]	0.9–3.1
InAs	3.1
In(AsSb)	3.1–5.4
PbS	4.3
InSb	5.4
PbTe	6.5
PbSe	8.5
Pb(SnTe)	6.5–28
Pb(SSe)	4.3–8.5
Pb(SnSe)	8.5→

[a] Includes temperature tuning between 77° K and 300° K.

[b] Lasing at 300° K has not been reported but should be possible with suitable material and diode technology. This would permit temperature tuning to ~ 0.94 μ.

The properties of lasers in InP (*193*), In(AsSb) (*194*), InSb (*195*), and In(AsP) (*196*) also were reported.

B. INFRARED LASERS IN OTHER COMPOUNDS

Diode lasers have been fabricated in PbSe (*197*), PbTe (*198*), PbS (*199*), PbSnTe (*200,201*), PbSSe (*202*), and PbSnSe (*203,204*). The longest wavelength laser to date ($\lambda = 28 \mu$) was obtained in PbSnTe operating at 12° K ($Pb_{0.13}Sn_{0.27}Te$) (*201*). Table 15 shows the operating characteristics of lasers made with varying Pb and Sn concentrations. A review of the materials preparation and operating characteristics of lasers in PbSnTe and PbSnSe has been published by Melngailis (*205*) who suggested that the ultimate wavelength limit of diode lasers has yet to be reached.

An interesting recent development is optical heterodyning between a CO_2 gas laser and a PbSnTe laser emitting at 10.6 μ (*206*). This is the first time that the coherence properties of an injection laser have been used directly, in a specific application. (In typical applications, the

TABLE 15

Operating Characteristics of $Pb_{1-x}Sn_xTe$ Diode
Lasers (201)

x	Emission wavelength (μ)		Minimum threshold current density (A/cm²)	
	12° K	77° K	12° K	77° K
0.15	11.7	9.5	250	3,000
0.17	13.0	9.9	55	10,000
0.19	14.5	11.0	130	9,000
0.20	15.1	11.2	275	12,500
0.21	16.8	12.0	175	7,000
0.22	17.2	—	175	>30,000
0.24	20.0	—	230	>30,000
0.27	28.0	—	125	>30,000

directionality of the diode laser beam rather than its coherence is of interest.)

XIV. Electron Beam Pumped Lasers

In principle, electron beam pumping (EBP) is the simplest technique for observing stimulated emission in the broadest possible range of compounds. In practice, the technique is limited by the need for rather elaborate electronics and costly, as well as bulky, power supplies if high output powers are desired. Furthermore, the question of sample degradation has been troublesome in some cases.

The sample size is generally on the order of 1 mm or less which is commensurate with the typical diameter of the electron beam. Three types of laser cavities are schematically shown in Figs. 57 and 58. Figure 57a shows the configuration where the Fabry–Perot cavity is perpendicular to the beam and only the surface region of the sample is excited. In the configuration shown in Fig. 57b, the cavity is parallel to the beam. In the third configuration (Fig. 58), described by Nicoll (207), lasing occurs in a crystal of rectangular cross section by total internal reflection. This configuration is useful in a material like CdS where the internal absorption of the light emitted is relatively small. The coherent light is emitted as a 360° disklike beam centered on the crystal with a divergence of about 5° perpendicular to the disk. In this

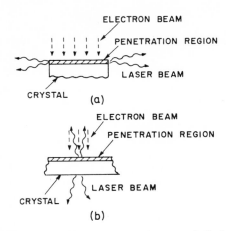

Fig. 57. Two lasing schemes in EBP lasers. In scheme (a) the lasing is transverse to the direction of the electron beam, the optical confinement being due to the high plasma density near the sample surface. In scheme (b) the electron beam also penetrates only a relatively shallow distance into the sample, but the Fabry–Perot cavity is paralleled to the electron beam. The optical losses are generally higher in scheme (b) than in scheme (a).

mode of operation lasing occurs at the lowest threshold current density. The second-lowest threshold is obtained with the configuration of Fig. 57a and the highest with that of Fig. 57b where the internal absorption is generally highest. The width of the recombination region increases with electron energy, with the upper limit set by radiation damage ($\gtrsim 290 \, \text{kV}$).

Table 16 lists all materials in which EBP lasers (but not $p–n$ junction lasers) have been reported.

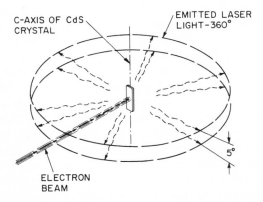

Fig. 58. Total internal reflection mode of EBP laser operation(*207*).

TABLE 16

Materials in Which Stimulated Emission Was Obtained Only by Either
Electron-Beam (EBP) or Optical Pumping (OP)

Material	Approximate lasing wavelength (μ)	Photon energy (eV)	Method of excitation[a]
ZnS	0.33	3.82	EBP
ZnO	0.37	3.30	EBP
ZnS-CdS	0.50–0.32	2.5–3.82	OP
ZnSe	0.46	2.7	EBP
CdS	0.5	2.50	EBP, OP
ZnTe	0.58	2.14	EBP
GaSe	0.59	2.09	EBP
CdS_xSe_{1-x}	0.69	1.8–2.5	EBP
CdSe	0.68–0.69	1.8	EBP, OP
CdTe	0.78	1.58	EBP
$CdSnP_2$	1.01	1.24	EBP
Cd_3P_2	2.1	0.59	OP
Te	0.36	0.34	EBP
$Cd_xHg_{1-x}Te$	3.8–4.1	0.3–0.33	OP

[a]See Section XIV (EBP), or Section XV (OP) for references.

The maximum theoretical power conversion efficiency has been
calculated by Klein(208). For example, in CdS it is ~27% while for
GaAs it is 25%.

A great deal of research dealing with EBP II-VI compounds has
been reported in the literature and a comprehensive review of this
work up to 1966 will be found in a paper by Reynolds(209). The highest power efficiency in an EBP laser was reported by Hurwitz(210) in
CdS at 110°K—26.5% external (35% internal) power efficiency at
4900 Å with the emission of 350 W. Ultraviolet lasers (λ = 3750 Å)
were reported by Nicoll(211) using ZnO operating at 77°K, and by
Hurwitz(212) in ZnS (3200 Å) at 4.2°K. Room temperature lasing in
the green was first obtained using CdS crystals by Nicoll(213) who
also demonstrated a sealed off cathode ray tube suitable for room temperature lasing of these crystals(214). Brewster(215) has reported
the highest peak power (200 kW) output at 300°K with CdS samples,
where no cavity was formed, in a system operating at 1400 A/cm².
The optical output of this system is useful in flash photography since
pulses of 3 nsec duration have been obtained. Single-mode operation
was reported at 300°K by Packard et al.(216) in CdS. Auger processes
in CdS EBP lasers were studied by Benoit a la Guillaume et al.(217).

The EBP lasers in ZnSe(218), ZnTe(219), Ga(AsP)(220), CdTe (221–223), Te(224), InAs(225), and GaSe(226) also have been reported. It is of interest to note that the first report of lasing at 77°K in a new class of compounds ($A^{II}B^{IV}C_2^{V}$) is by Berkovskii et al.(227) who studied CdSnP$_2$. The emission wavelength was $1.0139\,\mu$. The results of recent Russian EBP laser experiments in CdSe, PbS, PbTe, PbSe, and Pb$_x$Se$_{1-x}$ have been described by Kourbatov et al.(228).

The EBP GaAs lasers have been extensively studied. Cusano(229) studied the laser transitions in n- and p-type material. (Similar studies were later made using optical pumping, see Section XV.) Bogdankevich et al.(230,231) obtained the highest peak power to date — 100–200 W at 300°K with an efficiency of 11%, and 300–400 W at 77°K using n-type GaAs (about 2×10^{18} cm^{-3} was found to be the optimum doping level). Lavine and Adams(232) estimated that the internal quantum efficiency is about 0.9 at 77°K, a value in agreement with estimates reached from studies of injection lasers. The effect of inhomogeneous carrier distribution has been studied(233). The effect of dislocations on the homogeneity of the EBP emission was studied by Lavine et al. (234) who found (as in junction lasers, Section IX) that dislocation-free GaAs is not sufficient for homogeneous emission. Optical losses in EBP lasers were studied by Hunsperger(235).

At this time, practical applications of EBP lasers appear to be fairly remote, with the possible exception of specialized applications like flash photography. Nevertheless, electron-beam excitation is a power-ful experimental technique for studying a broad range of compounds. The low-electron-beam (< 25 kV) energy technique of Nicoll(236a) is particularly useful in this respect because of the relative simplicity of the equipment required. Figure 59 shows the experimental setup used by him. The electron gun is the same as that used in a 5TP4 tube. A special thoria cathode is used for demountable tube operation and a magnetic focusing coil is added for focusing lower-voltage electron beams. Deflection of the beam is obtained by tilting the whole gun structure with a micrometer by using a ball joint construction.

XV. Optical Pumping

Population inversion by optical excitation is similar to that by electron-beam excitation, with the additional simplicity that some materials can be pumped using injection lasers made from materials with higher bandgap energies. For example, Cd$_x$Hg$_{1-x}$Te(236b), InSb(237), and InAs(238) were pumped with GaAs lasers in which

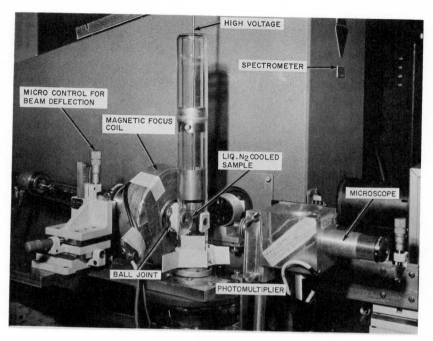

Fig. 59. Electron beam apparatus for laser studies. Courtesy F. H. Nicoll.

case the absorption occurs near the sample surface. Basov et al. (*239*) used a Raman shifted ruby laser to obtain the highest laser power reported to date from a semiconductor — 30,000 W of peak power (GaAs). Alternatively, more homogeneous excitation is possible using a pump source with a photon energy much smaller than the bandgap energy of the material of interest, and relying on two photon processes to generate free carriers. The latter scheme was used by Basov et al. (*240*) to pump GaAs using a Nd-doped glass laser. Konyukhov et al. (*241*) similarly used a ruby laser to pump CdS, and Brodin et al. (*242*) to excite $Zn_xCd_{1-x}S$.

Optical pumping is useful for investigations of laser transitions and time-dependent optical properties. The GaAs can be conveniently excited using either Ga(AsP) or (AlGa)As injection lasers. The sample is placed in close proximity to the diode laser and the whole assembly can be operated in a dewar. Furthermore, the short rise and fall times of the diode emission make it possible to study rapid rise and decay optical processes occurring in the sample under study (*243*). Figure 60 shows the pumping technique used by Rossi et al. (*59*). Figure 61 shows

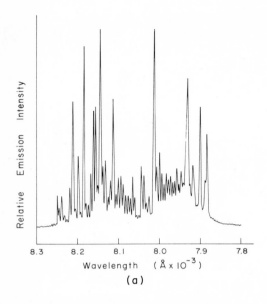

(a)

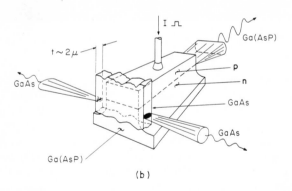

(b)

Fig. 60. Laser emission spectrum of a heavily doped ($\sim 10^{19}$) *n*-type GaAs platelet operated as illustrated in (b). The mode spacing of about 5 Å corresponds to a resonator cavity length from platelet edge to edge of 125 μ. The Te-doped GaAs was grown from Ga solution(*244*). Pump: Ga(AsP), 7200 Å; T $\approx$ 77° K.

the dependence of the laser photon energy on GaAs material doping (*244*). The flexibility possible with this excitation technique has been used to great advantage in measuring, in a direct manner, the shift in the quasi-Fermi level in the lasing GaAs material with increasing excitation(*293*). Some anomalies in the modes of OP lasers are dis-

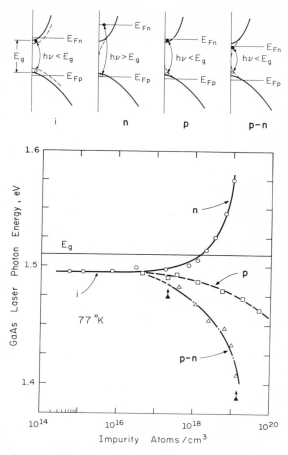

Fig. 61. Dependence of the laser photon energy on impurity concentration in an optically pumped GaAs laser. The *n*-type samples are Se, Sn, or Te doped; the *p*-type samples are Cd or Zn doped. The compensated material is Zn–Sn, Zn–Te, or Zn–Se doped. The laser photon energy becomes asymptotic to $h\nu \cong E_g - 0.015$ eV at low-impurity concentrations. The top diagrams illustrate the recombination processes appropriate to the various curves(*244*).

cussed in Ref.(*245*). Lasing in Ge- and Si-doped GaAs via the acceptor levels introduced by these amphoteric dopants has been studied (*246,244*). In the Si-doped GaAs prepared by LPE, simultaneous lasing was observed(*244*) via the deep and shallow acceptors due to Si(*60*). Thin CdSe platelets also have been made to lase in a similar setup(*247,248*).

The most recent new compound in which stimulated emission has been reported is Cd_3P_2 emitting at 2.1 μ ($\sim$ 0.59 eV) (*249*), and (InGa)P is a candidate in the direct bandgap portion of the alloy system ($\lesssim$ 2.1 eV).

XVI. Avalanche-Pumped Lasers

Basov et al. (*250*) suggested in 1959 that the carriers needed to obtain population inversion in semiconductors (and hence, stimulated emission) could be generated by impact ionization. This could be done either by applying sufficient high electric fields to homogeneous samples, or by using back-biased *p–n* junctions. Theoretical studies were made of the conditions required to achieve lasing in InSb by this technique. Haacke (*251*) considered the possibility of using crossed electric and magnetic fields. The fact that the plasma tends to "pinch" in InSb complicates the issue because the electron-hole density becomes very high in the pinched region. As a result, population inversion may occur locally, leaving the rest of the sample highly absorbing. The effect of the pinch phenomenon on the potential lasing behavior was considered by Steel (*252*), Schmidt (*253*), and Vladimirov (*254*).

One of the drawbacks of using impact ionization as the method of carrier generation is that the carriers formed are "hot" and, as a result, it is difficult to obtain a degenerate population. Thus, stimulated emission will occur *after* the electric field has dropped and the carriers have reached thermal equilibrium. Alternatively, the carriers may be generated in a high-field region and recombine in a low-field one.

In practice, only in GaAs and InP has stimulated emission been observed as a result of carrier generation by impact ionization. Weiser and Woods (*255*) used a $p^+p^-p^+$ structure where the electric field in the p^- region can be made sufficiently high for impact ionization. The electrons generated are pushed by the electric field into one of the (low electric field) p^+ regions where they recombine with the possibility of stimulated emission.

In the case of materials where Gunn domains are formed, stimulated emission occurs after the electric field has been reduced in a portion of the sample. Laser action at 77°K in bulk GaAs ionized by the passage of Gunn domains was first reported by Southgate (*256*). Subsequently, lasing to 300°K was observed at current densities somewhat higher (*257*). The samples used were *n* type and shaped in the form of dumbbells where the "bridge" was 0.1 × 0.1 × 0.5 mm. Operation was possible only with short pulses (10 nsec) because of heating. The photon energy is centered at about 1.48 eV at 77°K. The measured average power conversion efficiency at 77°K was only about 10^{-4}, hence considerably smaller than in $p–n$ junction lasers. A peak power of 3 W was obtained. The low power efficiency is inherent in the device because of strong internal absorption of the emitted light. It is estimated that values on the order of a few percent are theoretically possible under optimum conditions. A detailed study of the temperature dependence of the lasing and its time-dependent properties was made by Southgate(257). Stimulated emission has been observed also in InP(258), at lower threshold currents than in GaAs for unknown reasons.

XVII. Miscellaneous Diode Laser Configurations

A. Laser Operation with External Cavity Coupling

In applications which require the transmission of the laser output over large distances, it is desirable to obtain the smallest possible beam spread. A scheme for external mode locking of laser arrays, which involves the use of mirrors and a single lens, was described by Crowe and Ahearn(259). From a single diode, 0.2 W in a beam $0.12° \times 15°$ was obtained. This constitutes, of course, only a small fraction of the light emitted from the laser which normally is in a beam of $15° \times 15°$ or $20° \times 20°$. Similarly, laser diode arrays can be phase locked(259). The utilization of such schemes for reducing the beam spread by external means depends on a consideration of the alternative which consists of simply using a large lens placed close to the laser to collect as much of the light as possible. Factors such as cost, bulk of the optics needed, and size of the laser array all enter into making a choice in specific applications.

Factors affecting the laser properties when operated in external cavities were studied by Bachert and Raab(260), Broom and Mohn (261), Basov et al.(262), and Edmonds(263). Microwave self-modulation has been discussed in detail by Broom et al.(294) when lasers are coupled to an exterior cavity.

B. Laser Amplifiers

The basic idea of laser amplifiers consists of using two lasers, the first as an oscillator and the second, optically coupled but electrically independent, as an amplifier which may be modulated as required without affecting the oscillator. The optical coupling can be done using lenses(264–267) or two lasers very close to each other. The latter scheme was described by Kosonocky and Cornely(*268*) who at 77°K obtained a gain of about 150 in the amplifier. Figure 62 shows the laser amplifier used by these authors.

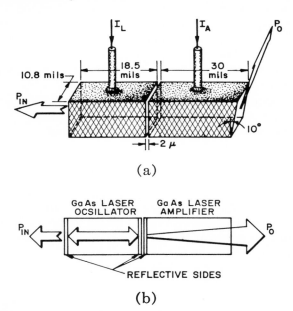

(a)

(b)

Fig. 62. Directly coupled laser amplifier. (a) Pictorial view, (b) schematic of the active laser regions. I_L is the oscillator current, and I_A is the amplifier current(*268*).

The theory of optically coupled lasers is treated in Refs.(*269,270*). The limitation in the optical amplification is the saturation of the gain above certain power levels, which is discussed in Refs.(*271–273*). The subject of laser amplifiers is of potential interest in computer applications(*274,275*). A nearly linear dependence of the gain in the amplifier on the current through it was found by Nishizawa and Tagusagawa(*28*) at 300°K and 77°K (below the saturation level) when using LPE homojunction laser structures for both the oscillator and the amplifier.

LIST OF SYMBOLS

A_d	diode area
CC	close confinement
d	width of active (lasing) region
D	reduced cavity width $= (2\pi d/\lambda_0)(n_2{}^2 - n_3{}^2)^{1/2}$
E	energy

EBP	electron beam pumped (laser)
E_g	bandgap energy
ΔE_g	bandgap energy discontinuity at heterojunction
$E_{g_1}, E_{g_2}, E_{g_3}$	bandgap energy in laser regions 1, 2, and 3 (ignoring bandtail "shrinking") (see Fig. 7)
F_c	quasi-Fermi level in conduction band
F_v	quasi-Fermi level in valence band
F_f	optical flux density (in W/cm of facet) at which catastrophic failure occurs
F_1, F_2	quantities defined in Eq. (10)
g	laser gain coefficient
I_{th}	threshold current
I	diode current
J	current density
J_f	current density in laser at which catastrophic failure occurs
J_m	operating current density for optimum power conversion efficiency
J_{th}	threshold current density for lasing
k	Boltzmann factor
K	thermal conductivity
l	distance between junction (i.e., heat source in laser) and heat sink
L, W	Fabry–Perot cavity length and width, respectively
L_m	cavity length for optimum power conversion efficiency
L_t	thermal diffusion length
LPE	liquid phase epitaxy
n_1, n_2, n_3	refractive index at the lasing photon energy in regions 1, 2, and 3, respectively (see Fig. 7)
n	electron density in the n-type material (region 3), also the donor density in the active region of SH-CC lasers
p	hole concentration in the active region
P_0	laser power output
p^+	hole concentration in region 1 adjoining the active region 2
PL	photoluminescence
R_s	diode series resistance
R	reflectivity of cleaved GaAs (0.32)
t	time
T	temperature
V	junction voltage
VPE	vapor phase epitaxy
x_m	laser geometrical factor used in computation of power conversion efficiency (see Section VI)

α_1, α_3	absorption coefficient at the lasing photon energy $h\nu_L$ in regions 1 and 3, respectively (see Fig. 7)
α_2	absorption coefficient in the active region 2, which must be negative for stimulated emission
α_{fc}	free-carrier absorption coefficient in the active region
α	averaged laser absorption coefficient
α'	$\alpha - \alpha_{fc}$
β	gain constant defined in Eq. (8a) ($g = \beta J^b$, where $b =$ constant)
δ	laser quality factor (see Section VI)
η	$(n_2 - n_1)/(n_2 - n_3)$; waveguide asymmetry factor
η_{ext}	laser differential external quantum efficiency
η_P	laser power conversion efficiency
η_{int}	internal quantum efficiency
θ	thermal resistance
κ	thermal diffusivity
λ	wavelength
λ_L	lasing wavelength
ν	frequency
$\Delta\nu$	radiative recombination linewidth
ρ	density of semiconductor; density of states in Eqs. (5) and (9)

ACKNOWLEDGMENTS

The author is grateful to H. Nelson, H. Lockwood, H. S. Sommers, Jr., L. R. Weisberg, D. Redfield, and J. I. Pankove for discussions of various aspects of the laser research described here; and to S. Dierk for technical assistance with the manuscript. Figure 32 was kindly provided by M. Coutts.

REFERENCES

1. R. N. Hall, G. E. Fenner, J. D. Kingsley, T. J. Soltys, and R. O. Carlson, *Phys. Rev. Letters*, **9**, 366 (1962).
2. M. I. Nathan, W. P. Dumke, G. Burns, F. H. Dill, Jr., and G. J. Lasher, *Appl. Phys. Letters*, **1**, 62 (1962).
3. T. M. Quist, R. H. Rediker, R. J. Keyes, W. E. Krag, B. Lax, A. L. McWhorter, and H. J. Zeiger, *Appl. Phys. Letters*, **1**, 91 (1962).
4. G. Burns and M. I. Nathan, *Proc. IEEE*, **52**, 770 (1964).
5. K. Unger, *Fortschr. Physik*, **13**, 701 (1965).
6. F. Stern, *Physics of III–V Compounds* (R. K. Willardson and A. C. Beer, eds.), Vol. 2, Academic Press, New York, 1966, Chap. 14.
7. H. Haken, *Festkörperprobleme IV* F. Sauter, ed., Vieweg, Braunschweig, 1965, p. 1.

8. M. I. Nathan, *Proc. IEEE*, **54**, 1276 (1966); *Appl. Optics*, **5**, 1514 (1966).
9. W. P. Dumke, *Advances in Lasers*, Vol. 2 (A. K. Levine, ed.), Marcel Dekker, New York, 1968,
10. M. H. Pilkuhn, *Phys. Status Solidi*, **25**, 9 (1968).
11. A. Yariv, *Quantum Electronics*, Wiley, New York, 1967.
12. A. H. Herzog, W. O. Groves, and M. G. Craford, *J. Appl. Phys.*, **40**, 1830 (1969).
13. H. Nelson and H. Kressel, *Appl. Phys. Letters*, **15**, 7 (1969).
14. H. C. Casey and M. B. Panish, *J. Appl. Phys.*, **40**, 4910 (1969).
15. J. J. Tietjen and L. R. Weisberg, *Appl. Phys. Letters*, **7**, 261 (1965).
16. G. J. Lasher and F. Stern, *Phys. Rev.*, **133**, A553 (1964).
17. W. P. Dumke, *Phys. Rev.*, **127**, 1559 (1962).
18. F. L. Vinetskii, V. S. Mashkevich, and G. Yu. Buryakovskii, *Soviet Phys.-Semicon.*, **1**, 42 (1967).
19. V. S. Mashkevich, *Soviet Phys.-Semicond.*, **1**, 48 (1967); M. J. Adams and P. T. Landsberg, *Proc. International Conference on Phys. Semiconductors, Moscow, 1968*, p. 619.
20. M. G. A. Bernard and G. Duraffourg, *Phys. Status Solidi*, **1**, 699 (1961).
21. W. P. Dumke, *Optical Masers*, Polytechnic Press, Brooklyn, New York, 1963, p. 461.
22. J. R. Biard, W. N. Carr, and B. S. Reed, *Trans. AIME*, **230**, 286 (1964).
23. M. H. Pilkuhn, *Proc. International Conference on Phys. Semiconductors, Moscow, 1968*, p. 523.
24. G. J. Lasher, *IBM J. Res. Develop.*, **7**, 58 (1963).
25. F. Stern, *Phys. Rev.*, **148**, 186 (1966).
26. B. J. Halperin and M. Lax, *Phys. Rev.*, **148**, 722 (1966).
27. H. C. Hwang, *Phys. Rev. B (Solid State)*, **1**, 4117 (1970).
28. J. Nishizawa and M. Tagusagawa, private communication, 1969.
29. G. C. Dousmanis, H. Nelson, and D. L. Staebler, *Appl. Phys. Letters*, **5**, 174 (1964).
30. N. N. Winogradoff and H. K. Kessler, *Solid State Commun.* **2**, 119 (1964).
31. G. E. Pikus, *Soviet Phys.-Solid State*, **7**, 2854 (1966).
32. H. Sato, *Japan. J. Appl. Phys.*, **7**, 409 (1967).
33. M. J. Adams, *Solid State Electron.*, **12**, 661 (1969).
34. A. N. Chakravarti, S. N. Biswas, and S. Rakshit, *Intern. J. Electron.*, **26**, 95 (1969) and references therein.
35. G. J. Burrell, T. S. Moss, and A. Hetherington, *Solid-State Electron.*, **12**, 787 (1969).
36. H. Kressel and H. Nelson, *RCA Rev.*, **30**, 106 (1969).
37. I. Hayashi, M. B. Panish, and P. Foy, *IEEE J. Quantum Electron*, **5**, 211 (1969).
38. K. L. Ashley and J. R. Biard, *IEEE Trans. Electron. Devices*, **14**, 429 (1967).
39. I. Hayashi and M. B. Panish, *J. Appl. Phys.*, **41**, 150 (1970).
40. M. B. Panish, I. Hayashi, and S. Sumski, *IEEE J. Quantum Electron.*, **5**, 210 (1969).
41. Zh. I. Alferov, M. V. Andreev, E. I. Koroklov, E. I. Portnoi, and D. N. Tretyakov, *Soviet Phys.-Semicond.*, **2**, 1289 (1969).
42. H. Kroemer, *Proc. IEEE*, **51**, 1782 (1963).
43. H. Kressel and F. Z. Hawrylo, *Proc. IEEE (Corr.)*, **56**, 1598 (1968).
44a. H. Nelson and G. C. Dousmanis, *Appl. Phys. Letters*, **4**, 192 (1964).
44b. H. Kressel and F. Z. Hawrylo, *J. Appl. Phys.*, **39**, 205 (1968).
45. M. H. Pilkuhn and H. Rupprecht, *J. Appl. Phys.*, **38**, 5 (1967).

46. R. O. Carlson, *J. Appl. Phys.*, **38**, 661 (1967).
47. J. J. Tietjen and J. A. Amick, *J. Electrochem. Soc.*, **113**, 724 (1966).
48. J. J. Tietjen, J. I. Pankove, I. J. Hegyi, and H. Nelson, *Trans. AIME*, **239**, 385 (1967).
49. H. Nelson, *RCA Rev.*, **24**, 603 (1963).
50. H. Nelson, *Proc. IEEE*, **55**, 1415 (1967).
51. H. Rupprecht, J. M. Woodall, and G. D. Pettit, *Appl. Phys. Letters*, **11**, 81 (1967).
52. H. Nelson, U.S. patent 3,565,702.
53. W. Susaki, T. Oku, and K. Works, *IEEE J. Quantum Electron.*, **3**, 332 (1967).
54a. H. Beneking and W. Vits, *IEEE J. Quantum Electron.*, **4**, 201 (1968).
54b. The effect of heat treatments on the laser parameters was studied by M. Takusagawa, J. Nishizawa, F. Endo, and Y. Goto, *Japan. J. Appl. Phys.*, **7**, 1301 (1968).
55. H. Kressel and F. Z. Hawrylo, *J. Appl. Phys.*, **41**, 1865 (1970).
56. R. D. Burnham, P. D. Dapkus, N. Holonyak, Jr., D. L. Keune, and H. R. Zwicker, *Solid-State Electron.*, **13**, 199 (1970).
57. H. Kressel, *J. Appl. Phys.*, **38**, 4383 (1967); H. Kressel, F. Z. Hawrylo, and P. LeFur, *J. Appl. Phys.*, **40**, 4059 (1969).
58. Zh. I. Alferov, D. Z. Garbuzov, E. P. Morozov, and D. N. Tretyakov, *Soviet Phys.-Semicond.*, **3**, 600 (1969).
59. J. A. Rossi, N. Holonyak, Jr., P. D. Dapkus, R. D. Burnham, and F. V. Williams, *J. Appl. Phys.*, **40**, 3289 (1969).
60. H. Kressel, J. U. Dunse, H. Nelson, and F. Z. Hawrylo, *J. Appl. Phys.*, **39**, 2006 (1968).
61. *Proc. of the First International Symposium on GaAs, Reading, England, 1966,* Institute of Physics and Physical Society, London, 1967.
62. *Proc. of the Second International Symposium on GaAs, Dallas, Texas, 1968,* Institute of Physics and Physical Society, London, 1969.
63. Ch. Deutsch, *Solid-State Electron.*, **11**, 877 (1968).
64. H. Nelson and H. Kressel, *High Pulse Power Injection Laser,* Final Report Contract No. DA28-043-AMC-02471(E), U.S. Army Electronics Command, Fort Monmouth, New Jersey, Oct. 1967.
65. J. C. Dyment and L. A. D'Asaro, *Appl. Phys. Letters*, **11**, 292 (1967).
66. W. Becke, *IEEE J. Quantum Electron.*, **4**, 364 (1968).
67. W. F. Kosonocky, R. H. Cornely, and I. J. Hegyi, *IEEE J. Quantum Electron.*, **4**, 176 (1968).
68. D. R. Muss, C. S. Duncan, and S. Scuro, *Trans. AIME*, **239**, 404 (1967).
69. A. L. McWhorter, H. J. Zeiger, and B. Lax, *J. Appl. Phys.*, **34**, 235 (1963).
70. A. Yariv and R. C. C. Leite, *Appl. Phys. Letters*, **2**, 55 (1963).
71. W. L. Bond, B. G. Cohen, R. C. C. Leite, and A. Yariv, *Appl. Phys. Letters*, **2**, 57 (1963).
72. A. L. McWhorter, *Solid State Electron.*, **6**, 417 (1963).
73. F. Stern, *Symp. Radiative Recombination in Semiconductors,* Dunod, Paris, 1964.
74. F. Stern, *Phys. Rev.*, **133**, A1653 (1964).
75. D. F. Nelson and J. McKenna, *J. Appl. Phys.*, **38**, 4057 (1967).
76a. T. H. Zachos and J. E. Ripper, *IEEE J. Quantum Electron.*, **5**, 29 (1969).
76b. T. L. Paoli, J. E. Ripper, and T. H. Zachos, *IEEE J. Quantum Electron.*, **5**, 271 (1969).
77. G. Diemer and B. Bolger, *Physica*, **29**, 600 (1963).
78. W. W. Anderson, *IEEE J. Quantum Electron.*, **1**, 228 (1965).
79. N. E. Byer and J. Butler, *IEEE J. Quantum Electron.*, **6**, 291 (1970).

80. I. Kudman and T. Seidel, *J. Appl. Phys.*, **33**, 771 (1962).
81. G. C. Dousmanis and D. L. Staebler, *J. Appl. Phys.*, **37**, 2278 (1966).
82. H. Kressel, H. Nelson, and F. Z. Hawrylo, *J. Appl. Phys.*, **41**, 2019 (1970).
83. D. E. Hill, *Phys. Rev.*, **133**, A866 (1964).
84. A. R. Goodwin and P. R. Selway, *IEEE J. Quantum Electron.*, **6**, 285 (1970).
85. Ch. Deutsch, *Phys. Letters*, **24A**, 467 (1967).
86. J. E. Ludman and K. M. Hergenrother, *Solid-State Electron.*, **9**, 863 (1966).
87. J. C. Dyment, *Appl. Phys. Letters*, **10**, 84 (1967).
88. J. C. Dyment and T. H. Zachos, *J. Appl. Phys.*, **39**, 2923 (1968).
89. T. H. Zachos, *Appl. Phys. Letters*, **12**, 318 (1968).
90. H. S. Sommers, Jr., *Solid-State Electron.*, **11**, 909 (1968).
91. C. H. Gooch, *IEEE J. Quantum Electron.*, **4**, 140 (1968).
92. J. I. Pankove and J. E. Berkeyheiser, *Rev. Sci. Instr.*, **39**, 884 (1968).
93. J. C. Dyment, J. E. Ripper, and T. H. Zachos, *J. Appl. Phys.*, **40**, 1802 (1969).
94. J. Vilms, L. Wandinger, and K. L. Klohn, *IEEE J. Quantum Electron.*, **2**, 80 (1966).
95. W. Engeler and M. Garfinkel, *Solid-State Electron.*, **8**, 585 (1965).
96. M. Ciftan and P. P. Debye, *Appl. Phys. Letters*, **6**, 120 (1965).
97. R. J. Keyes, *IBM J. Res. Develop.*, **9**, 303 (1965).
98. H. Knapp, O. Krumpholz, and S. Maslowski, *Z. Angew. Phys.*, **25**, 277 (1968).
99. W. E. Engeler and M. Garfinkel, *J. Appl. Phys.*, **35**, 1734 (1964).
100. M. H. Pilkuhn and G. T. Guettler, *IEEE J. Quantum Electron.*, **4**, 132 (1968).
101. R. F. Broom, *IEEE J. Quantum Electron.*, **4**, 135 (1968).
102. H. Kressel and H. P. Mierop, *J. Appl. Phys.*, **38**, 5419 (1967).
103. K. L. Konnerth and J. C. Marinace, *IEEE J. Quantum Electron.*, **4**, 173 (1968).
104. C. D. Dobson and F. S. Keeble, in Ref. (*61*).
105. Yu. I. Kruzhilin, V. I. Shveykin, N. V. Antonov, and Yu. I. Koloskov, *Proc. International Conference on Phys. Semiconductors, Moscow, 1968*, p. 541.
106. H. Kressel, N. E. Byer, H. Nelson, H. Lockwood, and H. S. Sommers, Jr., Interim Report No. 1, Contract F33615-69-C-1208, U.S. Air Force Avionics Lab., Dayton, Ohio, November 1969.
107. H. Kressel and N. E. Byer, *Proc. IEEE*, **57**, 25 (1969).
108. C. J. Nuese and H. Schade, private communication, 1970.
109. N. E. Byer, *IEEE J. Quantum Electron.*, **5**, 242 (1969).
110. H. Kressel, N. E. Byer, H. Lockwood, F. Z. Hawrylo, H. Nelson, M. S. Abrahams, and S. H. McFarlane, *Met. Trans.*, **1**, 635 (1970).
111. J. F. Black and E. D. Jungbluth, *J. Electrochem. Soc.*, **114**, 181, 188 (1967).
112. R. D. Gold and L. R. Weisberg, *Solid-State Electron.*, **7**, 811 (1964).
113. S. A. Steiner and R. L. Anderson, *Solid-State Electron.*, **11**, 65 (1968).
114. J. J. Loferski, H. Flicker, R. M. Esposito, and M. H. Wu, *Proc. Conf. on Radiation Effects in Semiconductors, Tokyo, 1966*.
115. H. Kressel, H. Nelson, S. H. McFarlane, M. S. Abrahams, P. LeFur, and C. J. Buiocchi, *J. Appl. Phys.*, **40**, 3587 (1969).
116. J. Hatz, *IEEE J. Quantum Electron.*, **3**, 643 (1967).
117. M. S. Abrahams, L. R. Weisberg, and J. J. Tietjen, *J. Appl. Phys.*, **40**, 3754 (1969).
118. G. H. Schwuttke and H. Rupprecht, *J. Appl. Phys.*, **37**, (1966).
119. S. Prussin, *J. Appl. Phys.*, **32**, 1876 (1961).
120. J. F. Black and P. Lublin, *J. Appl. Phys.*, **35**, 2462 (1964).
121. M. S. Abrahams, C. J. Buiocchi, and J. J. Tietjen, *J. Appl. Phys.*, **38**, 760 (1967).
122. H. Kressel, F. Z. Hawrylo, M. S. Abrahams, and C. J. Buiocchi, *J. Appl. Phys.*, **39**, 5139 (1968).

123. M. S. Abrahams, private communication, 1970.

124. C. J. Nuese, H. Schade, and D. Herrick, *Met. Trans.* **1**, 587 (1970).

125. A. Itoh and J. Nishizawa, *J. Phys. Soc. Japan*, to be published.

126. L. R. Weisberg and H. Schade, *J. Appl. Phys.*, **39**, 5149 (1968).

127. H. C. Hwang and J. C. Dyment in Ref. (*62*).

128a. P. R. Thornton, *The Physics of Electroluminescent Devices*, Spon., London, 1967.

128b. Zh. I. Alferov, V. M. Andreev, E. L. Portnoi, and M. K. Trukan, *Fiz. Tekh. Poluprov.*, **3**, 1328 (1969) [*Soviet Phys.-Semicond.*, **3**, 1107 (1970)].

129. J. I. Pankove, *IEEE J. Quantum Electron.*, **4**, 119 (1968).

130. R. Gill, *Proc. IEEE*, **58**, 949 (1970).

131. W. E. Ahearn and J. W. Crowe, *IEEE J. Quantum Electron.*, **6**, 377 (1970).

132. J. C. Marinace, private communication, 1969.

133. A. Zouridies, *RCA*, *Raritan*, private communication, 1969.

134. E. A. Ulmer, Jr., and I. Hayashi, *IEEE J. Quantum Electron.*, **6**, 297 (1970).

135. M. H. Pilkuhn, H. Rupprecht, and S. Blum, *Solid-State Electron.*, **7**, 905 (1964).

136. H. F. Quinn and W. O. Morton, paper presented at *IEEE Semiconductor Laser Conference, Mexico, 1969.*

137. W. Susaki, T. Oku, and T. Sogo, *IEEE J. Quantum Electron.*, **4**, 122 (1968).

138. K. Konnerth, *IEEE Trans. Electron. Devices*, **12**, 506 (1965).

139. C. D. Dobson, J. Franks, and F. S. Keeble, *IEEE J. Quantum Electron.*, **4**, 151 (1968).

140. G. E. Fenner, *Solid-State Electron.*, **10**, 753 (1967).

141. J. C. Dyment and J. E. Ripper, *IEEE J. Quantum Electron.*, **4**, 155 (1968).

142. J. E. Ripper, *IEEE J. Quantum Electron.*, **5**, 391 (1969).

143. J. E. Ripper and J. C. Dyment, *IEEE J. Quantum Electron.*, **5**, 396 (1969).

144. J. I. Pankove, *IEEE J. Quantum Electron.*, **4**, 161, 427 (1968).

145. T. Ikegami and Y. Suematsu, *IEEE J. Quantum Electron.*, **4**, 148 (1968).

146. J. Nishizawa, *IEEE J. Quantum Electron.*, **4**, 143 (1968).

147. S. Takamiya and J. Nishizawa, *Proc. IEEE (Corr.)*, **56**, 135 (1968).

148. J. E. Ripper, *IEEE J. Quantum Electron.*, **2**, 603 (1966); *ibid.*, **3**, 202 (1967).

149a. N. G. Basov, V. N. Morozov, V. V. Nikitin, and A. S. Semenov, *Soviet Phys.-Semicond.*, **1**, 1305 (1968); (and references therein).

149b. T. P. Lee and R. Roldan, *IEEE J. Quantum Electron.*, **6**, 339 (1970).

150. L. A. D'Asaro, J. M. Cherlow, and T. L. Paoli, *IEEE J. Quantum Electron.*, **4**, 164 (1968).

151. T. L. Paoli and J. E. Ripper, *Phys. Rev. Letters*, **22**, 1085 (1969).

152. R. F. Broom, *Electron. Letters*, **5**, 571 (1969).

153. T. L. Paoli and J. E. Ripper, *Appl. Phys. Letters*, **16**, 96 (1970).

154. J. E. Ripper, T. L. Paoli, and J. C. Dyment, *IEEE J. Quantum Electron.*, **6**, 300 (1970).

155. J. A. Armstrong and A. W. Smith, *Phys. Rev.*, **140**, A155 (1965).

156. R. Roldan, *Appl. Phys. Letters*, **11**, 346 (1967).

157. G. Guekos, and M. J. O. Strutt, *IEEE J. Quantum Electron.*, **5**, 129 (1969).

158. G. Guekos and M. J. O. Strutt, *Electron. Letters*, **4**, 408 (1968).

159. D. E. McCumber, *Phys. Rev.*, **141**, 306 (1966).

160. H. Haug, *Z. Phys.*, **200**, 57 (1967).

161. H. Haug, *Z. Phys.*, **206**, 163 (1967).

162. H. Haug and H. Haken, *Z. Physik*, **204**, 262 (1967).

163. J. E. Ripper and J. C. Dyment, *Appl. Phys. Letters*, **12**, 365 (1968).

164a. J. E. Ripper and T. L. Paoli, *Proc. IEEE*, **58**, 1457 (1970).

164b. M. S. Abrahams, L. R. Weisberg, C. J. Buiocchi, and J. Blanc, *J. Mater. Sci.*, **4**, 223 (1969).

165a. N. Holonyak, Jr., and S. F. Bevacqua, *Appl. Phys. Letters*, **1**, 82 (1962).

165b. J. I. Pankove, H. Nelson, J. J. Tietjen, I. J. Hegyi, and H. P. Maruska, *RCA Rev.*, **28**, 560 (1967).

166. N. Holonyak, Jr., *Trans. AIME*, **230**, 276 (1964).

167. M. Pilkuhn and H. Rupprecht, *J. Appl. Phys.*, **36**, 684 (1965).

168. N. G. Basov et al., *Soviet Phys.-Solid State*, **7**, 1539 (1965).

169. C. J. Nuese, G. E. Stillman, M. D. Sirkis, and N. Holonyak, Jr., *Solid-State Electron.*, **9**, 735 (1966).

170. M. B. Panish, *J. Phys. Chem. Solids*, **30**, 1083 (1969); G. A. Wolff, H. E. LaBelle, and B. N. Das, *Trans. AIME*, **242**, 436 (1968).

171. H. Kressel, H. Lockwood, H. Nelson, *IEEE J. Quantum Electron.*, **6**, 278 (1970).

172. W. Susaki, T. Sogo, and T. Oku, *IEEE J. Quantum Electron.*, **4**, 422 (1968).

173. H. Rupprecht, J. M. Woodall, G. D. Pettit, J. W. Crowe, and H. F. Quinn, *IEEE J. Quantum Electron.*, **4**, 35 (1968).

174. H. P. Maruska and J. I. Pankove, *Solid-State Electron.*, **10**, 917 (1967).

175. S. M. Ku and J. F. Black, *J. Appl. Phys.*, **37**, 3733 (1966).

176. I. Melngailis, *Appl. Phys. Letters*, **2**, 176 (1963).

177. B. M. Vul, A. P. Shotov, and V. S. Bogarev, *Soviet Phys.-Solid State*, **4**, 2689 (1963).

178a. C. Chipaux, *Symp. Radiative Recombination in Semiconductors*, Dunod, Paris, 1965, p. 217.

178b. B. Pistoulet and H. Mathieu, *Proc. International Conference on Phys. Semiconductors, Moscow, 1968*, p. 532.

179. T. Deutsch, R. C. Ellis, Jr., and O. M. Warschauer, *Phys. Status Solidi*, **3**, 1001 (1963).

180. R. J. Phelan, A. R. Calawa, R. H. Rediker, R. J. Keyes, and B. Lax, *Appl. Phys. Letters*, **3**, 143 (1963).

181. K. Weiser and R. S. Levitt, *Appl. Phys. Letters*, **2**, 178 (1963).

182. J. Melngailis, A. J. Strauss, and R. H. Rediker, *Proc. IEEE*, **51**, 1154 (1963).

183. F. B. Alexander et al., *Appl. Phys. Letters*, **4**, 13 (1964).

184. L. G. Bailey, *Trans. AIME*, **239**, 310 (1968).

185. I. Melngailis and R. H. Rediker, *J. Appl. Phys.*, **37**, 899 (1966).

186. T. L. Maslennikova, V. A. Kudryashov, and N. S. Baryshev, *Soviet Phys.-Semicond.*, **2**, 1414 (1969).

187. I. D. Anisimova and L. N. Kurbatov, *Soviet Phys.-Semicond.*, **2**, 989 (1969).

188. M. Rodot, P. Leroux-Hugon, J. Besson, and H. Lebloch, *l'Onde Electrique*, October 1965.

189. T. I. Galkina, N. B. Kornilova, and N. A. Penin, *Soviet Phys.-Solid State*, **8**, 1974 (1967).

190. C. Chipaux and R. Eymard, *Phys. Status Solidi*, **10**, 165 (1965).

191. I. V. Kryukova et al., *Soviet Phys.-Solid State*, **8**, 822 (1966).

192. Ya. E. Pokrovskii and K. J. Svistunova, *Soviet Phys.-Semicond.*, **1**, 118 (1967).

193. N. G. Basov et al., *Soviet Phys.-Solid State*, **8**, 2087 (1967).

194. N. G. Basov et al., *Soviet Phys.-Solid State*, **8**, 847 (1966).

195. A. P. Shotov, M. S. Mirgalovskaya, R. A. Muminov, and M. R. Raukhman, *Soviet Phys.-Semicond.*, **1**, 1193 (1968).

196. P. G. Eliseev et al., *Soviet Phys.-Solid State*, **8**, 1025 (1966).

197. J. F. Butler et al., *Solid-State Commun.*, **2**, 301 (1964); I. Chambouleyron, *Proc.*

International Conference on Phys. Semiconductors, Moscow, 1968, p. 546.

198. J. F. Butler, A. R. Calawa, R. J. Phelan, Jr., T. C. Harman, A. J. Strauss, and R. H. Rediker, *Appl. Phys. Letters*, **5**, 75 (1964).

199. J. F. Butler and A. R. Calawa, *J. Electrochem. Soc.*, **54**, 1056 (1965).

200. J. O. Dimmock, I. Melngailis, and A. J. Strauss, *Phys. Rev. Letters*, **16**, 1193 (1966).

201. J. F. Butler and T. C. Harman, *Appl. Phys. Letters*, **12**, 347 (1968).

202. L. N. Kurbatov et al., *Soviet Phys.-Semiconductors*, **2**, 1008 (1969).

203. T. C. Harman, A. R. Calawa, I. Melngailis, and J. O. Dimmock, *Appl. Phys. Letters*, **14**, 333 (1969).

204. A. R. Calawa, J. O. Dimmock, T. C. Harman, and I. Melngailis, *Phys. Rev. Letters*, **23**, 7 (1969).

205. I. Melngailis, *J. Phys. Radium*, **29**, C4-84 (Suppl. to No. 11), (1968).

206. E. D. Hinkley, T. C. Harman, and C. Freed, *Appl. Phys. Letters*, **13**, 49 (1968).

207. F. H. Nicoll, *Proc. IEEE*, **55**, 114 (1967).

208. C. A. Klein, *IEEE J. Quantum Electron.*, **5**, 186 (1968).

209. D. C. Reynolds, *Trans. AIME*, **239**, 300 (1967).

210. C. E. Hurwitz, *Appl. Phys. Letters*, **9**, 420 (1966).

211. F. H. Nicoll, *Appl. Phys. Letters*, **9**, 13 (1966).

212. C. E. Hurwitz, *Appl. Phys. Letters*, **9**, 116 (1966).

213. F. H. Nicoll, *Appl. Phys. Letters*, **10**, 69 (1967).

214. F. H. Nicoll, *RCA Rev.*, **29**, 379 (1968).

215. J. L. Brewster, *Appl. Phys. Letters*, **13**, 385 (1968).

216. J. R. Packard, W. C. Tait, and D. A. Campbell, *IEEE J. Quantum Electron.*, **5**, 44 (1969).

217. C. Benoit a la Guillaume, J. M. Debever, and F. Salvan, *Proc. International Conference on Phys. Semiconductors, Moscow, 1968*, p. 581.

218. O. V. Bogdankevich, M. M. Zverev, A. I. Krasilnikov, and A. N. Pechenov, *Phys. Status Solidi*, **19**, K5 (1967).

219. C. E. Hurwitz, *IEEE J. Quantum Electron.*, **3**, 333 (1967).

220. N. G. Basov, O. V. Bogdankevich, P. G. Eliseev, and B. M. Lavrushin, *Soviet Phys.-Solid State*, **8**, 1073 (1966).

221. G. Mandel and F. F. Morehead, *Appl. Phys. Letters*, **4**, 143 (1964).

222. V. S. Vavilov, E. L. Nolle, G. P. Golubev, V. S. Mashtakov, and E. I. Tsarapaeva, *Soviet Phys.-Solid State*, **9**, 657 (1967).

223. G. P. Golubev et al., *Soviet Phys.-Semicond.*, **3**, 240 (1969).

224. C. Benoit a la Guillaume and J. M. Debever, *Solid-State Commun.*, **3**, 19 (1965).

225. C. Benoit a la Guillaume and J. M. Debever, *Solid-State Commun.*, **2**, 145 (1964).

226. N. G. Basov et al., *Soviet Phys.-Doklady*, **10**, 329 (1965).

227. F. M. Berkovskii et al., *Soviet Phys.-Semicond.*, **2**, 1027 (1969).

228. L. N. Kourbatov, *Proc. International Conference on Phys. Semiconductors, Moscow, 1968*, p. 587.

229. D. A. Cusano, *Appl. Phys. Letters*, **7**, 151 (1965).

230. O. V. Bogdankevich, N. A. Borisov, I. V. Krjukova, and B. M. Lavrushin, *Sov. Phys.-Semicond.*, **2**, 845 (1969).

231. O. V. Bogdankevich, N. A. Borisov, I. V. Krjukova, and B. M. Lavrushin, *Proc. International Conference on Phys. Semiconductors, Moscow, 1968*, p. 575.

232. J. M. Lavine and A. Adams, Jr., *IEEE J. Quantum Electron.*, **5**, 195 (1968).

233. C. A. Klein and J. M. Lavine, *Appl. Phys. Letters*, **12**, 125 (1968).

234. J. M. Lavine, R. L. Mozzi, and A. Adams, Jr., *IEEE J. Quantum Electron.*, **5**, 422 (1969).

235. R. Hunsperger, *Solid-State Electron.*, **12**, 215 (1969).
236a. F. H. Nicoll, *Rev. Sci. Instr.*, **41**, 1175 (1970).
236b. I. Melngailis and A. J. Strauss, *Appl. Phys. Letters*, **8**, 179 (1966).
237. R. J. Phelan and R. H. Rediker, *Appl. Phys. Letters*, **6**, 70 (1965).
238. I. Melngailis, *IEEE J. Quantum Electron.*, **1**, 104 (1965).
239. N. G. Basov, A. Z. Grasyuk, and V. A. Katulin, *Soviet Phys.-Doklady*, **10**, 343 (1965).
240. N. G. Basov, A. Z. Grasyuk, I. G. Zabarev, and V. A. Katulin, *JETP Letters*, **1**, 118 (1965).
241. V. K. Konyukhov, L. A. Kulevskii, and A. M. Prokhorov, *IEEE J. Quantum Electron.*, **2**, 1 (1966).
242. M. S. Brodin, P. I. Budnik, N. I. Vitrikhovskii, and S. V. Zakrevskii, *Proc. International Conference on Physics of Semiconductors, Moscow, 1968*, p. 610.
243. D. L. Keune, J. A. Rossi, N. Holonyak, Jr., and P. D. Dapkus, *J. Appl. Phys.*, **40**, 1934 (1969).
244. P. D. Dapkus, N. Holonyak, Jr., J. A. Rossi, F. V. Williams, and D. A. High, *J. Appl. Phys.*, **40**, 3300 (1969).
245. N. Holonyak, Jr., M. R. Johnson, J. A. Rossi, and W. O. Groves, *Appl. Phys. Letters*, **12**, 151 (1968).
246. R. D. Burnham, P. D. Dapkus, N. Holonyak, Jr., and J. A. Rossi, *Appl. Phys. Letters*, **14**, 190 (1969).
247. G. E. Stillman, M. D. Sirkis, J. A. Rossi, M. R. Johnson, and N. Holonyak, Jr., *Appl. Phys. Letters*, **9**, 268 (1966).
248. N. Holonyak, Jr., M. R. Johnson, and D. L. Keune, *IEEE J. Quantum Electron.*, **4**, 199 (1968).
249. S. G. Bishop, W. J. Moore, and E. M. Swiggard, *Appl. Phys. Letters*, **15**, 12 (1969).
250. N. G. Basov, B. M. Vul, and Yu. M. Popov, *J. Exptl. Theoret. Phys. (USSR)*, **37**, 587 (1959); *Sov. Phys.-JETP*, **10**, 416 (1960).
251. G. Haacke, *Solid-State Commun.*, **2**, 317 (1964).
252. M. C. Steele, *RCA Rev.*, **27**, 263 (1966).
253. H. Schmidt, *Phys. Rev.*, **149**, 564 (1966).
254. V. V. Vladimirov, *Sov. Phys.-JETP*, **28**, 675 (1969).
255. K. Weiser and J. F. Woods, *Appl. Phys. Letters*, **7**, 225 (1965).
256. P. D. Southgate, *Appl. Phys. Letters*, **12**, 61 (1968).
257. P. D. Southgate, *IEEE J. Quantum Electron.*, **4**, 179 (1968).
258. P. D. Southgate and R. T. Mazzochi, *Phys. Letters*, **28A**, 216 (1968).
259. J. W. Crowe and W. E. Ahearn, *IEEE J. Quantum Electron.*, **4**, 169 (1968).
260. H. Bachert and S. Raab, *Phys. Status Solidi*, **29**, 175 (1968).
261. R. F. Broom and E. Mohn, *J. Appl. Phys.*, **39**, 4851 (1968).
262. N. G. Basov, O. V. Bogdankevich, A. N. Pechenov, A. S. Nasibov, and K. P. Fedoseev, *Sov. Phys.-JETP*, **28**, 900 (1969).
263. H. D. Edmonds, *Proc. IEEE (Corr.)*, **57**, 1307 (1969).
264. J. Crowe and R. Craig, *Appl. Phys. Letters*, **4**, 57 (1964).
265. M. Coupland, K. H. Hambleton, and C. Hilsum, *Phys. Letters*, **7**, 231 (1963).
266. R. Vuilleumier, *Proc. IEEE*, **55**, 1420 (1967).
267. J. W. Crowe and W. E. Ahearn, *IEEE J. Quantum Electron.*, **2**, 283 (1966).
268. W. F. Kosonocky and R. H. Cornely, *IEEE J. Quantum Electron.*, **4**, 125 (1968).
269. C. E. Kelley, *IEEE Trans. Electron. Devices*, **12**, 1 (1965).
270. Yu. P. Zakharov, V. V. Nikitin, A. S. Semenov, A. V. Uspenskii, and V. A. Shcheglov, *Sov. Phys.-Solid State*, **8**, 1660 (1967).
271. A. B. Fowler, *Appl. Phys. Letters*, **3**, 1 (1963).

272. F. Stern, *Proc. Physics of Quantum Electronics Conf.*, McGraw-Hill, New York, 1966.
273. A. A. Sheronov, *Soviet Phys.-Semicond.*, **3**, 314 (1969).
274. W. F. Kosonocky, *Symp. on Optical and Electro-Optical Information Processing Technology, Boston, Mass., November 9–10, 1964*, M.I.T. Press, Cambridge, Mass., 1965, pp. 269–304.
275. W. I. Kosonocky, *1968 Wescon Technical Papers, Session 16-Optics and Electro-Optics in Computers, August 20–23, 1968*, p. 1614.
276. H. F. Lockwood, H. Kressel, H. S. Sommers, Jr., and F. Z. Hawrylo, *Appl. Phys. Letters*, **17**, 499 (1970).
277. D. G. Herzog and H. Kressel, *Appl. Opt.*, **9**, 2249 (1970).
278. M. B. Panish, I. Hayashi, and S. Sumski, *Appl. Phys. Letters*, **16**, 326 (1970).
279. H. Kressel and F. Z. Hawrylo, *Appl. Phys. Letters*, **17**, 169 (1970).
280. M. J. Adams and M. Cross, *Phys. Letters*, **32A**, 207 (1970).
281. H. Kressel, F. Z. Hawrylo, and H. F. Lockwood, *IEEE Electron Devices Conf., Washington, 1970*.
282. J. K. Butler (to be published).
283. J. K. Butler, H. S. Sommers, Jr., and H. Kressel, *Appl. Phys. Letters*, **17**, 403 (1970).
284. C. E. Barnes, *Phys. Rev. B (Solid State)*, **1**, 4735 (1970).
285. I. Hayashi, M. B. Panish, P. W. Foy, and S. Sumski, *Appl. Phys. Letters*, **17**, 109 (1970).
286. H. Kressel, F. H. Lockwood, and F. Z. Hawrylo, *Appl. Phys. Letters*, **18**, 43 (1971).
287. Y. Unno, M. Yamamoto, and S. Iida, *Japan. J. Appl. Phys.*, **9**, 1181 (1970).
288. S. Iida, Y. Unno, and M. Yamamoto, *Japan. J. Appl. Phys.*, **9**, 424 (1970).
289. H. Yonezu, I. Sakuma, and Y. Nannichi, *Japan. J. Appl. Phys.*, **9**, 231 (1970).
290. F. Stern, private communication, 1970.
291. N. Patel and A. Yariv, *IEEE J. Quantum Electron.*, **6**, 383 (1970).
292. B. Ross and E. Snitzer, *IEEE J. Quantum Electron.*, **6**, 361 (1970).
293. P. D. Dapkus, N. Holonyak, Jr., R. D. Burham, and D. L. Keune, *Appl. Phys. Letters*, **16**, 93 (1970).
294. R. F. Broom, E. Mohn, C. Risch, and R. Salathé, *IEEE J. Quantum Electron.*, **6**, 328 (1970).
295. R. Ulbrich and M. H. Pilkuhn, *IEEE J. Quantum Electron.*, **6**, 314 (1970).
296. R. Ulbrich and M. H. Pilkuhn, *Appl. Phys. Letters*, **16**, 516 (1970).
297. G. Guekos and M. J. O. Strutt, *Electron. Letters*, **6**, 250 (1970).
298. M. Ettenberg, H. S. Sommers, Jr., H. Kressel, and H. Lockwood, 1970, to be published.
299. H. Kressel and F. Z. Hawrylo, to be published.

Chapter 2 · CO₂ LASERS

P. K. CHEO†

BELL TELEPHONE LABORATORIES, INC.
WHIPPANY, NEW JERSEY

I. Historical Review

Early in 1964, Patel et al.(*1*) first reported the observation of CW laser action in CO_2 gas at approximately $10\,\mu$. This discovery came at a time when the search for new laser transitions from ionic and molecular species was near its peak, and this particular laser did not attract any more attention than hundreds of other laser transitions

†*Present address:* United Aircraft Research Laboratories, East Hartford, Connecticut.

reported at that time. A few months later, Patel(2) published a detailed study of both the CW (~ 1 mW) and pulsed output power spectra of the CO_2 laser and also presented an interpretation(3) of previously reported results. Similar studies were made independently by a group of French scientists(4–6) whose interests were primarily in molecular spectroscopic research.

Within a two-year period, two great advances were made. The first was the use of a mixture of nitrogen and CO_2 as the active medium, which was suggested almost simultaneously by Legay and Legay-Sommaire(7) and by Patel(8). Patel(9) later demonstrated that resonant transfer of the vibrational energy from the N_2 metastable state ($v = 1$) to the upper CO_2 laser level ($00^0 1$) increased the CO_2 laser output from the milliwatts range up to about 10 W. The other advance was made by Moeller and Rigden(10), who showed that helium can enhance the laser output as well as nitrogen, although the mechanism was not well understood at that time. In the meantime, Patel et al.(11) obtained from a CO_2 laser a very impressive CW output power in excess of 100 W by using a flowing CO_2–N_2–He gas mixture. This was a significant achievement in view of the fact that good optical quality IR components were not available at that time. It was then evident that the efficiency and average output power of this laser are unique when compared with all other existing lasers. Recently(11a), CW laser output in excess of 11 kW and multijoule Q-switched laser pulse have been obtained from a compact CO_2 laser with a 1-m discharge length. A short sealed-off CO_2 laser (~ 1 ft in length) with a long-term stability better than one part in 10^{10} or a short term of only a few cycles at optical frequency has been achieved and can provide a single frequency and single mode output greater than 1 W with about 15% efficiency. Also short and fast-rise pulses on the order of nanoseconds can be generated by an electrooptic technique. Clearly this laser can be useful in basic research and in important communication, military, and industrial applications.

Recently (1966–1969), very intensive research and development activities on the CO_2 laser and related areas were carried out in industrial, government, and university laboratories. These can be categorized as follows:

1. Investigation of mechanisms responsible for high efficiency and high gain or power. This involves mainly the studies of relaxation (12–15) and electronic excitation(16–18) processes. In addition to the well-known resonant transfer(7,8), much information has been obtained concerning the low-lying CO_2 molecular vibrational levels

pertinent to the CO$_2$ laser and about the effects of foreign gases(*19,20*) on the population inversion. In this area a series of theoretical papers by Russian scientists(*21,22*) also appeared in the literature.

2. Measurement of CO$_2$ molecular lifetimes. This includes both the vibrational(*23–25*) and rotational(*26,27*) lifetimes as well as the rate constants(*14,15,23–28*) of kinetic equations for various collisional processes.

3. Studies of linewidth(*29–31*), line shape(*32,33*), rotational-level competition(*26,27,33*), gain characteristics(*19,34–39*), and saturation effects(*40–43*).

4. Work on active(*44–46*) and passive(*47–51*) Q switching, mode locking(*52,53,56*), modulation(*54,55*), and cavity dumping(*56*) of the CO$_2$ laser.

5. Development of CO$_2$ lasers(*57–63*) and laser amplifiers(*19,34, 35,43,64,65*) for both the flowing and nonflowing gas systems. In the related areas, effort was made in the studies of the characteristics (*66–68*), operating lifetime(*69–73*), and stability(*74,75*) of a sealed-off CO$_2$ laser, as well as the development of IR optical components(*76*) with high power capabilities.

6. Research and development work on various novel systems including thermal(*77–82*), chemical(*83*), high-speed transverse-gas-flow CO$_2$ laser systems(*84–87*), and high-pressure discharges(*88,89*).

7. Research and evaluation of CO$_2$ lasers for use in terrestrial and deep space communication(*90,91*), radar systems(*92*), and various industrial applications(*93*) such as cutting, drilling, welding, and hot plasma generation.

8. Studies of nonlinear optical phenomena(*94–98*) and coherent interactions between short CO$_2$ laser pulses and resonant media (*99–107*).

9. Deployment of CO$_2$ laser communication satellite ATS-F experiment (Applications Technology Satellite scheduled for launch in early 1973).

The above references are not a complete list of all the advances involving the CO$_2$ laser during the years 1966–1969. However, they represent most of the important contributions and achievements, and provide a broad perspective of the field. It should also be pointed out that several new areas of research—i.e., laser-induced fluorescence (*108*), infrared–infrared(*109*), and infrared–microwave(*110*) double resonance molecular spectroscopy—have recently become very active as a result of the advances with the CO$_2$ laser.

II. Introduction

To date, there are nearly 200 CO_2 laser oscillations resulting from vibration–rotation transitions among a number of low-lying ($E_v < 1$ eV) vibrational levels in the ground electronic $^1\Sigma$ state of the CO_2 molecule. These oscillations cover the spectral range from 9 to 18 μ. Among them, two of the strongest groups arise from the $00^0 1$–$10^0 0$ and $00^0 1$–$02^0 0$ bands with the band edges at 10.4 and 9.4 μ, respectively. With present-day optical technology in this IR wavelength range, one can easily construct a stable, single frequency, and single-mode CO_2 laser, operating on a number of the vibration–rotation transitions with a CW output power greater than 20 W ($\simeq 100$ W multimode output) or $\simeq 10$ kW Q-switched laser pulses from a 1-m-long gas discharge tube.

Mainly because its gain and efficiency are high, the CO_2 laser is by far the simplest of all gas lasers to fabricate and process. Neither the ultracleanliness required for He–Ne lasers nor the elaborate structural design required for high-power (~ 10 W) CW ion lasers is needed in the construction of CO_2 lasers. In fact, impurities often are helpful to the gain and power output of CO_2 lasers. The laser medium commonly consists of either a continuously flowing CO_2–N_2–He or a static CO_2–N_2–He–H_2–Xe gas mixture excited by either a dc or an rf electric discharge with water-cooled walls. For sealed-off CO_2 lasers, the power output can be as high as 60% of that obtainable from a flowing system but the operating life of the laser is limited to 1–2 khr. Because of the large number of impurities involved in this laser, the collisional processes in the active medium may be complicated. Investigation of these kinetic processes has led to some of the most significant advances in laser technology in recent years, particularly in the achievement of the highest efficiency ($\gtrsim 20\%$) and average power (more than 11 kW/m of active medium) ever obtained from a laser.

The CO_2 laser field has grown so enormously and is still expanding at such a fast rate that it is inappropriate to cover all aspects in one chapter. For this reason emphasis will be placed on the chemical physics and performance characteristics of CO_2 lasers and only incidental reference will be made to engineering techniques and systems involving the use of CO_2 lasers. Discussion of these topics will stress fundamental principles and concepts.

In general, the basic mechanisms of various types of CO_2 laser systems are fairly well understood. Since this laser involves a large number of collisional processes (i.e., electronic, atomic, and molecular) and in some cases involves chemical and fluid dynamic reactions, the exact analysis can be extremely complex. In many instances, therefore, assumptions are made to simplify the existing problem and treatments are limited to semiquantitative aspects. A number of review papers(*111–113*) are available in the literature but most of them are brief.

The remaining text of this chapter is divided as follows: Section III gives a brief review of CO_2 molecular structure and IR spectra with emphasis on those arising from low-lying vibration–rotation levels pertinent to CO_2 lasers. Also included in Section III are a summary of CO_2 laser transitions, the measurements of absorption and transmission spectroscopy, and determination of some molecular fine-structure constants by means of lasers. Section IV deals with excitation and relaxation processes leading to the achievement of population inversion in the CO_2 molecules. Data are presented on the inelastic collision cross section between CO_2 and N_2 with electrons, electron temperature in various gaseous discharges, and the collisional relaxation rates of 00^01, 10^00, 02^00, and 01^10 levels of CO_2 in various gas mixtures. The resonant process involving vibrational–vibrational energy exchange among CO_2 molecules and between CO_2 and other excited molecules also are detailed. These materials are extremely important for the understanding of CO_2 laser systems. Finally, Sections V and VI are devoted to various aspects involving both the properties and performance of conventional and novel laser systems. Conventional lasers refer to either the sealed-off or the slow longitudinal-flow (< 1 m/sec) lasers, whereas novel systems include transverse-subsonic-flow, supersonic-flow (or gasdynamic lasers), thermal, chemical, and high-pressure CO_2 lasers. A comprehensive discussion on gain and power output from various types of lasing media is given. Emphasis is placed on the mechanisms responsible for the observed results. For the conventional laser systems, the gain saturation and power output are analyzed by rate equations utilizing previously measured results on collisional and excitation cross sections and relaxation-rate constants. These calculations are found to provide reasonably good physical models. In the cases of high-speed transverse-flow lasers, agreement between experiment and analysis is not as good as for the conventional laser systems owing mainly to the lack of precise information about the rate constants involved in these high-speed flowing systems. Other topics such as linewidth, lineshape, rotational level and mode competitions, Q switching and mode locking, stabilization, and life of a sealed-off CO_2 laser are discussed in detail. Techniques for design and construction of various CO_2 laser devices; performance of various CO_2 laser components such as detectors, modulators, mirrors, beam splitters, interference filters, polarizers, and low-absorption IR transmitting materials for windows and so on; and satellite applications in space-qualified laser communication systems will be subjects of a subsequent article.

III. CO₂ Molecular Structure and Laser Spectroscopy

Carbon dioxide is a simple polyatomic molecule and, therefore, has been extensively studied(*114*). A brief review of the CO_2 molecular structure is presented in this section with only a minimum of mathematical detail. The main purpose is to build a clear picture of the laser medium and to make the subsequent presentation of the properties of CO_2 lasers more meaningful. Since our interest is in lasers, we shall consider only the IR spectra which arise from low-lying vibration–rotation levels in the ground electronic Σ_g^+ state. We shall first consider, as an approximation, the vibrational and rotational motion separately, and later consider briefly the mutual interactions of these motions. Readers who are already familiar with these topics and terminology may proceed onto Section III.D.

A. Normal Modes of Vibration

For ease in visualization, we shall first examine the vibrational structure classically, and later extend it to the quantum mechanical description. The CO_2 is a linear symmetric molecule which has an axis of symmetry, C_∞, and a plane of symmetry perpendicular to the C_∞ axis. There are three normal modes of vibration, ν_1, ν_2, and ν_3, which are associated with the species Σ_g^+, Σ_u^+, and π_u, respectively (Fig. 1). The designations of species for the CO_2 molecule are chosen in the usual way as for electronic states of homonuclear diatomic molecules. The species π_u corresponds to $\ell = 2$ representing a doubly degenerate vibration, usually indicated by ν_{2a} and ν_{2b}, which occurs with equal frequency both in the plane and perpendicular to the plane of the paper. As shown in Fig. 1, the S_i's are the symmetry coordinates(*115*) which in the case of CO_2 are identical to the normal coordinates. The potential V and kinetic T energies expressed in these coordinates are

$$2V = \sum_{ik} C_{ik} S_i S_k$$

$$= C_{11} S_1^2 + C_{22}(S_{2a}^2 + S_{2b}^2) + C_{33} S_3^2 \tag{1}$$

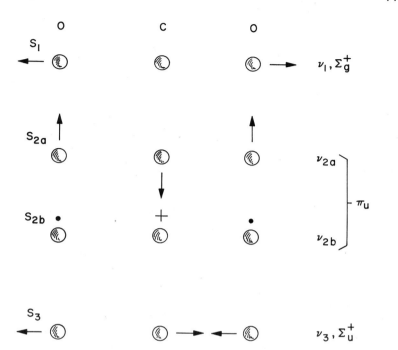

Fig. 1. Normal vibration of CO_2 molecule: symmetric (ν_1), bent (ν_2), and asymmetric (ν_3) modes.

$$2T = \sum_{ik} D_{ik}\dot{S}_i\dot{S}_k$$

$$= D_{11}\dot{S}_1{}^2 + D_{22}(\dot{S}_{2a}^2 + \dot{S}_{2b}^2) + D_{22}\dot{S}_3{}^2 \tag{2}$$

where S_{2a} and S_{2b} are the doubly degenerate and orthogonal symmetry coordinates of specie π_u, and C_{ik} and D_{ik} are related to the force constants a_{ik} in the internuclear or internal coordinate system(116) as

$$C_{11} = 2(a_{11} + a_{12})$$

$$C_{22} = \frac{4a_{33}(1+\mu)^2}{\ell^2} \tag{3}$$

$$C_{33} = 2(1+\mu)^2(a_{11} - a_{12})$$

where a_{11} is the force constant of C—O bond, a_{12} is the force constant that gives the interaction of the two bonds, a_{33} is the force constant for the bending mode of the molecule, $\mu = M_C/2M_O$, and ℓ is the equilibrium internuclear distance between C and O. From Eqs. (1) and (2), the secular determinant $|C_{ik} - \lambda D_{ik}| = 0$ resolves into three equations,

$$C_{11} - \lambda_1 D_{11} = 0$$
$$C_{22} - \lambda_2 D_{22} = 0 \qquad (4)$$
$$C_{33} - \lambda_3 D_{33} = 0$$

From Eqs. (3) and (4) the frequencies of the normal vibration are obtained as follows:

$$\omega_1{}^2 = \frac{a_{11} + a_{12}}{M_O}$$

$$\omega_2{}^2 = \left[\frac{2(M_C + 2M_O)}{M_C M_O} \right]\left(\frac{a_{33}}{\ell^2} \right) \qquad (5)$$

$$\omega_3{}^2 = \left[\frac{(M_C + 2M_O)}{M_C M_O} \right](a_{11} - a_{12})$$

For almost all linear symmetric molecules the force constants must be determined experimentally from the observed fundamental frequencies. Except a few cases, the number of force constants assumed over all symmetry types is larger than the number of normal modes and, therefore, the former cannot be determined from the latter. To overcome this difficulty, one usually uses isotopic molecules, for which the force constants are the same but the frequencies are different, thus providing additional equations. Another method is to make certain assumptions about the force fields in the molecule such that the number of force constants can be reduced. The most commonly used force field is the "central" force; however, for a linear triatomic molecule (symmetric or not) the assumption of central force would lead to zero frequency for the perpendicular (degenerate) vibration except in the higher order. Therefore the central force assumption is not always suitable. There are other force types, such as "valence" force which is often used by assuming a strong restoring force in the line of every valence bond if the distance of two atoms formed by this bond is changed, and also a restoring force opposing a change of angle between two valence bonds.

Application of valence forces to the CO_2 molecule yields the frequencies of three normal vibrations as

$$\omega_1{}^2 = \frac{K_1}{M_O}$$

$$\omega_2{}^2 = \left(1 + \frac{2M_O}{M_C} \right)\frac{2K_\delta}{M_O \ell^2} \qquad (6)$$

$$\omega_3{}^2 = \left(1 + \frac{2M_O}{M_C} \right)\frac{K_1}{M_O}$$

where K_1 and K_δ are the two valence-force constants. Note that Eqs. (6) and (5) are identical when

$$a_{11} = K_1 \qquad a_{12} = 0 \qquad a_{33} = K_\delta \qquad (7)$$

Table 1 gives the observed(*117–119*) frequencies for CO$_2$ and values for K_1 and K_δ/ℓ^2 obtained from ν_1 or ν_3. The two values of K_1 in Table 1 are fairly close indicating that the valence force is a reasonably good model. Using isotopic molecules, one can obtain additional equations for the force constants. In addition, information about the geometric structure and effects of Fermi resonance(*120*) also may be obtained, but these topics will not be discussed here.

TABLE 1

Fundamental Frequencies and Force Constants of CO$_2$ Molecule(*117–119*)

$\nu_1(cm^{-1})$	$\nu_2(cm^{-1})$	$\nu_3(cm^{-1})$	($\times 10^5$ dynes/cm)		
			K_1 from ν_1	K_1 from ν_3	K_δ/ℓ^2
1337	667	2349	16.8	14.2	0.57

It should be emphasized that the concept of normal vibrations rests on the assumption that the amplitudes of oscillations are infinitesimally small. Actually the amplitudes of the quantized oscillations, though small, are by no means infinitesimal and therefore the oscillations are more or less anharmonic. In other words, in addition to the quadratic terms, higher-order terms in the potential must be introduced into the wave equation. As a consequence, the energy is no longer a sum of independent terms corresponding to the different normal vibrations but contains cross terms corresponding to the vibrational quantum numbers of two or more normal vibrations. Classical analysis of anharmonic vibrations is rather lengthy and since it would not yield much more useful information relevant to this discussion, it will not be included here. However, the effects of anharmonicity on energy levels and on the determination of transition probabilities between CO$_2$ laser levels will be discussed later in a quantum mechanical treatment. It may be worth pointing out that the anharmonic corrections to the vibrational energy levels, in most cases, increase monotonically with increasing vibrational quantum numbers. Fortunately, CO$_2$ laser transitions involve the lowest vibrational levels, namely, $\nu_3 - \nu_1$ and $\nu_3 - 2\nu_2$, therefore the anharmonic corrections are not very large. It will be shown, however, that they are important in the calculation of transition probabilities. Certain IR bands which are forbidden in the harmonic approximation are partially allowed when anharmonic force constants are taken into account.

B. ENERGY LEVELS

In this section we shall present quantum mechanical results of the vibrational–rotational structure of the CO_2 molecule. For simplicity, we shall first consider the rotational and vibrational motions separately and then treat the mutual interaction of the two motions as a perturbation.

1. Rotational Energy Levels

In the electronic ground state of the CO_2 molecule, the angular momentum of the electrons about the internuclear axis is zero. Therefore, one can use the same treatment as for a diatomic molecule rotating about its equilibrium position. The energy levels are simply given by the well known formula as

$$\frac{E_r}{hc} = F(J) = BJ(J+1) = DJ^2(J+1)^2 + \cdots \tag{8}$$

where E_r is the rotational energy, $F(J)$ is the term value (in cm^{-1}), J is the rotational quantum number, and B is the rotational constant. The $DJ^2(J+1)^2$ and higher-order terms enter Eq. (8) because of the non-rigidity of the molecule and other effects which are small compared with the first term.

The population n_J of the various rotational levels can be described by the Boltzmann distribution as

$$n_J \simeq n_T\left(\frac{hcB}{kT}\right)g(J)\exp-\left\{F(J)\frac{hc}{kT}\right\} \tag{9}$$

where $g(J)$ is the statistical weight. For the CO_2 molecule or molecules belonging to point group $D_{\infty h}$, alternate rotational levels have different statistical weights. In the case of CO_2, spins of identical nuclei are zero. Therefore, the antisymmetric rotational levels are missing entirely. That is, for Σ_g^+ electronic states, the odd rotational levels are absent in accordance with Bose statistics. Therefore $g(J) = (2J+1)$. The thermal distribution of the rotational levels in the 00°1 upper vibrational state for $B = 0.38714$ cm^{-1} and $T = 400°$K is shown in Fig. 2. The J_{max} at

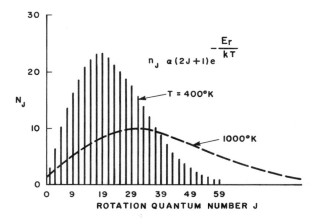

Fig. 2. Thermal distribution of rotational level population in the 00° 1 upper laser state.

which n_J is a maximum is obtained from Eq. (9) and is given by

$$J_{max} = \left(\frac{kT}{hcB}\right) - \frac{1}{2} \tag{10}$$

At $T = 400°\,K$, $J_{max} \simeq 19$, while at $T = 1000°\,K$, $J_{max} \simeq 29$, and the distribution of rotational level population spreads over a wider range of J values with correspondingly decreasing amplitude.

The rotational eigenfunction Ψ_r of CO_2 associated with these eigenstates are the surface harmonics such as

$$\Psi_r = \Theta_{JM}(\theta)e^{iM\phi} \tag{11}$$

where θ is the angle made by the axes of an inclined spinning top with respect to a spatially fixed z axis through the equilibrium position, and ϕ is its azimuth angle about the z axis; M is the magnetic quantum number which gives the components of J in the direction of the z axis in units of $h/2\pi$, and can have the values $J, J-1, \ldots, -J$; $\Theta_{JM}(\theta)$ represents the Jacobi (hypergeometric) polynomials. The selection rules for rotational transitions are $\Delta J = \pm 1$ for molecules having permanent dipole moment. Since the CO_2 molecule has no permanent dipole moment, transitions between rotational levels in a given vibrational state are forbidden.

2. Vibrational Energy Levels

In the harmonic approximation (displacements are small) the Schrödinger equation can be resolved into three uncoupled equations in terms of normal coordinates ξ_i:

$$\frac{1}{\Psi_i}\frac{d^2\Psi_i}{d\xi_i^2} + \frac{8\pi^2}{h^2}\left(E_i - \frac{1}{2}\lambda_i\xi_i^2\right) = 0 \qquad (12)$$

The solution of Eq. (12) is well known,

$$\Psi_i(\xi_i) = N_{v_i}H_{v_i}(\sqrt{\alpha_i}\xi_i)\exp\left(\frac{-\alpha_i}{2}\right)\xi_i^2 \qquad (13)$$

where H_{v_i} is the Hermite polynomials of degree v_i, $\alpha_i = 4\pi^2\nu_i/h$, and N_{v_i} is a normalization constant. The eigenvalues are

$$E_i = \hbar\omega_i(v_i + \tfrac{1}{2}) \qquad (14)$$

with $v_i = 0, 1, 2, \ldots$; the ω_i are given by Eq. (6). For the doubly degenerate vibration ν_2 where two of the ω's are the same, each of the mutually degenerate pair of vibrations gives its contribution $\omega_2/2$ to the zero-point energy, therefore,

$$E_2 = \hbar\omega_2(v_2 + 1) \qquad (14a)$$

and the corresponding eigenfunction is

$$\Psi_2 = N_{v2a}H_{v2a}(\sqrt{\alpha_2}\xi_{2a})H_{v2b}(\sqrt{\alpha_2}\xi_{2b})\exp(-\alpha_2/2)(\xi_{2a}^2 + \xi_{2b}^2) \quad (15)$$

In the harmonic approximation, the selection rules for both the IR and Raman transitions in a polyatomic molecule are the same as those for the diatomic molecule, that is, $\Delta v_i = \pm 1$ for each normal vibration. The occurrence of certain fundamentals in the IR or Raman spectrum depends on the presence of a change of either the dipole moment or the polarizability, respectively. Under symmetry operation, only the symmetric vibration ν_1 of the CO_2 molecule can change its quantum number v_1 by ± 1 owing to the Raman effect, whereas the bending ν_2 and asymmetric ν_3 vibrations of the CO_2 molecule can change their quantum numbers only by ± 1 owing to electric dipole interaction.

If the anharmonicity is taken into account, the total vibrational eigenfunction Ψ_v will include an additional term $\chi(\xi_1, \xi_{2a}, \xi_{ab}, \xi_3)$, which is small compared with the product $\Psi_1\Psi_{2a}\Psi_{2b}\Psi_3$. The term values for the vibrational energy as a result of anharmonicity are given by

$$G(v) = \sum_{i=1}^{3}\omega_i\left(v_i + \frac{d_i}{2}\right) + \sum_i\sum_k x_{ik}\left(v_i + \frac{d_i}{2}\right)\left(v_k + \frac{d_k}{2}\right) + \sum_i g_{ii}\ell_i^2 \quad (16)$$

where $d_i = 1$ for nondegenerate and 2 for doubly degenerate vibrations. The first summation in Eq. (16) is the dominant one; the double sum gives the higher-order correction terms which arise from cross coupling between different modes of vibration through the anharmonic force constants x_{ik}; the last summation is called the ℓ-type doubling, where g_{ii} represents small constants of the order of the x_{ik}. For non-degenerate vibrations, $\ell_i = 0$ and $g_{ii} = 0$. In the case of CO_2, only one degenerate vibration is excited. This mode of vibration is doubly degenerate on account of the equivalence of the two directions of the angular momentum ℓ. This double degeneracy can be removed by increasing rotation as a result of the Coriolis effect. Thus, for each J a splitting into two components occurs whose separation increases with increasing J.

For every vibrational state there exist a set of rotational levels, but with slightly different spacings for the different vibrational levels. It must be realized that for degenerate vibrational levels, J must be larger than or equal to ℓ, as $J = \ell, \ell + 1, \ell + 2, \ldots$; in other words, the first $\ell - 1$ rotational levels are missing in the degenerate vibrational state. Figure 3 gives the energy level diagram of a few low-lying vibrational states of the CO_2 molecule. Table 2 lists both the measured IR absorption and Raman spectra for transitions between levels corresponding to those shown in Fig. 3, as well as to a few neighboring levels. Two extremely strong absorption bands at 667.3 and 2349.3 cm^{-1} belong to the bending mode ν_2 (species π_u) and the asymmetric mode ν_3 (species Σ_u^+), respectively. The intense Raman spectrum consists of two lines at 1285.5 and 1388.3 cm^{-1} with an intensity ratio $1:0.6$, which correspond well with the ν_1 and $2\nu_2$ modes.

It should be pointed out that the lower laser levels 02⁰0 and 10⁰0 are almost in resonance, which leads to a perturbation of the energy levels, as recognized first by Fermi(121), owing to the anharmonic terms in the potential energy. As a result, the 10⁰0 level is shifted up and the 02⁰0 level is shifted down so that the actual separation of the two levels is much greater than expected. At the same time a mixing of the eigen-functions of the two levels occurs. The smaller the original energy difference of the two levels, the stronger the mixing. In addition, the magnitude of the perturbation depends on the value of the corresponding matrix element $E_{1,2}$ of the perturbation function V' as

$$E_{1,2} = \langle 10^00|V'|02^00\rangle. \tag{17}$$

where V' is given essentially by the anharmonic terms in the potential energy, and Ψ_{10^00}, Ψ_{02^00} are the zero approximation harmonic oscillator eigenfunctions of the two interacting vibrational levels. The perturbed energy E is given by

$$E = \tfrac{1}{2}(\delta_+) \pm \tfrac{1}{2}(4|E_{1,2}|^2 + \delta_-^2)^{1/2} \tag{18}$$

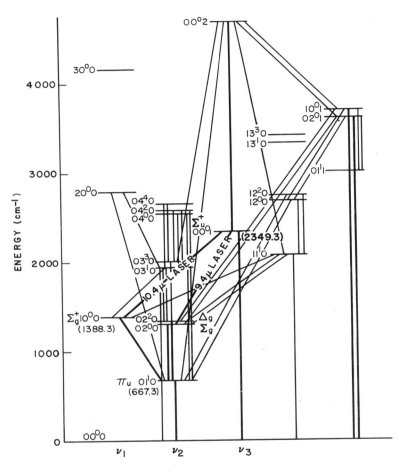

Fig. 3. Energy level diagram of low-lying vibrational levels of the CO_2 molecule.

when δ_+ and δ_- are the sum or difference of the two unperturbed eigenvalues of the 10^00 and 02^00 levels, respectively. It should be noted that 02^00 is only one of the two sublevels belonging to the $2\nu_2$ mode. The other sublevel, 02^20 (species Δg) also lies very close to the 10^00 level. Because of the symmetry operation, the rule requires that only two vibrational levels of the same species can interact with the other; therefore only the Σ_g^+ (02^00) sublevel of the 020 level can perturb the 10^00 level which has the species Σ_g^+. The energy level values of the two states 10^00 (1388.3 cm^{-1}) and 02^00 (1285.5 cm^{-1}) as given by Eq. (18) differ considerably from $2\nu_2$ ($\nu_2 = 667.3$ cm^{-1}). For the same

TABLE 2

Low Lying Infrared and Raman Bands of CO$_2$

Upper state		Lower state		ν_{vac} Observed(114) (cm^{-1})
$\nu_1\nu_2\nu_3$	Species	$\nu_1\nu_2\nu_3$	Species	
01^10	π_u	00^00	Σ_g^+	667.3
02^00	Σ_g^+	00^00	Σ_g^+	1285.5
10^00	Σ_g^+	00^00	Σ_g^+	1388.3
03^10	π_u	00^00	Σ_g^+	1937.5
00^01	Σ_u^+	00^00	Σ_g^+	2349.3
02^00	Σ_g^+	01^10	π_u	618.1
02^20	Δ_g	01^10	π_u	668.3
10^00	Σ_g^+	01^10	π_u	720.5
03^10	π_u	01^10	π_u	1264.8
04^00	Σ_g^+	01^10	π_u	1886
20^00	Σ_g^+	01^10	π_u	2137
03^10	π_u	02^20	Δ_g	596.8
03^10	π_u	02^00	Σ_g^+	647.6
00^01	Σ_u^+	10^00	Σ_g^+	960.8
00^01	Σ_u^+	02^00	Σ_g^+	1063.6
04^20	Δ_g	02^20	Δ_g	1242
04^20	Δ_g	02^00	Σ_g^+	1305.1
20^00	Σ_g^+	02^00	Σ_g^+	1528

reason, the splitting of the two sublevels $\ell_2 = 0$ and 2 of the 020 state is anomalously large (49.9 cm^{-1}). As a consequence of the strong perturbation (Fermi resonance), a strong mixing of the eigenfunctions of the two levels occurs so that the two observed levels can no longer be unambiguously designated as 10^00 and 02^00. Each actual level is a mixture of two. The corresponding eigenfunctions of these two mixed states $(10^00, 02^00)'$ and $(10^00, 02^00)''$ can be expressed in terms of a linear combination of Ψ_{10^00} and Ψ_{02^00} as

$$\Psi_{(10^00, 02^00)'} = a_+\Psi_{10^00} - a_-\Psi_{02^00}$$
$$\Psi_{(10^00, 02^00)''} = a_-\Psi_{10^00} + a_+\Psi_{02^00} \tag{19}$$

where

$$a_\pm = \left(\frac{(4|E_{1,2}|^2 + \delta_-^2)^{1/2} \pm \delta_-}{2(4|E_{1,2}|^2 + \delta_-^2)^{1/2}}\right)^{1/2} \tag{20}$$

These results are valid not only for the perturbation owing to the effect of one adjacent level but also for the integrated perturbing effect of a large number of vibrational levels of which each can contribute a term in $|E_{1,2}|$ through the anharmonic terms in the potential energy. We shall see the significance of these effects later in Section III.E and in Section IV.B on relaxation processes.

C. VIBRATIONAL–ROTATIONAL SPECTRA

The energy of a linear symmetric molecule can be obtained to a good approximation simply by adding the rotational $hcF(J)$ given by Eq. (8) and the vibrational energy $hcG(v)$ given by Eq. (16). Actually, the two types of motion occur simultaneously and give rise to the fine structure of IR and Raman bands. The coupling between the rotational and vibrational motions introduces additional terms in the energy expression which will be discussed later in Section III.F.3.

We shall now briefly consider the symmetry properties of the eigenfunctions and the selection rules for vibrational–rotational transitions. The rotational levels of linear molecules are positive or negative depending on whether the sign of the total eigenfunction

$$\Psi_T = \Psi_e \Psi_r \Psi_v \tag{21}$$

remains unchanged or changed upon an inversion. The Ψ_e is the eigenfunction for the ground electronic state $^1\Sigma$ and is totally symmetric. In all symmetric excited vibrational levels (species Σ^+), the even rotational levels are positive and the odd are negative. In antisymmetric vibrational levels (species Σ^-), the converse is true. In $\pi, \Delta, \cdots$ vibrational levels, for each J value there is a positive and a negative level of slightly different energy whose order alternates as: $+-, -+, +-, \cdots$ or $-+, +-, -+, \cdots$. This double degeneracy for π, Δ vibrational levels, as mentioned before, is a result of the equivalence of the two directions of the angular momentum ℓ, thus a splitting into two components for each J occurs whose separation increases with increasing J. As an example, for species π, the ℓ-type splitting would give a term $g_{22}\ell_2^2$ in Eq. (16).

The selection rules for the vibrational–rotational transitions in the IR spectrum are

$$\Delta l = 0, \pm 1, \quad \Sigma^+ \leftrightarrow \Sigma^-, \quad g \not\leftrightarrow g, \quad u \not\leftrightarrow u$$
$$\Delta J = 0, \pm 1, \quad (J=0) \not\leftrightarrow (J=0), \quad + \not\leftrightarrow -, \quad s \not\leftrightarrow a \tag{22}$$

where s and a designate the symmetric and antisymmetric Ψ_R and the symbols $\leftrightarrow$ and $\not\leftrightarrow$ represent "allowed" and "not allowed," respectively.

Specifically, two of the strongest CO_2 lasers arise from the vibrational–rotational transitions of the Σ_u^+–Σ_g^+ vibrational bands (00^01–10^00) and (00^01–02^00), at laser wavelengths near their band edge 10.4

and $9.4\,\mu$, respectively. Laser transitions have been obtained from both the P branch ($\Delta J = -1$) and the R branch ($\Delta J = +1$) of each band. The Q branch ($\Delta J = 0$) is not allowed because transitions occur between two Σ states for which $\ell = 0$. Figure 4 gives a detailed transition diagram for laser oscillations from both the P and R branches of the $00^{0}1$–$10^{0}0$ and $00^{0}1$–$02^{0}0$ bands.

D. CO₂ LASER TRANSITIONS

The most complete measurements on laser transitions were made by Frapard et al.(122) using a prism inside the cavity in order to avoid the strong competition between rotational levels, especially in a

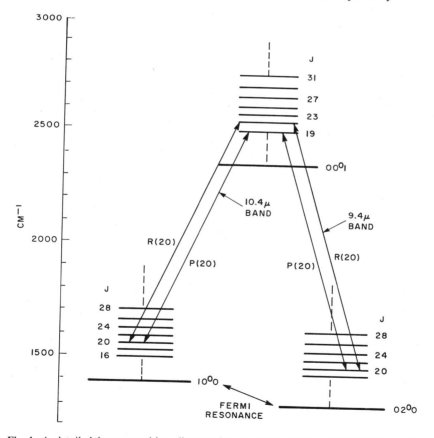

Fig. 4. A detailed laser transition diagram for the $00^{0}1$–$10^{0}0$ and $00^{0}1$–$02^{0}0$ bands, including rotational levels.

high-power CO_2 laser system (these competition effects will be discussed in great detail later). Their results along with the measurements by Patel(2) are given in Tables 3–6. Early investigations(1–11) of CO_2 laser spectroscopy were performed with laser systems which did not incorporate a dispersive element in the feedback interferometer. In these systems the larger the number of simultaneous laser transitions, the weaker the laser output. For example, Patel(2) observed a total of 14 simultaneously oscillating laser lines in the P branch of the

TABLE 3

Measured CO_2 Laser Wavelengths of the P Branch
of the $00^{0}1–10^{0}0$ Vibration–Rotation Transitions

Measured laser wavelength in vac. (μ)	Frequency (cm^{-1})	Transition[a,b] $(00^{0}1)–(10^{0}0)$
10.4410	957.76	$P(4)$
10.4585	956.16	$P(6)$
10.4765	954.52	$P(8)$
10.4945	952.88	$P(10)$
10.5135	951.16	$P(12)$
10.5326	949.43	$P(14)$
10.5518	947.70	$P(16)$
10.5713	945.96	$P(18)$
10.5912	944.18	$P(20)$
10.6118	942.35	$P(22)$
10.6324	940.52	$P(24)$
10.6534	938.67	$P(26)$
10.6748	936.78	$P(28)$
10.6965	934.88	$P(30)$
10.7194	932.89	$P(32)$
10.7415	930.96	$P(34)$
10.7648	928.95	$P(36)$
10.7880	926.95	$P(38)$
10.8120	924.90	$P(40)$
10.8360	922.85	$P(42)$
10.8605	920.77	$P(44)$
10.8855	918.65	$P(46)$
10.9110	916.51	$P(48)$
10.9360	914.41	$P(50)$
10.9630	912.16	$P(52)$
10.9900	909.92	$P(54)$
11.0165	907.73	$P(56)$

[a]See Ref. (2). A total of 14 laser transitions, $P(12)$ to $P(38)$.
[b]See Ref. (122). Additional 13 laser transitions, $P(4)$ to $P(10)$ and $P(40)$ to $P(56)$.

TABLE 4

Measured CO$_2$ Laser Wavelengths of the R Branch
of the $00^0 1$–$10^0 0$ Vibration–Rotation Transitions

Measured laser wavelength in vac. (μ)	Wave number (cm^{-1})	Transition[a,b] ($00^0 1$)–($10^0 0$)
10.3655	964.74	$R(4)$
10.3500	966.18	$R(6)$
10.3335	967.73	$R(8)$
10.3190	969.09	$R(10)$
10.3040	970.50	$R(12)$
10.2860	971.91	$R(14)$
10.2855	972.24	$R(16)$
10.2605	974.61	$R(18)$
10.2470	975.90	$R(20)$
10.2335	977.18	$R(22)$
10.2200	978.47	$R(24)$
10.2075	979.67	$R(26)$
10.1950	980.87	$R(28)$
10.1825	982.08	$R(30)$
10.1710	983.19	$R(32)$
10.1590	984.35	$R(34)$
10.1480	985.42	$R(36)$
10.1370	986.49	$R(38)$
10.1260	987.56	$R(40)$
10.1150	988.63	$R(42)$
10.1050	989.61	$R(44)$
10.0955	990.54	$R(46)$
10.0860	991.47	$R(48)$
10.0760	992.46	$R(50)$
10.0670	993.34	$R(52)$
10.0585	994.18	$R(54)$

[a]See Ref. (*124*). A total of 7 laser transitions, $R(14)$ to $R(26)$.
[b]See Ref. (*122*). Additional 19 laser transitions, $R(4)$ to $R(12)$ and $R(28)$ to $R(54)$.

$00^0 1$–$10^0 0$ band from a laser using only CO$_2$ as the active medium; the laser output was only in the milliwatt range. However, when a mixture of gases such as CO$_2$–N$_2$–He is used, the laser output increases more than four orders of magnitude as demonstrated by the work of Patel et al. (*11*). When the output was examined by a rapid scanning technique (*123*) only a few laser lines, namely, $P(18)$, $P(20)$, and $P(22)$, were found to oscillate at one time. The first observation of CO$_2$ laser oscillations in the R branch of the $00^0 1$–$10^0 0$ band was made by Howe (*124*) in a flowing CO$_2$–air laser system. Howe's laser oscillated

TABLE 5

Measured CO_2 Laser Wavelengths of the P Branch of
the $(00^0 1)$–$(02^0 0)$ Vibration–Rotation Transitions

Measured laser Wavelength in vac. (μ)	Wave number (cm^{-1})	Transition[a,b] $(00^0 1)$–$(10^0 0)$
9.4285	1060.61	$P(4)$
9.4425	1059.04	$P(6)$
9.4581	1057.30	$P(8)$
9.4735	1055.58	$P(10)$
9.4885	1053.91	$P(12)$
9.5045	1052.13	$P(14)$
9.5195	1050.47	$P(16)$
9.5360	1048.66	$P(18)$
9.5525	1046.85	$P(20)$
9.5690	1045.04	$P(22)$
9.5860	1043.19	$P(24)$
9.6035	1041.29	$P(26)$
9.5210	1039.34	$P(28)$
9.6391	1037.44	$P(30)$
9.6575	1035.46	$P(32)$
9.6760	1033.48	$P(34)$
9.6941	1031.56	$P(36)$
9.7140	1029.44	$P(38)$
9.7335	1027.38	$P(40)$
9.7535	1025.27	$P(42)$
9.7735	1023.17	$P(44)$
9.7940	1021.03	$P(46)$
9.8150	1018.85	$P(48)$
9.8360	1016.67	$P(50)$
9.8575	1014.46	$P(52)$
9.8790	1012.25	$P(54)$
9.9010	1010.00	$P(56)$
9.9230	1007.76	$P(58)$
9.9465	1005.38	$P(60)$

[a]See Ref. (6). A total 14 laser transitions, $P(10)$ to $P(36)$.
[b]See Ref. (122). Additional 15 laser transitions, $P(4)$ to $P(8)$ and $P(38)$ to $P(60)$.

in seven transitions and had a low power output. Laser oscillations in the P branch of the $00^0 1$–$02^0 0$ band, on the other hand, were first observed by Barchewitz et al.(5,6). In addition to these $00^0 1$–$10^0 0$ and $00^0 1$–$02^0 0$ bands, Frapard et al.(122) have reported CO_2 laser oscillations in the P branch of the $01^1 1$–$03^1 0$ band near 11 μ. These results are given in Table 7. Other laser transitions in the wavelength range from 11 to 18 μ also have been reported(125,126). A total of 16

TABLE 6

Measured CO$_2$ Laser Wavelengths of the R Branch of
the $(00^0 1)$–$(02^0 0)$ Vibration–Rotation Transitions

Measured laser wavelength in vac. (μ)	Wave number (cm^{-1})	Transition[a,b] $(00^0 1)$–$(10^0 0)$
9.3677	1067.50	$R(4)$
9.3555	1068.89	$R(6)$
9.3420	1070.43	$R(8)$
9.3295	1071.87	$R(10)$
9.3172	1073.28	$R(12)$
9.3055	1074.63	$R(14)$
9.2937	1076.00	$R(16)$
9.2825	1077.30	$R(18)$
9.2715	1078.57	$R(20)$
9.2605	1079.85	$R(22)$
9.2500	1081.08	$R(24)$
9.2397	1082.29	$R(26)$
9.2295	1083.48	$R(28)$
9.2197	1084.63	$R(30)$
9.2103	1085.74	$R(32)$
9.2010	1086.84	$R(34)$
9.1920	1087.90	$R(36)$
9.1830	1088.97	$R(38)$
9.1740	1090.04	$R(40)$
9.1660	1090.99	$R(42)$
9.1575	1092.00	$R(44)$
9.1490	1093.01	$R(46)$
9.1420	1093.85	$R(48)$
9.1340	1094.81	$R(50)$
9.1265	1095.71	$R(52)$

[a]See Ref. (4). A total 7 laser transitions, $R(16)$ to $R(28)$.
[b]See Ref. (122). Additional 18 laser transitions, $R(4)$ to $R(14)$ and $R(30)$ to $R(52)$.

laser transitions has been precisely measured(125) and assigned to the P branch of the $01^1 1$–$11^1 0$ band as given in Table 8. Thirteen additional laser transitions have been assigned(126) to $14^0 0$–$05^1 0$ (13.2 μ), $21^1 0$–$12^1 0$ (13.5 μ), $14^0 0$–$13^1 0$ (16.6 μ), $03^1 1$–$02^2 1$ (17.0 μ), and $24^0 0$–$23^1 0$ (17.4 μ) bands. Laser intensities from these bands are much weaker than those oscillating in the $00^0 1$–$10^0 0$ and $00^0 1$–$02^0 0$ bands and most of them oscillate only in the pulsed mode. Detailed information concerning these weaker CO$_2$ laser transitions can be found in Ref.(126). With present day technology, only a few additional laser lines can be added(127) to those that have already been observed.

TABLE 7

Measured(*122*) CO_2 Laser Wavelengths of the *P* Branch of the 01^01–
03^10 Vibration–Rotation Transitions

Wavelength in vac. (μ)	Wave number (cm^{-1})	Transition
10.9735	911.29	*P*(19)
10.9951	909.50	*P*(21)
11.0165	907.73	*P*(23)
11.0300	906.62	*P*(24)
11.0385	905.92	*P*(25)
11.0535	904.69	*P*(26)
11.0610	904.08	*P*(27)
11.0760	902.85	*P*(28)
11.0850	902.12	*P*(29)
11.1000	900.90	*P*(30)
11.1070	900.33	*P*(31)
11.1235	899.00	*P*(32)
11.1315	898.35	*P*(33)
11.1485	896.98	*P*(34)
11.1555	896.42	*P*(35)
11.1736	894.97	*P*(36)
11.1791	894.53	*P*(37)
11.1980	893.02	*P*(38)
11.2035	892.58	*P*(39)
11.2235	890.99	*P*(40)
11.2295	890.51	*P*(41)
11.2495	888.93	*P*(42)
11.2545	888.53	*P*(43)
11.2770	886.76	*P*(44)
11.2804	886.49	*P*(45)

E. Vibrational Transition Probabilities and Radiative Lifetimes

The radiative transition probabilities between two states m and n are given by the Einstein A coefficient

$$A_{mn} = \frac{1}{2J_m + 1}\left(\frac{64\pi^4}{3h\lambda^3}\right)S_{mn} \tag{23}$$

where S_{mn} is the line strength. It is given by the sum of the squares of all matrix elements,

$$S_{mn} = \sum_{m,n} |\langle J_mM_m|p|J_nM_n\rangle|^2 \tag{24}$$

TABLE 8

Measured(*125*) CO$_2$ Laser Wavelengths of the *P* Branch of the 01^11– 11^10 Vibration–Rotation Transitions

Wavelength in vac. (μ)	Wave number (cm^{-1})	Transition
10.9730	911.33	*P*(19)
10.9856	910.28	*P*(20)
10.9944	909.55	*P*(21)
11.0078	908.45	*P*(22)
11.0164	907.74	*P*(23)
11.0300	906.62	*P*(24)
11.0385	905.92	*P*(25)
11.0529	904.74	*P*(26)
11.0610	904.08	*P*(27)
11.0762	902.84	*P*(28)
11.0840	902.20	*P*(29)
11.0999	900.91	*P*(30)
11.1073	900.31	*P*(31)
11.1238	898.97	*P*(32)
11.1309	898.40	*P*(33)
11.1483	897.00	*P*(34)

where J and M are the angular momenta and the z components, respectively, and p is the dipole operator.

Statz et al.(*128*) have calculated the line strengths for a number of transitions among the low-lying vibrational levels of CO$_2$. Instead of the usual perturbation approach, they constructed the eigenfunction by a linear combination of a large number of levels which are connected by "matrix elements" with the levels of interest, i.e., 00^00, 01^10, 10^00, 02^00, 00^01. This involves the diagonalization of large Hamiltonian matrices, in some cases as large as 30×30. In the Hamiltonian, anharmonic forces were included and the potential energy contains terms up to fourth order. Contributions from these terms are significant in that some of the forbidden transitions in the harmonic approximation become partially allowed. The magnitude of the dipole moment as a function of various normal coordinates also enters in the determination of transition probabilities. Quadratic and higher-order terms in the normal coordinates are involved, especially for the forbidden transitions. However, they do not contribute to the calculated line strength significantly. Of all the levels considered, only the 00^01–00^00, 00^02–00^01, 01^10–00^00, and 02^20, 02^00–01^10 transitions are allowed in the absence of anharmonicities. Table 9 presents the Einstein coefficients

TABLE 9

Einstein A Coefficient(128) for Some Vibrational Transitions in CO_2

Transition		Transition probability (sec^{-1})	
		Untrapped	Trapped
$00^0 1 \rightarrow 00^0 0$	R	2.0×10^2	8.8
	P	2.1×10^2	10.1
$00^0 1 \rightarrow 02^0 0$	R	0.19	
	P	0.20	
$00^0 1 \rightarrow 10^0 0$	R	0.33	
	P	0.34	
$00^0 2 \rightarrow 00^0 1$	R	3.9×10^2	
	P	4.1×10^2	
	R	0.48	0.48
$01^1 0 \rightarrow 00^0 0$	P	0.46	0.46
	Q	0.94	
	R	0.22	
$02^0 0 \rightarrow 01^1 0$	P	0.26	
	Q	0.48	
	R	1.07	
$02^0 0 \rightarrow 01^1 0$	P	0.84	
	Q	1.89	
	R	0.20	
$10^0 0 \rightarrow 01^1 0$	P	0.23	
	Q	0.44	

for several transitions calculated(128) by choosing suitable values for anharmonic force constants to best fit the experimental data. Table 10 gives the radiative lifetimes calculated(128) from the A coefficients. The transition lifetime for the $00^0 1$–$10^0 0$ band, on the other hand, has been determined to be 4.7 sec from the absorption measurements ($29,30$) using the CO_2 laser (as discussed in the following subsection). From these results it is clear that radiative processes cannot account for the laser action since the radiative lifetimes of the lower laser levels are two to three orders of magnitude longer than that of the upper laser level. Relaxation of these lower laser levels $10^0 0$ and $02^0 0$ with relatively long radiative lifetimes by collisions with other molecules or atoms must play an important role in the CO_2 laser. A detailed discussion is given in Section IV. B.

TABLE 10

Radiative Lifetimes (*128*) of Some Vibrational Levels in CO$_2$

Level	Lifetime (sec)	
	Untrapped	Trapped
00^{0}1	2.4 × 10^{-3}	5.0 × 10^{-2}
00^{0}2	1.3 × 10^{-3}	
01^{1}0	1.1	2.8 (even J), 1.1 (odd J)
02^{0}0	1.0	
02^{2}0	0.26	
10^{0}0	1.1	

F. CO$_2$ LASER SPECTROSCOPIC MEASUREMENTS

The extremely narrow linewidth of CO$_2$ lasers as radiation sources allows spectroscopic studies of the CO$_2$ molecule with high precision. In this section we shall present both the absorption and transmission spectra of a CW CO$_2$ laser through and absorption cell filled with CO$_2$ gas and the measurements of the rotational constants of the CO$_2$ molecule by a heterodyne of two stable CW CO$_2$ lasers operating on any two by a heterodyne of two stable CW CO$_2$ lasers operating on any two neighboring transitions.

1. Absorption Spectra

The absorption coefficient has been measured (*29,30*) by using a single vibration–rotation laser line, 10.6 μ P(20) transition, through an absorption cell filled with pure CO$_2$ gas. For an inhomogeneously or Doppler broadened line, the absorption coefficient a at the line center is given by

$$a(\lambda_0) = -\frac{1}{8\pi} \frac{\lambda_0^3}{c\tau_r} \left(\frac{Mc^2}{2\pi kT}\right)^{1/2} \left[n_u - \frac{g_u}{g_\ell} n_\ell\right] \qquad (25)$$

where τ_r is the radiative lifetime of the vibration–rotation transition, k is the Boltzmann constant, T is gas temperature, M is the mass of the CO$_2$ molecule, n_u and n_ℓ are the upper- and lower-level population, respectively, and g_u and g_ℓ the corresponding statistical weights. For a collision-broadened case, a is given by

$$a(\lambda_0) = -\frac{1}{8\pi} \frac{\lambda_0^2}{\tau_r \nu_c} \left[n_u - \frac{g_u}{g_\ell} n_\ell\right] \qquad (26)$$

where ν_c is the collision frequency and all other symbols have their usual meanings. In the Doppler broadened region a is proportional to the gas pressure P because both n_u and n_ℓ are directly proportional to P. But in the case of collision broadening, a should be independent of P because ν_c is also proportional to P. The measured absorption coefficient(30) as a function of the CO_2 gas pressure is shown for the $P(20)$ line in Fig. 5. The combined Doppler and collision contour which best fits the data is represented by the solid curve. These measurements yield a collision frequency $\nu_c = 7.8 \pm 0.8 \times 10^9 \, \text{sec}^{-1}$, an optical broadening cross section of $5.7 \times 10^{-15} \, \text{cm}^2$ for collisions with CO_2 molecules and a transition lifetime $\tau_r = 4.7 \pm 0.5$ sec. The break in the absorption curve occurs at $P = 5.2$ Torr of CO_2, indicating that if the gas pressure in a CO_2 laser exceeds this value, the linewidth would be homogeneously broadened.

2. *Transmission Spectra*

The transmission spectra of the $00^0 1 - 10^0 0$ laser lines in both the P and R branches for $R(2)$ to $R(40)$ and $P(4)$ to $P(42)$ have been measured(129) by using a White absorption cell with a total path length of 10 m. From data on transmission vs. CO_2 gas pressure, one can determine the line intensity S_J and the half-width Δ_J. Figure 6 shows typical transmission data for both the P- and R-branch laser lines at a fixed CO_2 pressure of 300 Torr. The upper curve is the scan of laser frequencies in the absence of absorbing gas and the lower curve shows the attenuation of these lines. It is noteworthy that strong lines near the band center are absorbed more strongly than the weaker ones in the wings. This observation suggests that it may be more desirable for laser communication through atmosphere to use laser lines in the wings of the band at which the least absorption loss is assumed.

The spectral transmittance t at the peak of the collision-broadened line is given by

$$t = \exp\left[\frac{-S_J L}{\pi \Delta_J}\right] \tag{27}$$

where L is the total path length through the absorption cell. The transmittance was found to be independent of gas pressure for pressures above 50 Torr. Careful measurements of t were made(129) for a number of selected lines in the 10.4-μ band in the collision-broadened region. The values S_J/Δ_J (in cm^{-1}) at $300°$ K for these lines are deduced from the transmittance measurements. Results are summarized in Table 11, for which each of those values represents about 10 independent measurements. The dependence of linewidth Δ_J ($\text{cm}^{-1}\text{–atm}^{-1}$) at

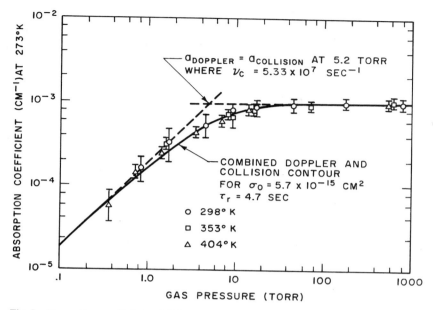

Fig. 5. Absorption coefficient of CO_2 gas for the 10.6-μ CO_2 laser as a function of CO_2 density. The absorption coefficients are scaled to 273° K, using the equilibrium temperature dependence for the lower state $J = 20$. The expected temperature dependence is verified by the fact that the data lie on a single curve, while the measured absorption coefficients in the high density limit varied by a factor of five for the temperature range shown. See text. [After Gerry and Leonard(*30*).]

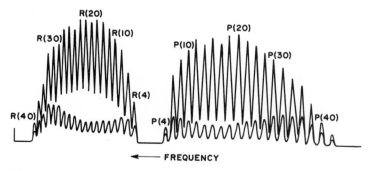

Fig. 6. The upper traces show CO_2 laser emission in the P and R branches of the 00°1– 10°0 band without attenuation. These spectra were obtained by rotating a diffraction grating as one of the cavity mirror. The lower traces show the same emission after attenuation by a 10-m path of pure CO_2 (300 Torr), equivalent to a path of approximately 30 km of air. [After Oppenheim and Devir(*129*).]

TABLE 11

Ratios of Laser Line Intensity S_J
to the Half-Width Δ_J for Selected
CO_2 Laser Transitions in the
10.4-μ Band at 300° K (129)

Transition	S_J/Δ_J (10^3 cm^{-1})
P(4)	2.00 ± 0.11
P(8)	3.72 ± 0.12
P(18)	5.84 ± 0.16
P(32)	2.83 ± 0.05
P(40)	1.35 ± 0.05
R(2)	1.54 ± 0.10
R(6)	3.53 ± 0.09
R(16)	5.87 ± 0.13
R(26)	4.75 ± 0.18
R(30)	3.34 ± 0.18
R(40)	1.30 ± 0.11

300° K on the rotational quantum number m for the 10.4-μ band CO_2 laser is shown in Fig. 7, where $m = -J$ for the P branch and $m = J + 1$ for the R branch. For the P(20) line, one obtains from Oppenheim and Devir's results (129) a value of ~ 3.3 MHz/Torr as the collision-broadened linewidth. Similar measurements were made earlier by Rossetti et al. (130), and are shown in Fig. 7.

3. Rotational Constants

The advantage of the narrow laser linewidth can be utilized also to obtain reliable information about the fine structure constants of the molecule. The coupling between the rotational and vibrational motion results in a slight modification of B and D values in Eq. (8) both of which vary with v_i. The change in B_v is owing to a change in the moment of inertia I during a vibration so that I does not equal I_e (value for the equilibrium position or the rigid rotator). The change in D_v is caused by the effects of centrifugal stretch; however, for practical purposes its dependence on v usually can be neglected. The modification of B takes on the form similar to that for diatomic molecules,

$$B_v = B_e - \sum_{i=1}^{3} \alpha_i \left(v_i + \frac{d_i}{2} \right) \tag{28}$$

where $d_i = 1$ for nondegenerate and 2 for doubly degenerate vibrations,

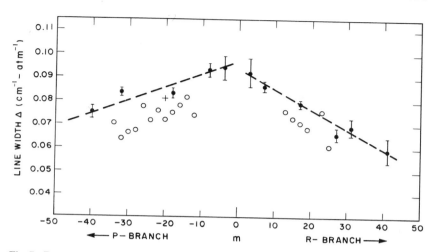

Fig. 7. Dependence of line width $\Delta_J(\text{cm}^{-1}\,\text{atm}^{-1})$ at 300°K on rotational quantum number m for the 10.4-μ band of CO₂. ●, Oppenheim et al.(*129*); ○, Rossetti et al.(*130*); +, McCubin et al.(*29*).

and α_1 are the fine structure constants. Other interactions, such as the Coriolis effect, would also contribute(*114*) a small correction to these constants; however, we shall not go into details here.

The B_v and D_v values recently(*131*) have been measured accurately using two stabilized CW CO₂ lasers. The experiment involves a mixing of stabilized vibration–rotation laser lines in a GaAs mixer. The beat frequencies, in the millimeter-wave region of 50–80 GHz, are measured for 37 pairs of transitions to an accuracy better than 1 MHz. An expression for beat frequencies, f, can be derived(*131*) from the term values $T(v, J)$ of vibrational–rotational energy levels which are the sum of Eqs. (8) and (16) as

$$f_P = T(v_u, J-2) - T(v_\ell, J-1) - T(v_u, J) - T(v_\ell, J+1)$$
$$f_R = T(v_u, J+2) - T(v_\ell, J+1) - T(v_u, J-2) - T(v_\ell, J-3)$$

$$(29)$$

where f_P and f_R are the beat frequencies in cm⁻¹ of two adjacent laser lines in the P and R branches, respectively, and J is the rotational quantum number (a running odd integer). The measured beat frequencies for 37 pairs of transitions were analyzed by a digital computer using a multiple regression method. The deduced rotational constants for 00^01, 10^00, and 02^00 levels are presented in Table 12, along with the best values obtained from the conventional spectroscopic data.

TABLE 12

Rotational Constant Measurements from Mixing CO_2 Lasers

| Constant | CO_2 Laser measurement[131] | | Previous best value[a,b] |
	(MHz)	(cm^{-1})	(cm^{-1})
B_{00^01}	11606.180 (±0.011)	387140.44 (±0.37) × 10^{-6}	387132 (±40) × 10^{-6}
D_{00^01}	39.728 (±0.063) × 10^{-4}	13.252 (±0.021) × 10^{-8}	13.45 (±0.50) × 10^{-8}
$B_{10^00}-B_{00^01}$	91.3584 (±0.0014)	3047.389 (±0.046) × 10^{-6}	3069 (±10) × 10^{-8}
$D_{10^00}-D_{00^01}$	−5.506 (±0.011) × 10^{-4}	−1.8366 (±0.0035) × 10^{-8}	−0.65 (±0.7) × 10^{-8}
$B_{02^00}-B_{00^01}$	100.1534 (±0.0019)	3340.757 (±0.063) × 10^{-6}	3344 (±10) × 10^{-6}
$D_{02^00}-D_{00^01}$	7.140 (±0.016) × 10^{-4}	2.3816 (±0.0053) × 10^{-8}	2.65 (±1.2) × 10^{-8}

[a] C. P. Courtoy, *Can. J. Phys.*, **35**, 608 (1957).
[b] R. Oberly and K. N. Rao, *J. Mol. Spect.*, **18**, 73 (1965).

IV. Population Inversion Mechanisms

In this section we shall be concerned mainly with the processes that produce population inversion between the $00^0 1$–$10^0 0$ and $00^0 1$–$02^0 0$ vibration–rotation levels of CO$_2$. According to the laws of thermo-dynamics, external excitation of some kind is required for all laser systems.† In the case of the CO$_2$ laser, collisional relaxation processes also play a vital role in the over-all inversion scheme. Collisional relaxation by itself is a rather complicated and thoroughly investigated field which involves a large number of kinetic processes. Before dis-cussing the collisional processes of relevant CO$_2$ levels, especially $10^0 0$, $02^0 0$, and $01^1 0$, we shall consider the processes which produce the $00^0 1$ upper laser level population.

A. Excitation Processes

It was suggested early by Patel[3] that population inversion may exist among certain vibration–rotation transitions in the $00^0 1$ and $10^0 0$ bands even though the population $n_v = \Sigma_J \, n_J$ of the upper vibrational level is slightly less than that of the lower laser level. The main argu-ment rests on the fact that within a given vibrational state, n_J is dis-tributed according to Eq. (9). Although this argument may lead to a small population inversion among some of the vibration–rotation transitions between two vibrational levels, it cannot possibly yield very high gain because only a small fraction of population density satisfies this condition known as "partial inversion." The two principal pumping mechanisms are now recognized as (1) direct electron impact and (2) resonance transfer of energy between N$_2$ ($v = 1$) and CO$_2$ ($00^0 0$).

1. *Electron Impact*

Measurements of rotational and vibrational excitation cross sections of CO$_2$ and N$_2$ by inelastic collisions of electrons at low energies (0–3 eV) have been reviewed by Phelps(*132*). Since then more accurate measurements have been made by Stamatovic and Schulz (*133*) for electron energy within 0.05 eV of threshold for all fundamen-tal modes in CO$_2$. As a result considerable knowledge has recently been added concerning excitation processes and cross-section mea-surements relevant to CO$_2$ lasers. Figure 8 shows the cross sections

---

†Recently a chemical CO$_2$ laser without external excitation sources has been reported by T. A. Cool [see Ref. (*216*)]. However this laser still requires external power to maintain a high-speed gas flow through the laser interacting region.

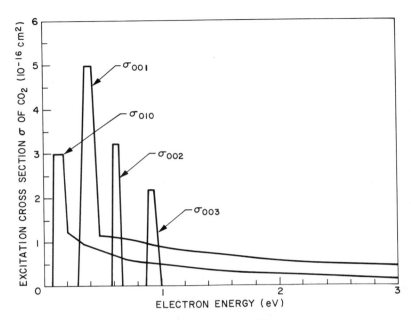

Fig. 8. Cross sections for vibrational excitation of CO_2 by electron impact. See text. [After Hake and Phelps(*17*).]

which were derived by Hake and Phelps(*17*) for the vibrational excitation of CO_2 molecules by electrons. Results show that there exists a set of four resonances at 0.08, 0.3, 0.6, 0.9 eV, two of which have high energy tails.† The resonances at 0.3, 0.6, and 0.9 eV have been observed by Boness and Schulz(*18*) who used high-resolution electron beam techniques. Hake and Phelps associate the 0.3, 0.6, and 0.9 eV energy loss processes with the three levels of the asymmetric mode (ν_3, $2\nu_3$, and $3\nu_3$) of the CO_2 molecule, and 0.08 eV is associated with the lowest bending mode (ν_2).

The unique feature of these results for CO_2 is that the resonances in each case occur very close to the threshold for onset of the particular energy loss process. This is quite different from the situation which is observed in diatomic gases, such as N_2, where the resonances occur at energies well above the ~ 0.3 eV threshold for $v = 1$ as shown in Fig. 9. These results indicate that the probability of excitation of N_2 ($v = 1$) by

†A. V. Phelps has pointed out (private communication) that the cross sections given in Fig. 8 may be significantly in error with regard to their energy dependence. Only the over-all value of the sum of the 0.3, 0.6, and 0.9 eV energy loss processes and the threshold energy dependence of the 0.08 eV process are considered reliable.

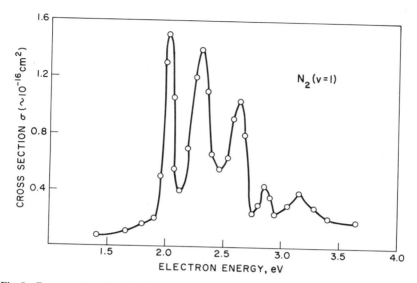

Fig. 9. Cross section for vibrational excitation of N_2 ($v = 1$) by electron impact. [After Schulz(*134*).]

electron impact at $kT_e > 1\,\mathrm{eV}$ depends less critically on electron temperature than does that of CO_2 (00°1). The total excitation cross section σ_T of N_2 also has been investigated by Schulz(*134*), who showed that σ_T reaches a maximum value $\sim 3 \times 10^{-16}\,\mathrm{cm}^2$ at electron energy $\sim 2.3\,\mathrm{eV}$, as shown in Fig. 10. This high probability was attributed to a "resonance" effect between the energetic electrons and the N_2 molecular potential, causing the formation of a short-lived negative-ion compound N_2^-. In the case of CO_2, the resonance theory is necessary to explain the shape and magnitude of the vibrational excitation cross section near the threshold (within 0.05 eV). But in the energy range far above threshold, the resonance theory cannot explain the behavior and magnitude of the excitation cross section. Thus one must attribute the excitation of the CO_2 (00°1) for electron energy in the range from 1 to 3 eV typically found(*16*) in a CO_2–N_2–He discharge to a "direct" process. Clearly the experimental conditions, such as gas pressure, mixture ratios, and electrical discharge current, must be adjusted so that the distribution of electron energy is most favorable for excitation of N_2 ($v = 1$) and CO_2 (00°1) but not CO_2 (01¹0). From the results of Figs. 8 and 9 and the electron temperature measurements(*16*), it is clear that excitation of CO_2 (00°1) level population by electrons is in the region dominated by the direct process and that a major portion of

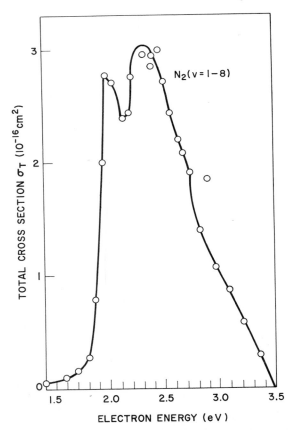

Fig. 10. Total effective cross section for vibrational excitation of N_2 ($v = 1$–8) by electron impact. [After Schulz(*134*).]

the electrons is used in the excitation of N_2 ($v = 1$), which in turn transfers its energy to excite CO_2 ($00^0 1$).

In the Born approximation, the vibrational excitation cross section σ is(*135*)

$$\sigma = \mu_{mn} \frac{3R_y}{8\pi a_0^2 \epsilon} \ln \frac{\epsilon^{1/2} + (\epsilon - E_i)^{1/2}}{\epsilon^{1/2} - (\epsilon - E_i)^{1/2}} \tag{30}$$

where $\mu_{mn} = \langle m|p|n \rangle$ are the dipole matrix elements from $01^1 0$ and $00^0 1$ to the ground state, ϵ is the average electron energy, R_y is the Rydberg constant, and a_0 is the radius of the Bohr orbit. For $\epsilon \gg E_i$, the ratio of the cross sections σ_1/σ_2 for two vibrational states evaluated at a particular electron energy ϵ is

$$\frac{\sigma_1}{\sigma_2} = \frac{f_1 E_2}{f_2 E_1} \frac{\ln(4\epsilon/E_1)}{\ln(4\epsilon/E_2)} \tag{31}$$

where f_1 and f_2 are the f numbers or the IR absorption intensities of the two vibrational transitions. They are

$$f_{mn} = \frac{E_\nu}{3R_y a_0^2 e^2} |\mu_{mn}|^2 \tag{32}$$

Using f values given by Penner(136), i.e., $f_{010} = 7.9 \times 10^{-6}$, $f_{001} = 1.14 \times 10^{-4}$ and $f_{100} = 0$, Eq. (31) yields the ratios $\sigma_{010}/\sigma_{001} \simeq 0.35$ for $\epsilon = 1.9$ eV which is in fairly good agreement with recent experimental measurements(133) for 15° scattering at 1.9 eV. From the preceding results, one concludes that in a CO_2-N_2-He laser, the average electron energy must be chosen far above the threshold value for excitation of the $00^0 1$ CO_2 level in order to avoid direct pumping of the $01^0 0$ level. In this case only a small fraction of the electrons is used to pump the ground state CO_2 molecules to the $00^0 1$ upper laser level; but for the most part, electrons in the discharge are used to excite the first vibrationally excited state of nitrogen. On the other hand, the average electron must be limited to < 3 eV in order to avoid excessive dissociation processes. If one assumes that the entire energy stored in the vibrationally excited N_2 (v) molecules (up to $v = 4$) can be transferred to the CO_2 $(00^0 1)$ level by a rapid resonant transfer process (to be discussed in detail later), it is possible to estimate the rate of excitation of the CO_2 $(00^0 1)$ level population from the excited N_2 molecules alone by using the total effective excitation cross section of N_2 (see Fig. 10) and data(137) on electron density n_e and drift velocity $\bar{v}_e$ in a positive column. At 1 Torr of N_2 pressure, the rate of excitation of N_2 is equal to $N_0 n_e \sigma_{N_2} \bar{v}_e \simeq 10^{18}$ sec^{-1} — cm^{-3} for electrons with energies in the 1–2 eV range. In a laser tube 1 m long and 2.5 cm i.d., this rate could produce a total $\sim 5 \times 10^{20}$ quanta/sec, corresponding to a maximum output power of the order 100 W.

2. Resonance Transfer

Excitation by resonance transfer (or collision of the second kind) was postulated as an important inversion mechanism in the first gas laser (He–Ne) by Javan et al.(138) using a helium metastable as the pump for Ne, the emitter. The importance of this process, however, was never completely convincing until the advent of the CO_2–N_2 laser (9). The selective excitation of CO_2 $(00^0 1)$ from the ground state by the first vibrationally excited N_2 ($v = 1$) occurs at a very rapid rate[25], $k_e = 1.9 \times 10^4$ Torr^{-1}-sec^{-1}, because of the extremely close coincidence between N_2 ($v = 1$) and CO_2 ($00^0 1$) as

$$N_2(v = 1) + CO_2(00^0 0) \xleftrightarrow{k'_e, k_e} N_2(v = 0) + CO_2(00^0 1) - 18 \text{ cm}^{-1}$$

(33)

and N_2 ($v = 1$) cannot otherwise decay to the ground state because N_2 has a zero permanent dipole moment. This resonance transfer, as given by Eq. (33), produces a mixed state in which the combined population of N_2 ($v = 1$) and CO_2 ($00^0 1$) are essentially in equilibrium. This, in effect, increases the effective lifetime of CO_2 ($00^0 1$) by almost a factor of two upon addition of a few Torr of N_2 pressure as measured by Cheo [24]. Relaxation of this mixed state takes place subsequently at a slower rate[24,25] ($k_{v_3} \simeq 110$ Torr^{-1}-sec^{-1}) as compared with the relaxation rate of v_3 in a pure CO_2 gas discharge ($k_{v_3} \simeq 385$ Torr^{-1}-sec^{-1}). We shall discuss the relaxation processes in detail later.

Vibrational energy transfer processes by collisions have been treated theoretically by Landau and Teller[139] and later extended by Schwartz, Slawsky, and Herzfeld (SSH)[140]. Because of the nonadiabatic nature of vibrational motions during molecular collisions, it has generally been assumed that only the strong and short-range repulsive forces are effective in causing the energy exchange. The SSH theory has been very successful in explaining a wide variety of collisional processes, in particular the $V \rightarrow T$ vibrational-to-translational energy exchange as we shall discuss later. However, an exception to this theory, as pointed out by Sharma and Brau[141], is the near-resonant vibrational energy transfer reaction described by Eq. (33). Recent experimental results obtained by laser-induced fluorescence[25,142] and shock-wave techniques[143] show that the cross section for this reaction has an anomalous inverse temperature dependence, in the temperature range below 1000°K, from that predicted by the SSH theory. Sharma and Brau[141] proposed a long-range dipole–quadrupole interaction by which this anomalous behavior can be explained. For simplicity, they assumed the expression for the dipole–quadrupole interaction with a spherical potential,

$$V = \frac{pq}{2r^4(t)}$$

(34)

where p is the dipole moment of CO_2, q is the quadrupole moment of N_2, and $r(t)$ is the classical trajectory describing the relative positions of the center mass of the molecules during the collision. These long-range forces are particularly important at low temperatures ($T <$

$1000° K$) and for processes of near-energy resonance. Using the interaction potential of the form given by Eq. (34), the probability $\mathscr{P}$ of vibrational energy transfer can be expressed as(*141*)

$$\mathscr{P} = \frac{1}{4h^2} |\mu_{10} q_{01}|^2 \left| \int_{-\infty}^{\infty} \frac{\exp(i\omega t/\hbar)}{r^4(t)} \, dt \right|^2 \qquad (35)$$

where $\omega = \Delta E/\hbar$ ($\Delta E = 18$ cm^{-1}), and $\mu_{10} = \langle 000|p|001 \rangle$ and $q_{01} = \langle 1|Q|0 \rangle$ are the dipole and quadrupole matrix elements between (000) and (001) states of CO_2 and between $v = 1$ and $v = 0$ states of N_2, respectively. In this approximation for rigid-sphere interaction, Sharma and Brau(*141*) obtained the formula for the total cross section of vibrational energy transfer in terms of the gas temperature, as

$$\mathscr{P}(T) = \frac{3\pi^3 m |\mu_{10} q_{01}|^2}{64 h^2 b^4 k T} \qquad (35a)$$

where m is the reduced mass and b the closest approach $\simeq 3.91 \times 10^{-8}$ cm (hard-sphere collision diameter). Since the rate constant k_{vv} for the $V \rightarrow V$ processes and the probability $\mathscr{P}$ are related(*15*) by

$$k_{vv} \propto \mathscr{P} T^{1/2} \qquad (36)$$

the theory(*141*) based on long-range forces provides an expression for the temperature dependence of the exchange rate as proportional to $T^{-1/2}$. This result is in excellent agreement with the measurements of Rosser et al.(*142*) in the temperature range from $300°$ to $1000° K$. At higher temperatures ($T > 1000° K$), experimental results(*141*) show that the transfer rate is approximately proportional to $T^{3/2}$, which indicates that the interaction is dominated by the short-range forces as predicted by the SSH theory.

It is a known fact that a substantial amount of CO ($\sim 10\%$ of the CO_2 pressure) is produced in a CO_2 gas discharge(*19*) as a result of dissociation of CO_2. Studies of CW and pulsed CO_2 laser amplifiers showed that CO increases the gain(*35*) and the effective lifetime(*24*) of the $00°1$ upper laser level substantially. These observations can be explained also by the same mechanism discussed above. The reaction

$$CO_2(00°1) + CO(v = 0) \xleftrightarrow{\ k_e(CO)\ } CO_2(00°0) + CO(v = 1) + 206 \text{ cm}^{-1} \qquad (37)$$

appears to be important in a CO_2 laser, especially for a sealed-off laser system, even though the energy resonance for this reaction is not as close as that of Eq. (33). The $k_e(CO)$ has been measured(*144*) to be 0.79×10^4 Torr^{-1}-sec^{-1}. Relaxation of this mixed state resulting from the resonant transfer process (37) takes place via collisions with CO or

CO_2 molecules in the ground state at a much slower rate (14), $k_{CO_2-CO} =$ 193 Torr^{-1}-sec^{-1}, which is a factor of about 2 less than $k_{CO_2-CO_2}$. It should also be pointed out here that helium, which is the other gas commonly used in a CO_2 laser, plays no important role, for all practical purposes, in the excitation or deactivation of the $00^0 1$ upper CO_2 laser level (24).

B. RELAXATION PROCESSES

To understand all the important aspects of population inversion in a CO_2 laser, one must not overlook relaxation processes which are comparable in magnitude to the resonance transfer processes for excitation of the upper CO_2 laser level. As mentioned before, the use of a CO_2–He mixture in a CO_2 laser results in an equal or slightly higher power output than that of a CO_2–N_2 mixture. This is primarily owing to efficient relaxation of the lower CO_2 laser level by collisions with helium (24,25) while the upper CO_2 laser level lifetime is left unaffected. The presence of many relaxation processes makes the CO_2 laser a much more complex system than most other existing lasers because the theory of a multilevel ($00^0 1$, $10^0 0$, $02^0 0$, $01^1 0$, $00^0 0$) laser system is difficult to treat. The mechanism is further complicated by the rotational relaxation and competition effects which are especially important when the laser is operated on a single rotational line. The concept of using relaxation processes for creating population inversions, in addition to excitation processes, is by no means new and has been introduced earlier by Gould (145). His concept has become successfully realized in the CO_2–N_2–He laser. To produce a high efficiency and high power laser, the lifetimes of the upper laser level, τ_u, and the lower level, τ_ℓ, must satisfy the condition,

$$\tau_u^r \gg \tau_u^c \gg \tau_\ell^c \tag{38}$$

where τ^r and τ^c represent the radiative and collisional (or effective) lifetimes, respectively. Furthermore, the laser levels should lie near the ground state so that the power conversion efficiency can be high and the relaxation of the lower laser level can be greatly increased by collisions with gas additives. The calculated radiative lifetimes of the lower laser levels $10^0 0$, $02^0 0$ and the lowest bending mode $01^1 0$ are very long (see Table 9). Therefore, relaxation via collisions are vital in the establishment of an inversion.

1. Gas Kinetics in a CO_2 Laser

Vibrational relaxation in CO_2 and mixtures of CO_2 has been extensively investigated by ultrasonic(15) and shock-wave techniques(143), and most recently by means of CO_2 lasers(14,24,25,142). Ultrasonic measurements(15) yield mainly the relaxation time of the lowest bending mode 01^10 of CO_2. As shown in Fig. 3, relaxation of this level involves a vibration-to-translation $(V \rightarrow T)$ energy exchange, as described by

$$CO_2(01^10) + M \xrightarrow{k_{v_2(M)}} CO_2(00^00) + M + 667 \text{ cm}^{-1} \qquad (39)$$

Other $(V \rightarrow T)$ processes as well as vibration–vibration $(V \rightarrow V)$ energy exchange are very important in the understanding of CO_2 lasers. For example, the Fermi resonance exchange between the two lower laser levels 10^00 and 02^00 is

$$CO_2(10^00, 02^00)' + M \xleftarrow{k_F} CO_2(10^00, 02^00)'' + M + 102 \text{ cm}^{-1} \qquad (40)$$

where $(10^00, 02^00)'$ and $(10^00, 02^00)''$ are two mixed states as discussed in Section III.B. Both of these levels can decay to the 01^10 level through resonant processes such as

$$CO_2(10^00, 02^00)' + CO_2(00^00) \xleftarrow{k_{(v_1,2v_2)}} 2CO_2(01^10) - 50 \text{ cm}^{-1} \quad (41a)$$

$$CO_2(10^00, 02^00)'' + CO_2(00^00) \xleftarrow{k_{(v_1,2v_2)}} 2CO_2(01^10) + 52 \text{ cm}^{-1}. \qquad (41b)$$

The upper laser level is coupled to the lower laser levels by collisions and by stimulated emission, as

$$CO_2(00^01) + M \xrightarrow{k_{v_3}} CO_2(n, m^\ell, 0) + M + E_{\text{trans}} \qquad (42)$$

and

$$CO_2(00^01) \xrightarrow{S_u \ell_{\text{laser}}} CO_2(10^00) \text{ or } CO_2(02^00) + h\nu \qquad (43)$$

Because a large amount of experimental information has been compiled on relaxation rates of $V \rightarrow T$ processes, the theory for the $V \rightarrow T$ process is fairly well established. On the other hand, the theory for the $V \rightarrow V$ process is less developed, and only very recently, with the advance of the CO_2 laser, has the experimental information on the $V \rightarrow V$ processes become available. In general, theoretical arguments indicate a faster rate for a $V \rightarrow V$ process than for a $V \rightarrow T$ process because the former involves a resonant exchange. It turns out, as we shall discuss in greater detail later, that the $V \rightarrow T$ rate for the lowest bending mode $01^1 0$ is one of the dominant factors in controlling the gain and power of CO_2 lasers.

Transfer of energy between two vibrational states or one vibrational and one translational degree of freedom occurs only through inter- or intramolecular collisions. The average number of collisions, Z, required to deactivate one vibrational quantum to its e^{-1} value is defined by

$$Z = \tau/\tau_c \tag{44}$$

where τ is the relaxation time or the effective lifetime of the state and τ_c is the time between successive collisions which is inversely proportional to gas pressure P. Therefore, $1/Z$ is the probability of deactivation of vibrational energy in one collision. Since Z depends only on the kinetics of the colliding molecules, it is a function of gas temperature and independent of P.

The relaxation rate of a collisional process is related to the exponential relaxation time τ by the simple expression

$$k = (P\tau)^{-1} \tag{45}$$

where k is the rate constant of a given relaxation process. This fact is used widely in experimental investigations of relaxation time of excited vibrational states of CO_2 in a pure CO_2 gas mixture such as in Eqs. (39)–(42), etc.

In a mixture involving CO_2 and a foreign gas M, relaxation of a vibrationally excited CO_2 molecule can take place through the binary processes by collisions with either a ground state of CO_2 and/or with M. The relaxation rate for the binary collisions is given by

$$k = x k_{CO_2-CO_2} + (x-1) k_{CO_2-M} \tag{46}$$

where x is the mole fraction of CO_2, and $k_{CO_2-CO_2}$ and k_{CO_2-M} are the respective rates for CO_2–CO_2 and CO_2–M collisions. This equation is useful in determining the relaxation rate constants of $01^1 0$, $00^0 1$, and $02^0 0$ or $10^0 0$ levels by collisions with foreign gases such as H_2, He, H_2O, CO, and N_2 which are commonly used in CO_2 lasers.

In the cases of CO_2–N_2 and CO_2–CO mixtures, where resonant transfer is important, mixed states are formed as described by Eqs. (33) and (37). The formation of these mixed states is very rapid (typically $\sim 10^{-5}$ sec in a working CO_2 laser). Thereafter the combined population CO_2 $(00^0 1)$ and N_2 $(v = 1)$ or CO $(v = 1)$ remains essentially in equilibrium. The relaxation of the mixed state in the case of a CO_2–N_2 mixture occurs at a much slower rate by means of the processes,

$$CO_2(00^0 1) + N_2(v = 0) \xrightarrow{\ \tau_{CO_2-N_2}\ } CO_2(n,m^l,0) + N_2(v = 0) \tag{47a}$$

$$N_2(v = 1) + CO_2(00^00) \xrightarrow{\tau_{N_2-CO_2}} N_2(v = 0) + CO_2(n,m^\ell,0) \quad (47b)$$

The relaxation rates of Eqs. (47) can be computed from the measurements of the effective lifetime τ of CO_2 $(00^0 1)$ in a CO_2–N_2 mixture, by the equation,

$$\frac{1}{\tau} = \frac{1}{x_1 + rx_2}\left[\frac{x_1^2}{\tau_{CO_2-CO_2}} + x_1 x_2\left(\frac{1}{\tau_{CO_2-N_2}} + \frac{r}{\tau_{N_2-CO_2}}\right)\right] \quad (48)$$

where $r = \exp[-(E_{N_2} - E_{CO_2})/kT]$ and x_1 and x_2 are the mole fractions of the CO_2 and N_2 molecules, respectively. The fraction of excitation CO_2 $(00^0 1)$ is $x_1/(x_1 + k_e x_2/k_e')$, where k_e and k_e' are the rates for the reaction as described by Eq. (33). Therefore the number of collisions between CO_2 $(00^0 1)$ and CO_2 in the ground state is less than that in pure CO_2 gas by the factor $x_1^2/(x_1 + k_e x_2/k_e')$. Since the observed rates of the processes in Eqs. (47) are proportional to $x_1 x_2$, they cannot be separated experimentally. Also in Eq. (48) a term owing to $N_2(v = 1)$–$N_2(v = 0)$ collisions has been neglected, since the relaxation time for this process is very long(*146*).

2. Relaxation of the 01¹0 Level

Among all processes discussed above, the $V \rightarrow T$ process as described by Eq. (39) is the most widely studied in both theory and experiment. Before presenting various experimental measurements of the relaxation time of CO_2 levels and rate constants for the above mentioned collisional processes, we shall briefly discuss some theoretical developments and mechanisms on $V \rightarrow T$ processes. Several books(*147–149*) on this subject are available for more details. Landau and Teller (*139*) were the first to treat the theory of energy transfer in a collision. Even though the treatment is classical, it contributes significant insight into the physical mechanisms. They showed that the probability of energy transfer depends on the interaction potential between colliding molecules and on their relative velocities. If the relative change in strength of an external force is small during τ_c the process is adiabatic, otherwise it is nonadiabatic. According to Ehrenfest's principle, only the nonadiabatic process can cause a quantum transition. Since the repulsive intermolecular forces are of shorter range than all others and thus can produce a sufficiently strong interaction to cause a transition, they are the only ones that need be considered.

Based on that assumption and also assuming that the velocities of the colliding molecules obey Maxwell's distribution, Landau and Teller (*139*) obtained an expression for the probability that the vibrational energy E_v of one of the colliding pair is transferred to the translational mode of the other. It is given by

$$\mathscr{P} \propto \exp[-(3E_v/kT)^{1/3}] \tag{49}$$

A quantum mechanical treatment of this problem was formulated by Schwartz et al.(*140*) about 16 years later, and they obtained some numerical results for Z from a priori calculation. In this treatment, they fitted the widely used Lennard–Jones potential(*150*), in which the dominant term, r^{-12}, is the strong repulsive part of the interaction between the two colliding particles, with an exponential form ($V \propto e^{-r}$), thus simplifying the mathematical analysis considerably. They obtained the probability $\mathscr{P}$ for a $V \to T$ process as given by

$$\mathscr{P} = 0.39 V_{10}^2 \left(\frac{8\pi^3 \mu E_v}{\alpha^2 h^2}\right) \sigma^{3/2} \left[1 - \exp\left(\frac{-2\sigma}{3}\right)\right]^{-1} \exp(-\sigma) \tag{50}$$

where $V_{10} = \langle 01^1 0 | V | 00^0 0 \rangle$, μ is the reduced mass of the colliding pair, and

$$\sigma = 3\left(\frac{2\pi^4 E_v^2 \mu}{\alpha^2 h^2 kT}\right)^{1/3} + \frac{E_v}{2kT} \tag{51}$$

The quantum mechanical results show a temperature dependence similar to that derived classically. Numerical results of these calculations are in better agreement with experimental data(*15*). Figure 11 shows both the experimental (ultrasonic) measurements and the SSH theoretical results for relaxation of $01^1 0$ CO_2 level by collisions with unexcited CO_2 molecules and He atoms. As we shall see later, relaxation of this level is very important for the CO_2 laser because population inversion densities associated with the $00^0 1 – 10^0 0$ and $00^0 1 – 02^0 0$ laser transitions are essentially limited by the relaxation rate of the $01^1 0$ level(*14*). Other gases such as H_2, H_2O, and to some extent CO (data on CO are not as abundant as for H_2 and H_2O) are known to enhance the relaxation rate of the lowest bending mode of CO_2. Table 13 gives a list of the $01^1 0$ relaxation rates k_{v_2} for CO_2 mixed with various gases.
    Several considerations should be pointed out.

1. H_2 and H_2O are very effective in relaxing the $01^1 0$ level, but they also relax the upper CO_2 laser level ($00^0 1$) very effectively. Therefore, these gases normally are not used intentionally in high-power CO_2 laser systems. Improvements in performance with H_2 or H_2O additives have been reported with the sealed-off CO_2 laser(*20,67*). Even then, the concentration of these gases is extremely small compared with other constituents and must be carefully controlled in order to obtain consistent results.

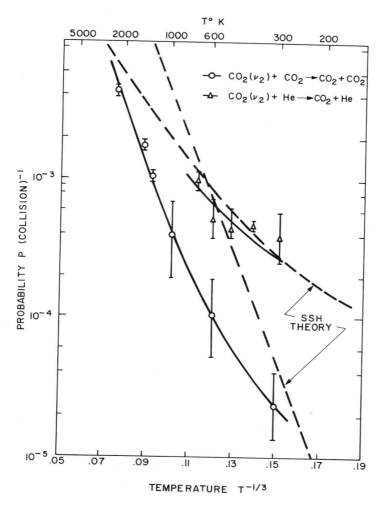

Fig. 11. Probability for vibrational–translational $(V \rightarrow T)$ energy exchange between the lowest bending mode, $01^1 0$, and the ground state of the CO_2 molecule. [After Taylor and Bitterman(*15*).]

2. Helium, even though less efficient in deactivating the $01^1 0$ level than H_2 and H_2O, yields the highest gain(*19*) compared with any other single gas additives. This is mainly because the upper CO_2 laser level is essentially unaffected by the admittance of a large amount of helium into the laser ($\gtrsim 10$ Torr), whereas the $10^0 0$ level relaxation time is reduced drastically(*24*). At such high pressures the relaxation time of

TABLE 13

Relaxation Rate Constant k_{ν_2} for CO_2 in Various
Gas Mixtures at $T = 300°\,K$

Mixture	k_{ν_2} (Torr^{-1} sec^{-1})	References
CO_2	194	15
CO_2–H_2	6.5×10^4	15
CO_2–He	3.27×10^3	15
CO_2–H_2O	4.5×10^5	15
CO_2–CO	2.5×10^4	148^a
CO_2–N_2	6.5×10^2	15^a

[a]Ultrasonic data on CO and N_2 are not as accurate as other data given here.

the 01^10 level is about 2×10^{-5} sec and is in fact shorter than $\tau_{CO_2-H_2O}$ (5×10^{-5} sec at 0.1 Torr of H_2O) in a typical working CO_2–H_2O laser.

3. In studies of the transient gain response of pulsed CO_2–H_2 and CO_2–H_2O laser amplifiers, Cheo(14) observed a substantial absorption in the afterglow period as shown in Fig. 12. When a minute amount of

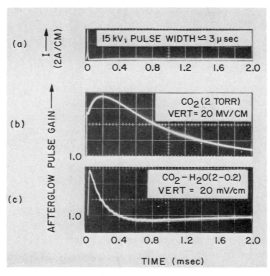

Fig. 12. Afterglow pulse gain measurements. (a) Discharge current pulse used to produce the optical gain pulse in curves b and c. (b) The 10.6-μ gain pulse in a 22-mm-bore laser amplifier tube filled with 2-Torr of CO_2. (c) The afterglow gain and absorption pulses at 10.6 μ in a 22-mm-bore laser amplifier tube filled with 2 Torr of CO_2 and 0.2 Torr of H_2O.

H_2O vapor (0.1–0.2 Torr) is introduced into the CO_2 laser amplifier, the afterglow gain decays rapidly at a rate more than one order of magnitude faster than that of pure CO_2. However, the absorption pulse following the initial gain has a very long tail with a time constant typically of the order of 1 msec. The peak absorption increases with increasing H_2O pressure at a constant excitation. This has led to the conclusion that a bottleneck is formed at the 01^10 level in the latter afterglow periods and relaxation of this long lifetime level is achieved primarily by the wall diffusion process. As we shall see later in discussions of deactivation of the 00^01 and 10^00 and 02^00 levels, relaxation of the lower CO_2 laser levels is essentially limited by the depletion rate of the 01^10 level. The accumulation of population at the 01^10 can occur through resonant intermolecular process as given by

$$CO_2\,(00^01) + H_2O\,(000) \rightarrow CO_2\,(01^10) + H_2O\,(010) + 87\ cm^{-1} \qquad (52)$$

and the 010 level of H_2O molecules can subsequently transfer its energy to CO_2 through the process

$$H_2O\,(010) + CO_2\,(00^00) \leftrightarrow H_2O\,(000) + CO_2\,(10^00) + 207\ cm^{-1} \qquad (53)$$

Available experimental data(15) show that the probability $\mathscr{P}$ for deactivation of the 01^10 level of CO_2 by collisions with H_2O has an opposite temperature dependence from that predicted by the SSH theory, a condition similar to that for the resonance transfer process observed in a CO_2–N_2 mixture. To date, there exists no satisfactory explanation of this anomalous behavior. If $V \rightarrow V$ resonant processes as described by Eqs. (52) and (53) play a significant role, assumptions made in the formulation of SSH theory would be inadequate. It is imperative that further experimental studies be carried out to verify the significance of these $V \rightarrow V$ processes by which H_2O interacts with the ν_1 and ν_2 modes of the CO_2 molecule.

4. Finally it should be stressed again that dissociation plays an important role in CO_2 lasers because CO can relax the 01^10 level of the CO_2 molecule very effectively as well as excite the 00^01 level by a resonant transfer, as discussed earlier. The CO has not been among the most widely studied additives to CO_2, nevertheless, it is always present in a CO_2 gas discharge. In studies of early CO_2 lasers, when pure CO_2 was the main gas used in sealed-off tubes, the presence of CO or certain impurities in a CO_2 gas discharge was indispensible for causing the population inversion between the vibrational levels of the CO_2 molecules. This fact has been substantiated by the pulsed-gain measurements(14) of relaxation times of the 00^01 and 10^00 levels of the CO_2 molecule.

3. *Relaxation of* $00^0 1$ *Level*

The relaxation times of the $00^0 1$ CO_2 upper laser level in various gas mixtures commonly used in CO_2 laser systems have been measured by an induced fluorescence technique(25) with a Q-switched CO_2 laser and by the afterglow pulse gain technique(14). The former was introduced by Hocker et al.(23) to study the relaxation time of the $00^0 1$ level via CO_2–CO_2 collisions. In this experiment, a small passive cell containing CO_2 gas or mixtures of CO_2 with other gases is placed inside a rotating-mirror Q-switched CO_2 laser cavity. During the Q-switching phase (typically the pulse width is 0.2 to 0.5 μsec in duration) the $10^0 0 \rightarrow 00^0 1$ transition is nearly saturated, and a small fraction ($\sim 10^{-3}$) of the CO_2 molecules in the passive cell is pumped to the $00^0 1$ level. The relaxation rate is determined by observing the spontaneous decay of the $00^0 1 \rightarrow 00^0 0$ emission intensity (4.3 μ) from the side window as a function of CO_2 gas pressure. The relaxation times measured by this method were limited to a minimum of a few μsec by the laser pulse width and to a maximum of about 1 msec by the frequency of the rotating mirror. Because the fluorescence is very weak, measurements are made usually at fairly high gas pressures. The advantage of this method, of course, is that it provides unambiguous measurements of gas composition because the interaction in the cell involves no discharges.

The afterglow pulse gain technique, on the other hand, can yield information concerning the relaxation times of both the upper $00^0 1$ and the lower $10^0 0$ or $02^0 0$ CO_2 laser levels simultaneously. This technique is applicable to most high-power CW laser systems satisfying the condition described by Eq. (38). In a pulsed CO_2 laser amplifier with a very short electrical excitation pulse (~ 1 μsec), the transient gain response to a CW CO_2 laser first rises exponentially with a time constant τ_r and then decays exponentially with a time constant τ_d (see for example, Fig. 12b). Analysis(24) indicates that τ_r and τ_d correspond to the relaxation times of the lower and the upper laser levels, respectively. This technique not only permits a direct measurement of effective lifetimes of both the upper and lower laser levels, at a significantly higher signal-to-noise ratio than from fluorescence measurements, but also yields information concerning the energy transfer processes which are relevant to the population inversion between $00^0 1$–$10^0 0$ or $00^0 1$–$02^0 0$ levels in various mixtures containing CO_2. However, it must be emphasized that relaxation time or rate measurements by this method were made in a pulsed laser amplifier with a suitable electrical excitation for optimum gain, therefore a certain amount of impurities, such as CO and O$_2$ (and in some cases NO, OH, etc.) is present and can cause complications in the interpretation of results. Nevertheless, information obtained in this way is much more relevant to CO$_2$ laser performance because the experimental conditions are similar to those for operating CO$_2$ lasers. For this reason we shall discuss here mainly the pulse-gain measurements and compare them with measurements obtained by other methods whenever appropriate data are available.

In a series of papers, Cheo[14,24] has reported the relaxation times of the 00^{0}1 and 10^{0}0 CO$_2$ laser levels by measuring the exponential rise and decay time constants of the afterglow gain pulse in CO$_2$, CO$_2$–H$_2$, CO$_2$–He, CO$_2$–H$_2$O, CO$_2$–N$_2$, CO$_2$–CO, and CO$_2$–Xe non-flowing gas mixtures. Some measurements for CO$_2$–He mixtures were made also in a flowing gas laser amplifier. In the cases of CO$_2$, CO$_2$–H$_2$, and CO$_2$–H$_2$O, the product of the 00^{0}1 level relaxation time constant τ_{00^01} and the additive gas pressure P for a fixed CO$_2$ concentration was found to remain at a constant value over a certain range of P. Thus the rate constant k_{ν_3} for relaxation of the 00^{0}1 level by collisions with these gases can be determined directly from the slopes of the linear plot of $1/\tau_{00^01}$ vs. P curves. These measured values of k_{ν_3} are listed in Table 14 along with fluorescence measurements[25] of k_{ν_3}. In general, k_{ν_3} values obtained from the two different methods are in fairly good agreement. For mixtures of CO$_2$–N$_2$ and CO$_2$–CO, τ_{00^01} increases by almost a factor of 2 upon addition of a few Torr of CO and N$_2$ pressure[14] as a result of the rapid resonant transfer process.

TABLE 14

Relaxation Rate Constant k_{ν_3} of CO$_2$ in Various Gas Mixtures X at $T = 300°$ K

Mixture	P_{CO_2} (Torr)	P_X range (Torr)	k_{ν_3} (Torr^{-1} sec^{-1})	
			Pulse gain[14]	Fluorescence[a]
CO$_2$	1–8	[b]	385	350
CO$_2$–H$_2$	3	0.5–3.0	4.5×10^3	3.8×10^3
CO$_2$–He	1	1–8	0–50	85
CO$_2$–H$_2$O	2	0.05	4.2×10^4	2.4×10^4
CO$_2$–N$_2$	1	1–7	115[c]	106
CO$_2$–CO	2	1–5	193[c]	—
CO$_2$–Xe	2	0.1–1.0	0–50	30

[a]Data by Moore et al.[25,141] at much higher gas pressure than pulse gain measurements.

[b]Other species such as CO–O$_2$ are also present owing to dissociation of CO$_2$.

[c]Obtained for largest P_X values.

Relaxation of the mixed states, as described by Eqs. (33) and (37), is at a much slower rate through processes (47b) than that by CO_2–CO_2 collisions. These rate constants k_{v3} for CO_2–N_2 and CO_2–CO collisions are computed from Eq. (48) for mixture ratios $1:7$ and $2:5$, respectively.

From Table 14 we see that both H_2 and H_2O relax the upper CO_2 laser level severely. These results, combined with the bottleneck formation at the 01^10 and 10^00 levels as described by Eqs. (52) and (53), can account for the extremely critical pressure dependence on gain and power output encountered with CO_2–H_2 and CO_2–H_2O laser systems. In the cases of CO_2–He and CO_2–Xe mixtures, no significant change in τ_{00^01} from that of pure CO_2 was observed for $0 < P_{He} < 8$ Torr and $0 < P_{Xe} < 1$ Torr. In most cases, the relaxation of CO_2 (v_3) takes place through the $V \rightarrow V$ coupling between v_3 and v_2 (or v_1) modes as

$$CO_2 \ (v_3) + M \longrightarrow CO_2 \ (v_2) + M + 416 \ cm^{-1} \qquad (54)$$

However, it has been pointed out by Yardley and Moore(151) that in the case of noble gases the most favorable collisions for the deactivation of v_3 occur when only a small amount of energy on the order of kT is exchanged with the translational mode. This argument is based on the fact that the measured(151) probability for deactivation of CO_2 (v_3) by noble gases has a peak for those whose reduced mass μ, and hence velocity of the rare-gas collision partners, corresponds to a resonance between the collision time τ_c and the frequency difference between the two levels involved, or

$$\tau_c \simeq \frac{h}{\Delta E} \qquad (55)$$

For $\tau_c \simeq 1.6 \ \text{Å}/(8 \ kT/\pi\mu)^{1/2}$, one obtains the energy difference ΔE of a transition between the initial and final states within $20 \ cm^{-1} \leqslant E \leqslant 300 \ cm^{-1}$. Based on these arguments, deactivation of CO_2 (v_3) by collisions with noble gases is attributed to the intramolecular energy exchange, for example,

$$CO_2 \ (v_3) + M \longrightarrow CO_2 \ (04^00) + M - 199 \ cm^{-1} \qquad (56)$$

and

$$CO_2 \ (v_3) + M \longrightarrow CO_2 \ (11^10) + M + 272 \ cm^{-1} \qquad (57)$$

It has been shown(151) that the highest transition probability for the intramolecular $V \rightarrow V$ energy transfer, when the difference of energy levels lies within a few hundred cm^{-1}, is owing mainly to a mixing of vibrational states through anharmonicity and Coriolis coupling terms. Therefore this process cannot be very efficient when compared with other collisional relaxation processes among polyatomic molecules. For this reason the population density at the upper CO_2 laser level is relatively unaffected upon addition of a large amount of He or Xe.

Diffusion also plays an important role in the deactivation of the excited CO_2 molecules. This has been studied by Kovacs et al.(152), who measured the wall depletion rate of the excited 00^01 level of CO_2 by monitoring the decay of the spontaneous 4.3-μ radiation arising from the 00^01 level of CO_2 molecules in an induced fluorescent cell excited by a rotating-mirror Q-switched CO_2 laser pulse. The measured diffusion coefficient for the 00^01 level was found to be different from the self-diffusion coefficient obtained from viscosity measurements. Several absorption cells of different diameters and wall materials were used in this experiment. The relaxation rate of the 00^01 level population is determined by both the collisional relaxation rate within the volume and the destruction rate of the wall. Thus

$$k_{\text{total}} = k_{\text{wall}} + k_{\text{volume}} \tag{58}$$

where $k_{\text{wall}} = \gamma^2 \mathscr{D}/r_0^2$ and $k_{\text{volume}} = k_{\nu_3}$ as given in Table 14. The symbol γ represents the roots of the boundary equation which is derived from the continuity equation at the wall, $\mathscr{D}$ is the diffusion coefficient, and r_0 is the radius of the absorption cell. For a cylindrical tube, the boundary equation is given by(152)

$$\gamma J_1(\gamma) - \frac{\bar{v} r_0}{2\mathscr{D}}\left(\frac{1-\beta}{1+\beta}\right) J_0(\gamma) = 0 \tag{59}$$

where β is called the wall reflection coefficient and $\bar{v}$ the average thermal velocity of the molecules. For CO_2 pressures greater than 1 Torr, k_{total} is dominated by k_{ν_3}, whereas for $P_{CO_2} < 1$ Torr, $k_{\text{total}} \simeq k_{\text{wall}}$, which increases rapidly and approaches a constant value $(\bar{v}/r_0) \times (1-\beta/1+\beta)$. Figure 13 shows the plot of k_{wall} vs. CO_2 pressure. The dashed curve represents the P dependence for the volume-quenching case, which is obtained by extrapolation from data taken above 1 Torr where diffusion is negligible. The measured diffusion constant $\mathscr{D}$ is 0.7 ± 0.01 cm²/sec which differs substantially from the self-diffusion constant(148) 0.119 ± 0.001 cm²/sec. Experiments with four different wall surfaces yield the same result, 0.22 ± 0.08, for the wall deexcitation probability $(1-\beta)$. This result is attributed(152) to a common type of comtamination absorbed by the surfaces of the absorption cell.

4. Relaxation of the 10^00 and 02^00 Levels

As mentioned before, because of Fermi resonance, relaxation of $(10^00, 02^00)'$ and $(10^00, 02^00)''$ takes place via processes (41) at a

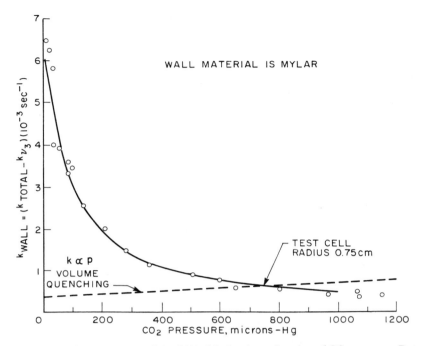

Fig. 13. Diffusion-decay rate of the $00°1$ CO_2 level as a function of CO_2 pressure. Data (open circles) were obtained from a cell with a 1.5-cm i.d. Mylar wall material. The dashed curve is the volume quanching rate as a function of CO_2 pressure. [After Kovacs et al. (*152*).]

rate $k_{(\nu_1,2\nu_2)}$ several orders of magnitude as fast (*153*) as a second-order transition, such as $10°0 \rightarrow 01^10$, which would require a Raman-like perturbation ($10°0 \rightarrow 00°0 \rightarrow 01^10$). To date no direct rate measurements have been made of either the Fermi resonance, Eq. (40), or the $V \rightarrow V$ processes, Eqs. (41). Pulse-gain measurements (*14*) provide only an effective relaxation rate of the lower laser level; because of the strong collisional coupling between the lower laser level and the bending mode (01^10 level), the measured rate constants $k^{\text{eff}}_{(\nu_1,2\nu_2)}$ are actually the trapped values owing to the accumulation of population at the 01^10 level. This is evident when the pulse-gain results as given in Table 15 are compared with the relaxation rates of the 01^10 level as given in Table 13. The measurements (Table 15), except for those taken with a pure CO_2 fill in which a considerable amount ($> 10\%$) of CO is present owing to dissociation of CO_2, are reasonably close to the k_{ν_2} values. This indicates that the effective relaxation rates of the lower laser levels are essentially limited by that of the 01^10 level.

TABLE 15

Effective Rate Constant $k^{eff}_{(\nu_1,2\nu_2)}$ of CO_2 in Various Gas Mixtures X

Mixture	P_{CO_2} (Torr)	P_X range (Torr)	$k^{eff}_{(\nu_1,2\nu_2)}$ (Torr^{-1} sec^{-1}) Pulse gain(14)
CO_2	1–8	a	2.2×10^3
CO_2–H_2	3	0.5–3.0	3.3×10^4
CO_2–He	1	1–8	4.7×10^3
CO_2–H_2O	2	0.05	1.2×10^6
CO_2–N_2	1	1–7	26
CO_2–CO	2	1–5	4.1×10^3
CO_2–Xe	2	0.1–1.0	5×10^3

^{a}Other species such as CO–O_2 are also present owing to dissociation of CO_2.

In the case of CO_2–CO_2 collisions, pulse-gain measurements(14) yield a $k^{eff}_{(\nu_1,2\nu_2)}$ value of 2.2×10^3 Torr^{-1}-sec^{-1} which is about 10 times that for a k_{ν_2} value obtained by ultrasonic methods(15). This discrepancy could be owing to dissociation of CO_2 into CO and O_2. From ultrasonic data(15), O_2 does not relax the CO_2 (01^10) level very efficiently, but very little data on CO is available. According to pulse-gain measurements(14), the relaxation rate for the 10^00 and 02^00 levels is 4.1×10^3 Torr^{-1}-sec^{-1}, obtained by collisions with CO. Therefore, the assumption of dissociation is a reasonable one and is consistent with other observations(35), such as enhancement of CO_2 laser gain and of power output by addition of CO, and also spectroscopic analysis of the side-light emission from a CO_2 laser discharge(19). Another interesting observation is that, in the case of a pulsed CO_2–N_2 laser, the effective lifetime of the lower laser level increases with increasing N_2 gas pressure similarly to the increased lifetime observed for τ_{00^01} in the same gas mixture as a result of resonance transfer between CO_2 (00^01) and N_2 ($v = 1$). It has been suggested(15) that since Ar also is known to be less efficient than CO_2 in deactivating mode ν_2, the same mechanism may be involved for these two cases.

Recently Rhodes et al.(28), using double-beam resonance techniques, observed a fairly fast rate of the order of 4×10^5 Torr^{-1}-sec^{-1} in a CO_2 absorption cell by monitoring the decrease in absorption of a CW 10.6-μ radiation immediately following a 9.6-μ rotating-mirror Q-switched CO_2 laser pulse. From their experiment, it is not clear what reaction corresponds to the observed rate. They attribute this rate to the relaxation of the mixed state (10^00, 02^00) to the 01^10 level via Eq. (41). Herzfeld(153) has made extensive computations of the vibrational relaxation rates in the presence of Fermi resonance, using an exponential repulsive interaction energy between the two colliding particles and eigenfunctions of the states $(10^00, 02^00)'$ and $(10^00, 02^00)''$ as given by Eq. (19). Mixing of nonresonant states was also taken into consideration, but symmetry rules were observed in this numerical analysis. Several of these numerical results, which are relevant to the CO_2 laser system, are tabulated in Table 16. Calculations of vibrational energy exchange rates, using Herzfeld's method, recently have been carried out by Gordietz et al.(154) for mixtures of CO_2–He, CO_2–N_2, and CO_2–CO. The calculated results are given in Table 17. Several significant points emerge from these calculations. (i) The numerical calculation of the Z number for the $01^10 \rightarrow 00^00$ is in excellent agreement with the experimental results given in Table 13.

TABLE 16

Z Number(153) for Various Relaxation Processes in Pure CO_2

Transition	300° K	600° K	1000° K	2000° K
$01^10 \rightarrow 00^00$	58,400	2,900	430	43
$(10^00, 02^00)' \rightarrow 00^00$	3.2×10^8	3.7×10^6	2.0×10^5	5,600
$(10^00, 02^00)'' \rightarrow 00^00$	8.7×10^8	8.2×10^6	3.8×10^5	8,900
$(10^00, 02^00)' \rightarrow 01^10$	29,300	1,550	250	29
$(10^00, 02^00)'' \rightarrow 01^10$	75,500	3,330	480	49
$02^20 \rightarrow 01^10$	29,200	1,450	215	22
$(10^00, 02^00)' \leftrightarrow (10^00, 02^00)''$	290	145	85	40
$(10^00, 02^00)' \leftrightarrow 02^20$	140	70	42	21
$(10^00, 02^00)'' \rightarrow 02^20$	163	81	49	24
$(11^10, 03^10)' \rightarrow (10^00, 02^00)'$	22,300	1,141	180	19
$(11^10, 03^10)' \rightarrow (10^00, 02^00)''$	56,400	3,900	760	100
$(11^10, 03^10)'' \rightarrow (10^00, 02^00)'$	3.5×10^6	1.2×10^5	14,300	880
$(11^10, 03^10)'' \rightarrow (10^00, 02^00)''$	29,580	2,040	300	27
$(11^10, 03^10)' \rightarrow 02^20$	38,100	2,180	390	110
$03^30 \rightarrow 02^20$	19,500	970	140	14
$(11^10, 03^10) \rightarrow$ sum over all	5,800	320	52	< 10
$00^01 \rightarrow 00^00$	8.6×10^{12}	1.6×10^{10}	2.2×10^8	1.2×10^6
$0^01 \rightarrow (10^00, 02^00)', (00^00)$	2.6×10^9	4.7×10^7	3.5×10^6	1.5×10^5
$0^01 \rightarrow (10^00, 02^00)'', (00^00)$	5.9×10^8	1.35×10^7	1.2×10^6	64,000
$0^01 \rightarrow (00^00), (10^00, 02^00)'$	6.8×10^9	1.2×10^8	9×10^6	3.9×10^5
$0^01 \rightarrow (00^00), (10^00, 02^00)''$	1.6×10^8	3.2×10^7	3.2×10^6	1.7×10^5
$0^01 \rightarrow (10^00, 02^00)', (01^10)$	1.1×10^8	1.25×10^7	3.3×10^6	6.9×10^5
$0^01 \rightarrow (10^00, 02^00)'', (01^10)$	1.4×10^7	2.6×10^6	8.1×10^5	2.4×10^5
$0^01 \rightarrow 01^10, (10^00, 02^00)'$	4.8×10^8	5.4×10^7	1.5×10^7	2.9×10^6
$0^01 \rightarrow 01^10, (10^00, 02^00)''$	6.4×10^7	1.1×10^7	3.8×10^6	1.05×10^6
$0^01 \rightarrow$ sum over all	2.7×10^6	4.9×10^5	1.25×10^5	21,000

TABLE 17

Calculated(154) Vibrational Energy Transfer Rate Constants
among CO$_2$, N$_2$, and CO Molecules

k (Torr^{-1} sec^{-1})	300° K	600° K	1000° K
N$_2$ ($v = 1$) → CO$_2(\nu_3)$	1.0×10^4	2.1×10^4	3.9×10^4
CO ($v = 1$) → CO$_2(\nu_3)$	1.38×10^3	7.3×10^3	2.6×10^4
k_{ν_3} (CO$_2$–N$_2$) for Eq. (44)	10	63	2.8×10^4
k_{ν_3} (He) for Eq. (38)	62	6.9×10^2	2.7×10^3
k_{ν_2} (CO)	4.5×10^3	6.5×10^4	5.8×10^5
CO ($v = 1$) → CO$_2(\nu_1) + $ CO$_2(\nu_2)$	9.5×10^2	3×10^3	8.2×10^3
CO ($v = 1$) → CO$_2(\nu_2)$	6	46	2.8×10^2

(ii) Only 290 collisions (corresponding to $\sim 4 \times 10^4$ Torr^{-1}-sec^{-1}) are required to achieve a mixed state among 10^00 and 02^00, the lower CO$_2$ laser levels; the formation of this state through Eq. (40) is at a rate about 300 times as fast as that derived from Eq. (39). (iii) Relaxation rates of the lower laser levels to the lowest bending mode (01^10) are of the same order of magnitude as that for the transition from $01^10 \rightarrow 00^00$. These results are about one to two orders of magnitude lower than that reported by Rhodes et al.(28) but are very close to the pulse-gain measurements of $k^{\text{eff}}_{(\nu_1, 2\nu_2)}$ by Cheo(14). (iv) The transfer rate between ($10^00, 02^00$) and 02^20 is very rapid even though a direct mixing between 10^00 and 02^20 is not allowed because of the symmetry rule. Furthermore, the transfer rate between 02^20 and 01^10 is of the same order of magnitude as that for $(100, 02^00) \rightarrow 01^10$; therefore the 02^20 level may be important also for the populations of 10^00, 02^00, and 01^10 to reach equilibrium(155).

It is clear from the above discussions that the vibrational energy exchange and collisional relaxation in CO$_2$ laser systems are very complex and further studies in this area are needed. However, for a typical CO$_2$ laser using a CO$_2$–N$_2$–He mixture, the four most important relaxation processes are given by Eqs. (39)–(42). The first three processes lead to the depletion of the population of the lower laser level which in general occurs at a rate ($\geqslant 10^2$) much greater than the process described by Eq. (42).

5. Rotational Relaxation

All the preceding discussions pertained to the relaxation of low-lying vibrational levels of the CO$_2$ molecule while assuming that rotational relaxation within a given vibration state occurs at an extremely fast rate. Using viscosity data for CO_2, a simple calculation shows that the collision time for hard-sphere collisions is about 100 nsec at 1 Torr. This corresponds to a rate constant of 10^7 $Torr^{-1} sec^{-1}$ if rotational thermalization occurs in one collision. Previous relaxation-time measurements by the pulse gain(14,24) and by rotating-mirror Q-switched laser techniques(25,28) have been limited to about 0.3 μsec or greater because of either the duration of the electrical excitation in a pulsed discharge or the length of the rotating-mirror Q-switched CO_2 laser. Recent development(56) of a GaAs electrooptic switch for cavity dumping a 10.6-μ Q-switched pulse of about 10 kW and 20 nsec duration has made direct measurement of the rotational thermalization time of the $00^0 1$ upper CO_2 laser level possible. A single-frequency $P(20)$ 10.6-μ cavity dumped pulse (for details see Section V.E) was used to measure the rotational relaxation rate constants k_{rot} of the CO_2 $(J, 00^0 1)$ level by collisions with CO_2(26), N_2, or He(27). This was accomplished by monitoring the transient change in the CW gain of a 9.6-μ $P(J)$ laser amplifier, where $J = 17, 19, \ldots, 39$, induced by passage of a 20-nsec 10.6-μ $P(20)$ laser pulse through a pure CO_2, CO_2–N_2, or CO_2–He laser amplifier. Since the $P(20)$ 10.6-μ and $P(20)$ 9.6-μ transitions share a common upper level, $J = 19$ (see Fig. 4), the CW gain of the 9.6-μ CO_2 laser amplifier will be suddenly reduced by the passage of the 10.6-μ Q-switched pulse through the same medium. The recovery time of the gain at 9.6 μ is a direct measure of the rotational relaxation time of the $J = 19$ upper laser level. More significantly, one can obtain the selection rules for rotational transition if the Q-switched laser is fixed on $P(20)$ of the 10.6-μ band but the 9.6-μ CW laser transition is scanned over different J values.

Figure 14a shows the 10.6-μ, $P(20)$ Q-switched pulse. Figure 14b, c are the 9.6-μ CW gain responses for the $P(20)$ and $P(30)$ transitions, respectively, when the medium is suddenly perturbed by the Q-switched 10.6-μ $P(20)$ laser pulse. Both traces b and c of Fig. 14 are obtained for a CO_2 pressure of 1.1 Torr. In Fig. 15, the measured rotational relaxation rate $1/\tau_{rot}$, or the recovery rate of the gain, is plotted as a function of CO_2, N_2, and He gas pressures. Rate constants k_{rot}, for CO_2–CO_2, CO_2–N_2, and CO_2–He collisions, are obtained from the slopes of these linear plots of $1/\tau_{rot}$ vs. CO_2, N_2, and He gas pressures and are given in Table 18. For pure CO_2, the value of $Z = \tau/\tau_{coll}$ (the number of gas kinetic collisions necessary to cause rotational thermalization) is equal to 0.6. This implies that collision-induced rotational mixing occurs more rapidly than simple hard-sphere collisions. Acoustical absorption experiments(156) show that $Z_{CO_2-CO_2} = 1.5$ for ground state CO_2 molecules. This result is the relaxation rate

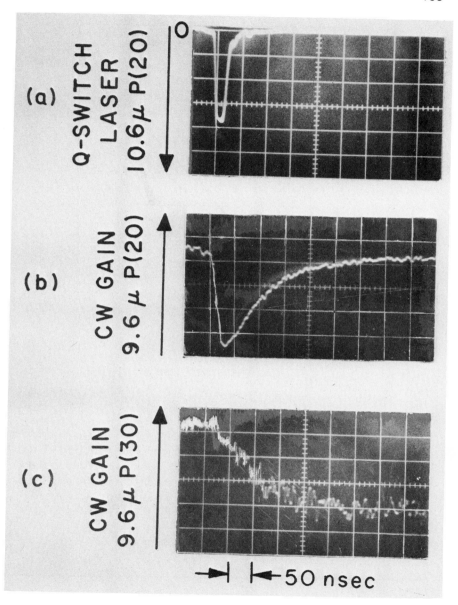

Fig. 14. Data on rotational relaxation in the 00°1 CO_2 level. Trace (a) shows the cavity-dumped $P(20)$ 10.6-μ laser pulse as measured by a Ge:Cu(Sb) detector. Traces (b, c) show the responses of CW gain of the $P(20)$ and $P(30)$ transitions in the 00°1–02°0 band, respectively, after the medium has been perturbed by the passage of a cavity-dumped $P(20)$ 10.6-μ laser pulse.

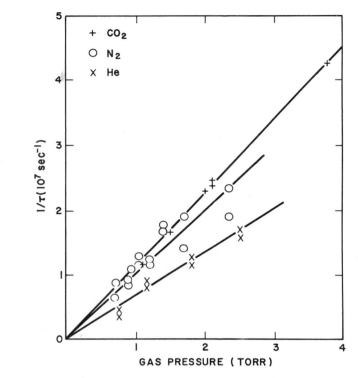

Fig. 15. Relaxation rate $1/\tau_{\mathrm{rot}}$ of the $J = 19$ rotational level of the $00°1$ CO_2 vibrational band as a function of CO_2, N_2, and He gas pressure.

of the average rotational energy when the entire distribution of the rotational levels is involved. On the other hand, the measurements made by using the double laser beam resonance technique(26,27)

TABLE 18

Rotational Relaxation Rate Constant $k_{\mathrm{CO_2-X}}^{\mathrm{rot}}$ of CO_2
($00°1, J = 19$) for X = CO_2, He, and N_2(26,27)

X	CO_2	He	N_2
$k_{\mathrm{CO_2-X}}^{\mathrm{rot}}$ (10^7 sec^{-1} Torr^{-1})	1.1 ± 0.2	0.7 ± 0.1	1.0 ± 0.2

involve a single perturbed J level in an excited vibrational state. The measured value($29,30$) of the optical-broadening collision frequency by CO_2–CO_2 collisions yields a rate constant of approximately 1.0×10^7 Torr^{-1} sec^{-1}, in good agreement with that obtained by the direct method described above, indicating that the 10.6-μ linewidth is owing primarily to collisions causing rotational mixing. The relative effectiveness of CO_2, N_2, and He for optical broadening of the CO_2 absorption line also has been measured(157). From these measurements(156), one obtains $k_{CO_2-CO_2}/k_{CO_2-N_2} = 1.33$ and $k_{CO_2-CO_2}/k_{CO_2-He} = 1.69$, consistent with values given in Table 18.

Figure 14c shows the decay of CW gain for the 9.6-μ $P(30)$ line when perturbed by the 10.6-μ $P(20)$ Q-switched laser. Similar results were obtained also for transitions from $P(18)$ to $P(38)$. The observed decrease in gain for the $P(30)$ line is owing to the collisional transfer of population from the $J = 29$ to the $J = 19$ rotational level. If rotational relaxation is a stepwise process, i.e., $\Delta J = \pm 2$, one would expect that the levels closest to $J = 19$ are most strongly coupled, so that there would be an increase in the time delay between the transient gain perturbation and the Q-switched laser pulse for either an increasing or decreasing value of J from $J = 19$. Experimentally, one finds that the observed waveforms for all transitions from $P(18)$ to $P(38)$ in the 00^01–02^00 band are very similar, with the exception of the $P(20)$. The decay times for all J's have almost the same value as the gain recovery time of the $P(20)$ line. This suggests that the rotational relaxation is J-independent, and only one collision is necessary to thermalize all rotational levels. From these measurements, it is clear that, in a CO_2 laser, all these tightly coupled rotational levels compete strongly for lasing. As a result, oscillation often switches from one line to the next, especially in a high-power laser system without a dispersive element. Further discussion on rotational level competition appears in Section V.C.

In summary, extensive investigations, both experimental and theoretical, have been carried out on various excitation and relaxation processes occurring in CO_2 discharges. Mechanisms responsible for population inversion of the two strongest bands, namely 00^01–10^00 and 00^01–02^00, are well understood; however, no attention has been given to other CO_2 laser transitions because they are weak and very little work has been done on them. Laser oscillations from those weak bands exist only under pulsed excitation. Measurements of collision cross sections and rate constants of the important processes allow reliable estimations on the performance of various CO_2 laser systems.

V. Conventional Laser Systems

Conventional CO_2 laser systems comprise either a sealed-off gas discharge or a low-speed longitudinal flow of CO_2 in a mixture along the discharge column. The gain and power characteristics of this type of laser have been investigated extensively, and the mechanisms are well understood. In this section we present the properties of these laser systems, such as the small-signal gain, gain saturation, linewidth and lineshape, CW and Q-switched laser output, and stabilization and life of a sealed-off CO_2 laser. A brief discussion of some nonlinear phenomena which have direct effects on the CO_2 laser is also included. Interpretation of experimental results is provided in terms of known physical processes and, in some cases, results are compared with theoretical models utilizing available information on lifetimes and collision cross sections of excitation and relaxation of the laser levels.

More recently concentrated effort has been devoted to laser systems involving high-speed (sub- or supersonic) transverse and longitudinal flow. Power output from these systems has exceeded that from conventional lasers by more than one order of magnitude, but the laser efficiency in general is lower. Experimental conditions for these high-flow systems are substantially different from the normal ones and involve either a high-speed exhaust channel at a flow rate greater than the relaxation rates of the laser levels or sudden expansion through a supersonic nozzle. We shall treat these subjects as well as some other novel systems in Section VI.

A. SMALL-SIGNAL GAIN

The theory of the CO_2 laser, especially for the conventional types, is fairly well established. Historically it was the experimental investigation of various gain characteristics that led to insight into mechanism and provided the stimulus for theoretical development. Investigations of CO_2 laser amplifiers usually yield more information than do studies of laser oscillators because in the former the gain is directly proportional to the population inversion of the laser levels whereas the oscillators are inherently nonlinear systems involving positive feedback. Furthermore, one can extract a large amount of information from an amplifier with negative gain value. An oscillator, on the other hand, provides no information when the medium gain is less than the oscillator loss.

A small signal I propagating through an amplifying medium with an unsaturated gain coefficient α_0 will be amplified with a rate of increase

$$\frac{dI}{dL} = \alpha_0 I \tag{60}$$

where L is the length of the medium. From the principle of energy conservation, the increase in intensity per unit length, dI/dL, is related to the population inversion density $n_u - (g_u/g_\ell)n_\ell$ by

$$\frac{dI}{dL} = h\nu\mathscr{W}\left(n_u - \frac{g_u}{g_\ell}n_\ell\right) \tag{61}$$

where n_u and n_ℓ are the population densities of the upper and lower laser levels, g_u and g_ℓ are the level degeneracies, and $\mathscr{W}$ is the induced emission rate which is related to the spontaneous transition lifetime τ_r by (158)

$$\mathscr{W} = \frac{\lambda^2 I}{8\pi h\nu\tau_r} g(\nu) \tag{62}$$

where $g(\nu)$ is the intensity distribution function describing the line-shape. Combining Eqs. (60)–(62) one obtains

$$\alpha_0 = \frac{\lambda^2}{8\pi\tau_r} g(\nu)\left(n_u - \frac{g_u}{g_\ell}n_\ell\right) \tag{63}$$

Equation (63) is equivalent to Eqs. (25) and (26) provided that a proper expression for $g(\nu)$ is chosen. In the case of homogeneous broadening,

$$g(\nu - \nu_0) = \Delta\nu/2\pi[(\nu - \nu_0)^2 + \tfrac{1}{4}(\Delta\nu)^2] \tag{64}$$

where $\Delta\nu$ is the full width of the Lorentzian line at half-intensity and ν_0 is the center frequency of laser transition. At low gas pressure, the Doppler lineshape yields a better description which can be approximated by a Gaussian distribution $g(\nu_0)$ over ν_0 for a gas in thermal equilibrium as

$$g(\nu_0) = (\sqrt{\pi}\Delta\nu_0)^{-1}\exp\left[-\left(\frac{\nu_0 - \nu_0'}{\Delta\nu_0}\right)^2\right] \tag{65}$$

where $\Delta\nu_0$ is the half-width of the Doppler-broadened line and ν_0' is the line center at zero velocity.

A number of parameters such as excitation current, wall temperature, gas mixtures and mixing ratios, tube bore, and flow rates can affect the gain in a CO_2 laser in addition to pressure broadening or Doppler effects. In what follows we shall present experimental investigations of these effects which provide not only quantitative information about the CO_2 laser but also guidelines to achieve the optimum performance of this laser.

1. *Current and Temperature Dependence*

Parametric studies on the unsaturated gain coefficient of both the nonflowing(*19*) and flowing(*34,35*) CO_2 laser amplifiers have been made by measuring the power ratio of the net increase in the output to the input for a fixed length of amplifying medium. Figure 16 shows the effects of excitation power (discharge currents) on gain for a number of

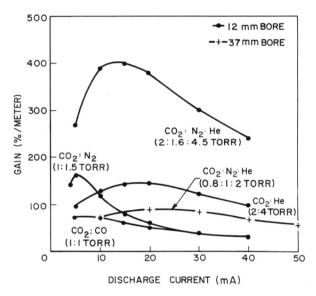

Fig. 16. Gain versus discharge current for various flowing CO_2 gas media at optimum mixture ratios and a constant CO_2 flow rate of 150 cm³/min in 12- and 37-mm-bore amplifier tubes.

flowing gas mixtures. In general, optimum gain occurs at higher discharge currents for systems containing helium. Similar behavior is observed also for the nonflowing laser systems. At optimum discharge current, the gain of a nonflowing CO_2–He laser amplifier decreases with increasing wall temperature, as shown in Fig. 17. In general, the gain depends more critically on wall temperature for systems without helium but the temperature dependence is found to be less critical for flowing systems than for nonflowing laser amplifiers. One of the important effects of helium is to reduce the gas temperature in a discharge, owing to its high thermal conductivity. Similarly, gas temperature in a conventional laser also can be reduced by slowing the gas flow. Another important role played by helium is to increase the relaxa-

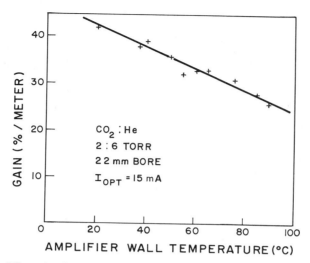

Fig. 17. Effect of wall temperature on optimum gain of a CO₂–He laser amplifier.

tion rate of the lower laser level. As shown in Fig. 16, an increase in discharge current leads to an increase in pumping rate, thus contributing to the initial increase of gain; however, it also leads to an increase in gas temperature, causing a rapid depopulation of CO₂ laser levels. With the addition of He the gas temperature decreases so that the depletion rate of the upper laser level population by collisions is decreased; furthermore the relaxation of the lower laser level is increased by CO₂–He collisions. Therefore, systems containing He usually can be pumped at higher discharge currents and thus yield a larger inversion density.

2. Pressure and Bore Dependence

For both flowing and nonflowing laser amplifiers, gain increases rapidly with increasing He pressure from 0 to about 3 Torr and thereafter gain becomes relatively insensitive to a further increase of helium pressure. However, gain depends critically on CO₂ partial pressure as shown in Fig. 18. This behavior can be easily explained from the measurements of volume quenching rate and wall destruction rate through diffusion of the vibrationally excited CO₂ molecules as shown in Fig. 13. It can be seen from Fig. 13 that the total decay rate of the $00^0 1$ level, k_{total}, has a minimum value at $P_{CO_2} \simeq 1$ Torr, at which a maximum gain is obtained for the pure CO₂ and CO₂–N₂ laser

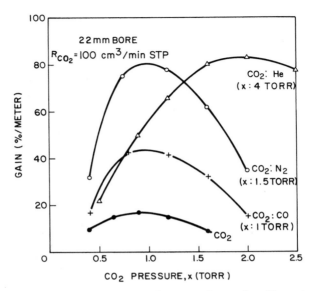

Fig. 18. Gain versus CO_2 pressure for various gas mixtures in a 22-mm-bore amplifier tube. The CO_2 flow rate is 100 cm³/min.

amplifiers of approximately the same bore size (~ 20 mm). This is expected because diffusion of CO_2 ($00^0 1$) through N_2 should have about the same coefficient as that measured in pure CO_2 gas (0.07 ± 0.01 cm²/sec), since the rate of vibrational energy exchange between these two molecules is extremely high(25). The rate of deactivation of N_2 ($v = 1$) at room temperature is known to be very low(15). Furthermore, relaxation rates of both the upper and lower CO_2 laser levels by collisions with N_2 are much lower(14) than those with CO_2 (see Tables 14 and 15). For a CO_2–He mixture, one can operate the laser at a higher CO_2 partial pressure by taking advantage of the fact that He relaxes the lower CO_2 laser level very effectively, especially at higher He pressure, while leaving the upper laser level population unaffected (14). Other effects of He, such as a shift in the electron energy distribution and a decrease of gas temperature in a CO_2 gas discharge, also are involved. In summary, the most important role played by He in a CO_2 laser is to remove the "bottleneck" $01^1 0$ level which lies in the depletion path of the lower laser level population.

The dependence of optimum gain on tube bore(19,35) is shown on Fig. 19. For a flowing-gas laser amplifier the gain varies inversely as the bore, a weaker dependence is exhibited by the nonflowing gas systems. In general, the optimum gain and power output occur at a *PD*

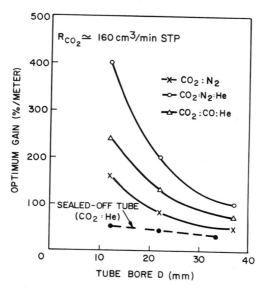

Fig. 19. Optimum gain of various flowing CO_2 gas media at CO_2 flow rate of 160 cm³/ min as a function of tube bore. The dashed curve is the maximum gain of a nonflowing laser amplifier versus tube bore.

product (19,35) of about 3 Torr-cm, where P is the CO_2 partial pressure and D is the bore of the tube. The highest gain reported for the non-flowing laser amplifier (19) is 1.7 dB/m from tubes with inner diameter less than 20 mm and with a CO_2–He gas mixture, whereas a maximum gain of 7.8 dB/m can be obtained from a flowing CO_2–N_2–He laser amplifier (35) with a 12-mm bore. The radial gain profiles (19) across a nonflow amplifier tube diameter for CO_2, CO_2–N_2, and CO_2–He are shown in Fig. 20. For CO_2 and CO_2–He discharges, the radial gain profiles have essentially the same shape with a fairly constant value near the center of the discharges and decreasing gradually toward the wall of the amplifier tube. In the case of CO_2–N_2, the flat portion of the gain extends further toward the wall. These radial gain profiles for various gas mixtures can be explained (154) by the volume destruc-tion of vibrationally excited CO_2 molecules, by the variation of the relaxation rate of the upper laser levels with gas mixtures, and by the nonuniformity of gas temperature as a function of the tube radius.

3. Effects of Gas Flow

The effects of gas flow on gain are shown in Fig. 21. These mea-surements (35) were made in water-cooled laser amplifiers with a dc

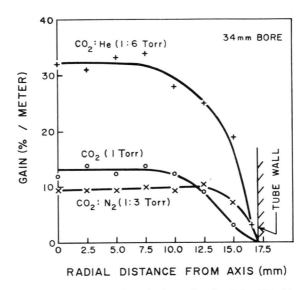

Fig. 20. Small-signal gain as a function of tube radius for CO_2, CO_2-N_2, and CO_2-He mixtures in a 34-mm-bore amplifier.

electrical discharge and a longitudinal flow of gas mixture either from anode to cathode or vice versa. In a conventional laser with a low-speed longitudinal gas flow, the transit time of gas flow is much too long compared to the diffusion and collisional relaxation times; therefore, the gain enhancement by the gas flow in this case is owing simply to a reduction of the gas temperature by convection and by exhaustion of the hot gas at a higher temperature. This is consistent with the observation that the gain and power output of a flowing laser depend less critically on the discharge tube wall temperature. An enhancement of gain by a factor of nearly two is observed(35) merely by maintaining a continuous flow of gas through the amplifier at a very low flow rate ($\leqslant 20$ cm³/min). A related phenomenon often observed in a freshly filled nonflow laser is that the output power increases slowly and reaches its maximum value in a time varying from a few minutes to several hundred hours depending on the gas mixture used in the discharge tube. In a sealed-off CO_2 gas discharge, a significant concentration of CO can be accumulated in the tube after a few minutes of operation(19). Dissociation and recombination processes continue until a quasi-equilibrium state is reached. Clearly other processes such as energy transfer(146) between CO and N_2 and a possible formation of complex molecules also could take place, thus making

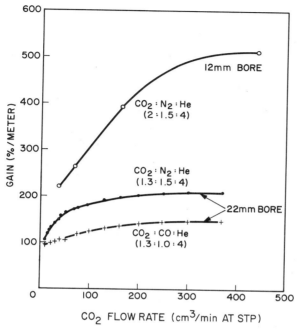

Fig. 21. Effect of gas flow on gain of CO$_2$–N$_2$–He and CO$_2$–CO–He laser amplifiers in 12- and 22-mm-bore tubes. Gas pressure and mixture ratio are near optimum in each case.

the nonflowing CO$_2$ laser a very complicated system to analyze. However, a continuous flow of gas at a very low rate (< 20 cm^3/min) is sufficient to remove the accumulation of CO which causes these complications in sealed-off lasers, and to produce a substantial gain enhancement by maintaining an optimum balance among various components of the gas mixture. Evidence of the accumulation of CO was obtained from the studies of spontaneous light emission of the CO angstrom band at 4835 and 5198 Å in a CO$_2$ laser discharge as a function of gas flow rate. Results(35) shown in Fig. 22 indicate that the relative spectral intensity of CO, resulting from dissociation of CO$_2$, in a CO$_2$ gas discharge decreases rapidly to $\leqslant 20\%$ of that in a sealed-off CO$_2$ gas discharge as the CO$_2$ flow rate is increased from zero to about 20 cm^3/min STP. On the other hand, the relative spectral intensity of CO in a CO gas discharge is relatively insensitive to the flow rate.

The small-signal gain coefficient has also been measured(87) in a very high speed (500 m/sec) gas flowing laser amplifier. In this experiment a uniformly mixed CO$_2$–N$_2$–He mixture [with only the nitrogen

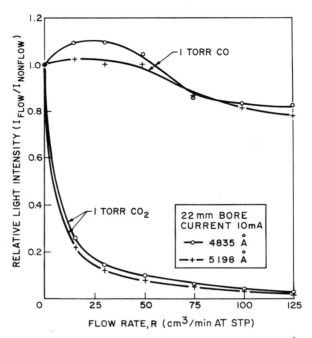

Fig. 22. Ratio of spontaneous light intensities ($I_{flow}/I_{nonflow}$) of 4835 Å and 5198 Å. Angstrom system of CO versus flow rate for both CO_2 and CO gas discharges in a 22-mm-bore tube.

being excited (pumped) prior to mixing] was injected into an amplifier column with an optical axis along the direction of flow. There was no glow discharge in the column and CO_2 molecules were pumped entirely by the N_2 excited in a separate chamber. Because the resonant transfer time is short compared with the fluid mixing time, the optical gain reached a maximum value at a short distance ($\simeq 2$ cm) downstream from the points of injection. For a 2.54-cm-bore amplifier tube with an active length of 18.5 cm, an average gain of 4.2 dB/m was obtained. Although the average gain value in this device was not unusually high, and was essentially limited by the development of a boundary layer downstream, the experiment demonstrated that high gain can be realized in a high-speed tranverse-flowing laser in which the gain should be independent of the length. In fact this system has been investigated recently by Targ and Tiffany(*159*) who have measured the small-signal gain as well as the gain-saturation parameter (to be discussed in Section VI) in an amplifier with an optical axis transverse to the flow.

4. Gain Distribution

From Eq. (63), one can derive the gain coefficients α for $P[(J-1) \rightarrow J]$ and $R[(J+1) \rightarrow J]$ branch vibration–rotation transitions between the $10^0 0$ and $00^0 1$ vibrational levels. Assuming that all rotational levels are in thermal equilibrium with the translational gas temperature T, and that the linewidth is Doppler broadened, one obtains(38) the gain coefficients α_P and α_R as given by

$$\alpha_P = AJT^{-3/2}\left\{\frac{n_1 B_1}{n_2 B_2} \exp\left[-F_1(J-1)\frac{hc}{kT}\right] - \exp\left[-F_2(J)\frac{hc}{kT}\right]\right\} \quad (66)$$

$$\alpha_R = A(J+1)T^{-3/2}\left\{\frac{n_1 B_1}{n_2 B_2} \exp\left[-F_1(J+1)\frac{hc}{kT}\right] - \exp\left[-F_2(J)\frac{hc}{kT}\right]\right\} \quad (67)$$

where 1 and 2 denote the upper $00^0 1$ and the lower $10^0 0$ laser levels, respectively. The parameter A is given by

$$A = \frac{8\pi^3}{3k}\left(\frac{Mc^2}{2\pi k}\right)^{1/2} n_2 B_2 S_{12} \quad (68)$$

where S_{12} is the line strength of the transition in question and is related to the transition lifetime τ_r by

$$\frac{1}{\tau_r} = \frac{1}{2J_1+1}\frac{64\pi^4}{3h\lambda^3}S_{12} \quad (69)$$

From a numerical analysis for the gain distribution with J, Patel(3) showed that population inversion can exist among these vibration–rotation transitions even though $n_1 < n_2$ where n_1 and n_2 are total population densities in the upper $00^0 1$ and the lower $10^0 0$ laser levels, respectively. The relative population inversion for the P branch is much larger than that for the R branch. In an experiment in which the small-signal gain coefficients α_P and α_R were measured for each individual transition, Djeu et al.(38) have obtained the actual gain profiles as shown in Fig. 23. The total gas pressure used in the experiment was about 5 Torr ($CO_2 = 0.65$, $N_2 = 1.4$, He = 2.9 Torr) so that the gain coefficient derived for Doppler broadening is appropriate(30). These measured gain values are somewhat smaller than previously measured values(35), largely owing to the use of low gas pressures which are far below the optimum value for the tube bore used in the experiment. They provide, however, a lower bound of the ratio of population densities for the $00^0 1$ and $10^0 0$ levels in a typical CO_2–N_2–He laser with a longitudinal gas flow. The ratio $n_{00^0 1}/n_{10^0 0}$ equals 2.27 for this laser medium in a 1-in.-i.d. tube with a dc discharge current of 15 mA. Taking the measured(29,30) value of 4.7 ± 0.5 sec for the

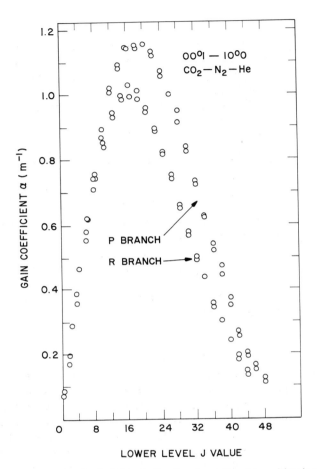

Fig. 23. Gain distribution for individual vibrational–rotational transition in the P and R branches of the $00^\circ 1$–$10^\circ 0$ band. Amplifier tube bore is 2.54 cm; gas mixture and flow velocity are: CO_2 (0.65 Torr), N_2 (1.4 Torr), and He (2.9 Torr); $v = 192$ cm/sec. [After Djeu et al.(*38*).]

spontaneous radiative lifetime of the $00^\circ 1 \rightarrow 10^\circ 0$ transition, one obtains (*38*) the absolute population densities, in this case

$$n_{00^\circ 1} = 3.27 \times 10^{15} \, \text{cm}^{-3} \quad \text{and} \quad n_{10^\circ 0} = 1.44 \times 10^{15} \, \text{cm}^{-3} \tag{70}$$

This corresponds to a lower bound of 17 and 8% of the total density of CO_2 molecules excited, respectively, into the $00^\circ 1$ and $10^\circ 0$ vibrational levels in a conventional flowing CO_2–N_2–He laser system.

5. Analysis—A Thermodynamic Approach

Various mechanisms leading to population inversion in the 00⁰1–10⁰0 band have been presented in Section IV. A large amount of quantitative information on gain and its dependence on various parameters also has been presented. On the other hand, a number of theoretical analyses have been made(22,25,42,46,154,160,161) in the attempt to explain the gain and power characteristics of the CO$_2$ laser. Because of the large number of complex processes involved, an exact treatment is very difficult, and a number of assumptions must be made in order to obtain semiquantitative calculations. In general this problem can be approached by using either the usual rate equations or a thermodynamic model. Using the former approach, Moore et al.(25) have treated the problem by grouping a number of vibrational levels which are strongly coupled with the 00⁰1 level into one combined upper laser level and similarly for the lower laser level. The other approach introduced by Gordietz et al.(22,154) is different from the usual rate equation analysis and will be discussed somewhat in detail here. This analysis has been further extended by Tulip(160) to include the effects of stimulated emission.

Gordietz et al.(22,154) have calculated the gain and population inversion of the vibrational levels in the CO$_2$ laser by a set of energy balance equations among various vibrational modes. In this model, a simple assumption is made that each vibrational mode can be characterized by a temperature, so that in each mode the Boltzmann distribution is established. This assumption is valid as long as the rate of exchange of the vibrational quanta within each mode is faster than excitation and transfer rates into other vibrational modes or translational motion. In this way a set of rate equations can be replaced by a reduced number of energy balance equations for the vibrational modes involved. Physical processes that are taken into account in the balance equations include (1) the electronic excitation of the N$_2$ ($v = 1$) and CO ($v = 1$) levels which subsequently transfer their energies to a CO$_2$ (00⁰1) level by the resonant exchange process and (2) the transfer of vibrational energy of CO$_2$ (00⁰1) to the bending mode and finally into the translational motion of molecules by collisions (for details see Section IV.B). In this analysis, effects of CO, arising from the dissociation of CO$_2$, on various vibrational modes of the CO$_2$ molecule also are emphasized. The direct transfer of the vibrational energy from ν_3 and ν_1 modes to the translational energy and both the spontaneous and stimulated processes have been ignored. The coupled energy balance equations for three vibrational modes of CO$_2$ and for the vibrationally

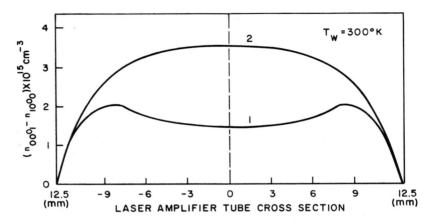

Fig. 24. The calculated radial profile of the population inversion density for (1) CO_2–N_2 (2–2 Torr) and (2) CO_2–N_2–He mixture (2–2–6 Torr), for average electron density $\bar{n}_e = 6.5 \times 10^9$ cm^{-3}. [After Gordietz et al.(154).]

excited N_2 and CO are too lengthy to present here. We shall simply present some of the numerical results and compare them with measured gain coefficients. Figure 24 shows the calculated(154) radial profiles of population inversion density for CO_2–N_2 and CO_2–N_2–He mixtures. These results are in excellent agreement with the experiments (Fig. 20). In the case of CO_2–N_2, the gain at the center of the tube can be even slightly less than that near the edge owing mainly to the high gas temperature on the axis of the discharge column. With the addition of He, the population inversion increases near the center of the tube as a result of a large (about a factor of two) reduction of gas temperature, which has a significant influence on the relaxation rates of vibrational levels (see Tables 16 and 17). Figures 25 and 26 show, respectively, the calculated current and pressure dependence on gain. These results are derived from the above theoretical model using rate constants given in Table 17 by assuming an average electron energy of 3 eV in a discharge tube of 12.5-mm bore. Again these computed values are in reasonably good agreement with experimental measurements (see Figs. 16 and 18) in that the optimum gain occurs at higher currents and high gas pressure for systems containing He. Furthermore the calculated inversion number density is of the same order of magnitude compared with the values(38) deduced from the measurements of the gain distribution profile for each individual transition as given by Eq. (70).

The inclusion of the stimulated emission term in the energy balance equations by an analysis similar to that of Gordietz et al.(154) has been

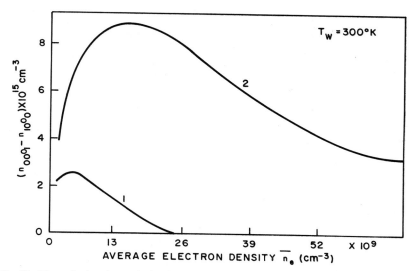

Fig. 25. The calculated population inversion density as a function of electron density for (1) CO_2-N_2 mixture (2-2 Torr) and (2) CO_2-N_2-He mixture (2-2-6 Torr). [After Gordietz et al.(*154*).]

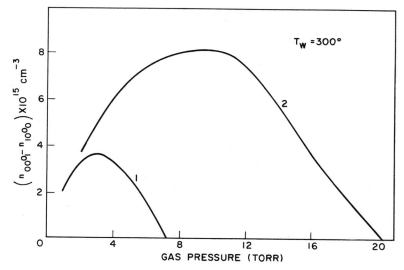

Fig. 26. The calculated population inversion density as a function of total gas pressure for (1) CO_2-N_2 mixture (1-1), $\bar{n}_e = 6.5 \times 10^9$ cm^{-3}, and (2) CO_2-N_2-He mixture (1-1-3), $\bar{n}_e = 2.6 \times 10^{10}$ cm^{-3}. [After Gordietz et al.(*154*).]

made recently by Tulip(160). In this analysis, assumptions are made
to equate the vibrational temperatures of the antisymmetric ν_3 mode
with the excited state of N_2 ($v = 1$), and those of the symmetric ν_1 and
the bending ν_2 modes with the translational temperature. This in effect
is equivalent to the assumption in the rate equation approach(25) of
grouping a number of levels into a single laser level. Further simplifica-
tions are made by ignoring a number of relaxation and excitation, as
well as dissociation and diffusion, processes. With these assumptions
and approximations, the set of energy balance equations is reduced to
only two equations, one of which describes the energy conservation
of the ν_3 mode, E_3, and the other of the combined ν_1 and ν_2 modes,
$E_1 + E_2$. They can be expressed as(160)

$$\frac{dE_3}{dt} = n_0 n_e h\nu_3 X_T - \xi'(T_3, T') + \xi_3(T_2, T_3) - \mathscr{W} h\nu\left(n_u - \frac{g_u}{g_\ell} n_\ell\right) \tag{71}$$

and

$$\frac{d}{dt}(E_1 + E_2) = \xi_2(T, T_2) - (1 - \Delta E)\xi_3(T_2, T_3) + h\nu_2 \mathscr{W}\left(n_u - \frac{g_u}{g_\ell} n_\ell\right) \tag{72}$$

In Eq. (71), the first term in the right-hand side represents electronic
excitation, where n_0 and n_e are the number densities of CO_2 and electrons, respectively. The total excitation probability X_T of both the ν_3
mode and N_2 ($v = 1$) is defined by $X(\nu_3) + (n_{N_2}/n_{CO_2})X(N_2)$. The second
and third terms represent the rates of energy transfer from the vibrationally excited nitrogen N_2 ($v = 1$) characterized by a temperature T'
to $CO_2(\nu_3)$ and from $CO_2(\nu_3)$ to $CO_2(\nu_2)$ levels, respectively. The
functional form of $\xi_i(T_j, T_i)$ is(161)

$$\xi_i(T_j, T_i) = \frac{1}{\tau_{ij}}[E_i(T_j) - E_i] \tag{73}$$

where $1/\tau_{ij}$ is the transfer rate at a given gas pressure, and

$$E_i(T_j) = \frac{n_0 h\nu_i}{\left[\exp\left(\dfrac{h\nu_i}{kT_j}\right) - 1\right]} \tag{74}$$

The last term in the right-hand side of Eq. (71) is the stimulated emis-
sion rate, in which $\mathscr{W}$ is given by Eq. (62). In Eq. (72), $\Delta E \equiv h\nu_3 -$
$(h\nu_1 + h\nu_2)$, and other symbols have their usual meanings. In the steady
state condition that $dE_i/dt = 0$, solutions for Eqs. (71) and (72) have
been obtained numerically(160). By using the measured vibrational
lifetimes(24) of the upper ($\tau_u = 4 \times 10^{-3}$ sec) and lower ($\tau_\ell = 2 \times 10^{-4}$

sec) laser levels and by assuming a line-center value(30) $g(\nu_0) =$ 6.3×10^{-9} which corresponds to a collisional linewidth of 75 MHz and a Doppler width of 50 MHz, this analysis has led to some interesting results on gain saturation, which will be discussed in detail in the next section. The calculated(160) small-signal gain coefficient is 4.8 dB/m at an optimum excitation $(n_e X_T = 200)$ corresponding to an electron density of 4×10^9 cm³. This result and the calculated(160) gain-vs.-excitation currents (data are not included here) are in good agreement with the measured values (Fig. 16), suggesting that this thermodynamic description of vibrational thermal equilibrium in the CO₂ laser is valid even in the presence of a high stimulated emission rate.

B. GAIN SATURATION

One of the interesting characteristics of the CO₂ laser is the extremely high gain saturation parameter which is the basic requirement for a powerful laser system. In the early studies of gain characteristics, Cheo and Cooper(19) observed that the gain remains unsaturated over a 30-dB input power range from a few milliwatts to several watts for a beam cross section ~6 mm diam. Subsequently a number of investigators$(40,41)$ have reported values of the gain saturation parameter ranging from 22 to 100 W/cm². Recent work(159) on a transverse gas flow amplifier yielded a value as high as 250 W/cm². Clearly the gain saturation of a CO₂ laser is not unique and depends much on operating conditions of the laser. Work by Tulip(160) showed that the gain saturation parameter increases with increasing excitation rate. Another detailed study(42) of the effects of diffusion showed that the gain saturation of a sealed-off CO₂ laser amplifier decreases monotonically from 97 to 25 W/cm² as the average input beam radius increases from 0.9 to 2.5 mm in an 18-mm-bore amplifier tube with electrical discharges.

1. A Two-Level Model

According to Eq. (60), a small input signal, of intensity I_0, propagating through an amplifier with an unsaturated gain coefficient α_0, will be amplified according to the expression

$$I = I_0 \exp(\alpha_0 L) \tag{75}$$

where L is the length of the gain medium. However, the growth of radiation is limited by saturation of the medium. Gain saturation as a function of radiation intensity I has been analyzed(*162*) for both homogeneously and inhomogeneously broadened media. In the homogeneous case, the gain coefficient α at the line center decreases with increasing I according to the relation

$$\alpha = \frac{\alpha_0}{(1 + I/I_s)} \tag{76}$$

whereas for inhomogeneous broadening,

$$\alpha = \frac{\alpha_0}{(1 + I/I_s)^{1/2}} \tag{77}$$

where α_0 is the small-signal gain and I_s the saturation parameter, which is defined by that value at which α reduces to one-half of the α_0 value. In terms of the relaxation rates of a two-level system, I_s at the center of the Lorentzian line is(*162*)

$$I_s = 8\pi^2 h\nu\Delta\nu\tau_r \Big/ \lambda^2 \left[k_u^{-1} + \frac{g_u}{g_\ell} k_\ell^{-1} \right] \tag{78}$$

where τ_r, is the radiative lifetime of the $001 \rightarrow 100$ transition, k_u and k_ℓ are the relaxation rates of the upper and the lower CO_2 laser levels, $\Delta\nu$ is the Lorentzian half-width of the transition, λ is the wavelength, and g_u and g_ℓ are the degeneracies of the two levels. Since $k_u \ll k_\ell$ (g_ℓ/g_u), Eq. (78) can be approximated by

$$I_s = \frac{8\pi^2 h\nu\Delta\nu\tau_r}{\lambda^2} k_u \tag{79}$$

A comparison of Eq. (79) with the induced transition rate $\mathcal{W}$ as given by Eq. (62) shows that saturation occurs when the rate of induced transition becomes comparable to the relaxation rate of the upper laser level. The induced transitions depopulate the upper laser level population at a rate which is eventually limited by the relaxation processes in the medium. Therefore, the saturation parameter can be enhanced by increasing the relaxation or the induced transition rate and consequently the small-signal gain coefficient is reduced. This, at first, seems surprising, but the paradox is explained when one considers that the criterion for a laser system to reach saturation is for the rate of excitation to the upper laser level to be as fast as the stimulated emission rate from the excited state. By substituting values $\tau_r = 5$ sec, $\tau_u = 1$ msec, and $\Delta\nu = 50$ MHz into Eq. (79) one obtains a saturation parameter of ≈ 0.3 W/cm² , which is about two orders of magnitude smaller than the measured values. Christensen et al.(*42*) pointed out that the main source of discrepancy between the theory and experiment is the fact that the CO_2 laser is a multilevel system with both homogeneous and inhomogeneous broadening. We shall disregard for the moment a large variation of the measured saturation parameter reported by a number of workers, ranging from 22 to 250 W/cm². To reconcile this discrepancy one must replace the relaxation rate of the upper laser level k_u in Eq. (79) by the weighted relaxation rates $\left(\Sigma_i^m n_i k_i\right)/n_u$ of all the other closely coupled levels to the upper laser level. As discussed in the next section, this, in effect, will increase the relaxation rate k_u by m approximately a factor of which is the number of strongly coupled levels within the 00⁰1 band, or the number of ways that the molecule can rapidly leave the lasing level into other tightly coupled levels, each of which can relax at the rate k_u. This modification has brought the theory in closer agreement with experiments.

2. Multilevel System

Since CO_2 laser levels involve a large number of strongly coupled rotational levels via collisions, it is apparent that a multilevel description is called for, especially when operating conditions are such that level competition is involved. As pointed out above, a theory based on the description of a simple two-level model for a CO_2 laser system cannot adequately account for the gain saturation in a CO_2 laser amplifier. Here we shall present a summary of the analysis for a multilevel system, treated by Christensen et al.(42). In their treatment, assumptions were made that the laser action occurs only between one upper and one lower rotational level, each of which is coupled strongly with a set of m rotational levels, and that the lasing levels are homogeneously broadened. The rate equations for this system are of the form,

$$\frac{dn_J{}^u}{dt} = -k_J{}^u n_J{}^u - \sum_i K_{iJ}^u + \sum_i K_{Ji}^u n_i{}^u - \mathscr{W}\left(n_J{}^u - \frac{g_u}{g_\ell} n_{J'}^\ell\right) + \Gamma_J{}^u$$

$$\frac{dn_{J'}^\ell}{dt} = -k_{J'}^\ell n_{J'}^\ell - \sum_i K_{iJ'}^\ell + \sum_i K_{J'i}^\ell n_i{}^\ell + \mathscr{W}\left(n_J{}^u - \frac{g_u}{g_\ell} n_{J'}^\ell\right) + \Gamma_{J'}^\ell \qquad (80)$$

$$\frac{dn_j}{dt} = -k_j n_j - \sum_i K_{ij} n_j + \sum_i K_{ji} n_i + \Gamma_j \qquad \begin{array}{l}(j \text{ represents all other} \\ \text{rotational levels})\end{array}$$

where $k_J{}^u$ and $k_{J'}^\ell$ are the relaxation rates of the upper and lower laser levels; K_{iJ}^u and $K_{iJ'}^\ell$ are the relaxation rates among the upper and lower levels; n_j's are the population densities of the remaining rotational levels in either the upper or lower vibrational states; Γ_j represents the pumping rate from ground to the jth level; and $\mathscr{W}$ is the induced transition rate as given by Eq. (62).

In the steady state condition, all the time derivatives are equal to zero. Equation (80) reduces to a set of algebraic equations which can be solved for $n_J{}^u$ and $n_{J'}^{\ell}$ in terms of rate constants and for $\mathscr{W}(I)$ by means of Cramer's rule. These solutions combined with Eq. (63) yield the gain coefficient as a function of radiation intensity I in a form similar to that of a two-level system, but the expression contains terms involving k_u's and K_{ij}'s in a very complicated way. We shall not go into detail here. In the assumption that the coupling among the rotational levels is very strong compared with the vibrational relaxation rate, $k_j \ll K_{ij}$, one obtains a simplified expression (42) for the saturation parameter I_s for a homogeneously broadened medium:

$$I_s = 8\pi\tau_r h\nu/g(\nu)\lambda^2 \left[n_J{}^u \Big/ \sum_i k_i{}^u n_i{}^u + g_u n_{J'}^{\ell} \Big/ g_{\ell} \sum_i k_i{}^{\ell} n_i{}^{\ell} \right] \quad (81)$$

By comparing Eqs. (81) and (78), it may be noted that the behavior of the gain saturation parameter I_s of a multilevel system is similar to that of a two-level system if one replaces the terms in Eq. (81) by

$$\frac{1}{n_J{}^u} \sum_i^m k_i{}^u n_i{}^u = k_u{}^{\text{eff}} \quad \text{and} \quad \frac{1}{n_{J'}^{\ell}} \sum_i^m k_i{}^{\ell} n_i{}^{\ell} = k_{\ell}{}^{\text{eff}} \quad (82)$$

If all k_i's and n_i's are assumed to be equal, then $k^{\text{eff}} = mk$, where m is the number of coupled levels in either the upper or the lower vibrational states. To bring the theory closer to experiment, a total of 50 levels must be actively involved in competing for population in a CO_2 laser. Actually n_i is a function of J and gas temperature as given by Eq. (9), therefore Eqs. (82) can be rewritten as

$$k_u{}^{\text{eff}} = k_u \frac{kT}{2hc} \frac{\sum_i^m \exp(-h\nu_i/kTB_i)}{(2J+1)\exp[-BJ(J+1)hc/kT]} \quad (83)$$

Again the assumption was made that all k's are equal. These relaxation rates are collisional in nature when diffusion is negligible, but become enhanced if the particles can diffuse into and out of the beam within a time comparable to the inverse collision relaxation rates. Effects of diffusion will be discussed in detail below. Based on collision effects alone, particles can leave the lasing level and move either into all the levels not explicitly included in the rate equations (80), directly at a rate k_u, or by a cross relaxation into any one of the m levels which subsequently relax to the lower vibration state at the rate k_u. Therefore a particle has m effective ways of leaving the system of upper levels with an effective relaxation rate k_u^{eff} approximately equal to mk_u. This causes an increase in the saturation parameter. As pointed out before, the increased power density obtainable from such a system is achieved because the rate at which excitation of CO_2 molecules into the upper level also is increased accordingly.

3. Effects of Excitation and Diffusion

The measured values of the gain saturation I_s reported by a number of investigators (40–42,159,160) vary over a wide range. The parameters which have the greatest effect on I_s are the excitation currents and the laser beam cross section. Figure 27 shows both the theoretical and measured results for I_s as a function of the excitation parameter (discharge currents). The measurements were made (160) in a laser amplifier tube of 6-cm bore with a flowing CO_2–N_2–He gas (1:2:3 Torr), excited by a dc electrical discharge. The laser beam, passed along the

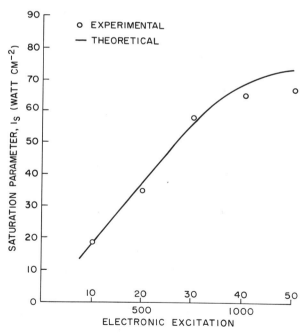

Fig. 27. The measured and calculated gain saturation parameter I_s as a function of excitation. The upper scale of the abscissa is the discharge currents in mA and the lower scale of the abscissa is the excitation parameter $n_e X_T$. [After Tulip(160).]

axis of this tube, had a Gaussian profile with radius of 3.9 mm at $1/e$ of maximum intensity. Detectors with 1-mm aperture size were placed in the near field of the oscillator. The curve in Fig. 27 represents the theoretical results using the thermodynamic approach(160). At low excitation current, the gain of the amplifier obeys a homogeneous saturation function as described by Eq. (76). At high excitation levels ($n_e X \sim 1000$ or currents > 50 mA), theory and experiment begin to deviate, indicating that the theory based on thermal equilibrium may not be valid under these circumstances. Figure 28 shows both the theoretical and experimental results(42) of I_s as a function of average beam radius r_0 of the laser input to the amplifier. In this case the measurements were made in a laser amplifier with i.d. of 1.8 cm and 70 cm of active length. The amplifier was filled with a sealed-off mixture of CO_2–N_2–He–H_2 (1.4:1.7:7.0:0.2 Torr) and dc excited at a constant

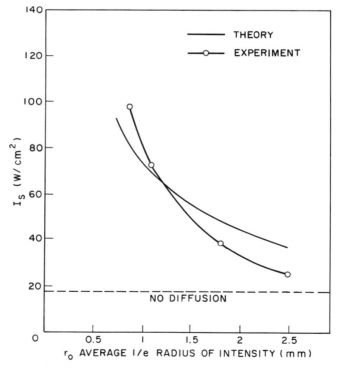

Fig. 28. The measured and calculated gain saturation parameter I_s as a function of average beam radius r_0. A multilevel system (for $n = 35$, $\Delta \nu = 43$ MHz) is assumed. [After Christensen et al.(42).]

discharge current of 26 mA for all the measurements. The measured saturation parameter varied from 25 to 97 W/cm^2 for four different average beam radii r_0 between 0.9 and 2.5 mm (measured to the $1/e$ value of the intensity). The theoretical results in Fig. 28 were obtained (42) by inclusion in the rate equations (80) of a term $\mathscr{D}'\nabla^2 n_J$, where $\mathscr{D}'$ is defined heuristically by the same analogy as that of Eq. (82), namely

$$\mathscr{D}_u' = \frac{\mathscr{D}}{n_u} \sum_i n_i^u, \quad \text{etc.} \tag{84}$$

where $\mathscr{D}$ is the diffusion coefficient of the excited CO_2 molecules. In this way, a set of coupled equations including diffusion effects can be reduced to two equations similar to a simple two-level system,

$$\mathscr{D}'\nabla^2 n_u - k_u^{\text{eff}} n_u - \mathscr{W}\left(n_u - \frac{g_u}{g_\ell} n_\ell\right) = -\Gamma_u$$

and

$$\mathscr{D}'\nabla^2 n_\ell - k_\ell^{\text{eff}} n_\ell + \mathscr{W}\left(n_u - \frac{g_u}{g_\ell} n_\ell\right) = -\Gamma_\ell \tag{85}$$

An analytic solution of Eqs. (85) for the input beam of a uniform intensity profile has been obtained (42) as a function of r_0 and results are plotted in Fig. 28 along with the measurements. Only qualitative agreement was found between theory and experiment, primarily owing to assumptions introduced in order to simplify the calculation. It should be pointed out that the actual beam intensity in a laser is Gaussian in the case of the TEM₀₀ mode; however, solutions of Eqs. (85) for a Gaussian beam is quite difficult. Other work(163) on beam distortion caused by gain saturation is available in the literature.

From these results, it is clear that excitation and diffusion play an important role in determining gain saturation in a CO_2 laser. The depletion rate of the upper laser level population is proportional to the stimulated emission rate but is inversely proportional to the excitation rate. At a fixed excitation rate, the laser field depletes the upper laser level population and causes an excess population in the lower level. The excited CO_2 molecules from outside the beam region diffuse into the space occupied by the laser beam, and molecules in the lower laser level diffuse out of the interaction region. Diffusion effects become more significant when the diffusion rate of molecules is comparable to the relaxation rate of each of these coupled levels. Within the relaxation time τ_u (~ 1 msec), the excited and deexcited CO_2 molecules can traverse optical beams of the order of a few millimeters in radius. Therefore it is easy to understand why the gain saturation parameter increases with decreasing beam size as shown in Fig. 28.

C. OTHER CHARACTERISTICS — ANOMALOUS AND NONLINEAR PHENOMENA

1. *Linewidth and Lineshape*

From absorption(*29,30*) and transmission(*129,130*) measurements, one infers a much wider oscillation linewidth as a result of collisional broadening in a CO_2 laser (typically operating at a gas pressure above 5 Torr) than the usual Doppler width owing to merely the thermal motion of the gas molecules ($\sim$ 60 MHz at a gas temperature of 400° K). A direct measurement(*31*) of the linewidth has been made in a laser amplifier by careful examination of the time response of the gain to a fast-rise step-input pulse. The homogeneous linewidth of this laser can be deduced also from the measurements of the depth of Lamb dip as a function of gas pressure. The former gives a direct measure of the homogeneous relaxation time T_2 and hence the bandwidth of the laser amplifier; the latter yields information not only on the linewidth, but also on the lineshape function $g(\nu)$ and on level competition between adjacent rotational transitions. In the following discussion we shall present experimental measurements on the linewidth and lineshape of the CO_2 laser and point out some interesting aspects of Lamb-dip spectroscopy.

Direct linewidth measurements were made(*31*) by analyzing the time response of the amplifier gain for a step pulse with subnanosecond risetime. Figure 29 shows the oscilloscope traces of both the step input (top trace) and the gain response (bottom trace) of a multipath CO_2 laser amplifier. It may be noted that the shape of the gain pulse is considerably distorted owing to the limitation of the finite bandwidth of the amplifying medium. For an impulse function U_0, the electric field response of the amplifier $\mathscr{E}(t,z)$ with a homogeneously broadened line at the line center is(*31*)

$$\mathscr{E}(t,z) = U_0 + \frac{1}{c}\left(\frac{Az}{ct-z}\right)^{1/2} I_1\left[2\left(\frac{Az}{c}\right)^{1/2}\left(\frac{t-z}{c}\right)^{1/2}\right]$$

$$\times \exp\left[-\frac{\left(\frac{t-z}{c}\right)}{T_2}\right] \qquad t > \frac{z}{c} \qquad\qquad (86)$$

$$\mathscr{E}(t,z) = 0 \qquad\qquad\qquad t < \frac{z}{c}$$

where $A = \alpha_0 c/2T_2$, α_0 is the small-signal gain coefficient, and I_1 the modified Bessel function of the first kind. The step response $G(t,z)$

can be obtained by squaring $\mathscr{E}(t, z)$ as described by Eq. (86), and integrating over the time period $0 \leqslant \tau \leqslant t$, as

$$G(t, z) = \left| \int_0^t \mathscr{E}(t - \tau, z) \, d\tau \right|^2 \tag{87}$$

Figure 30 shows some typical plots of the step response function $G(t, z)$ obtained by computer evaluation for a fixed gain coefficient

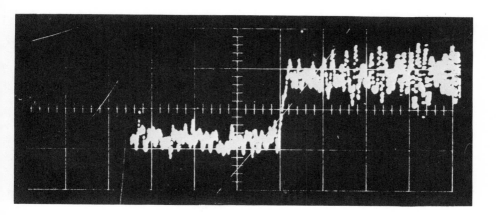

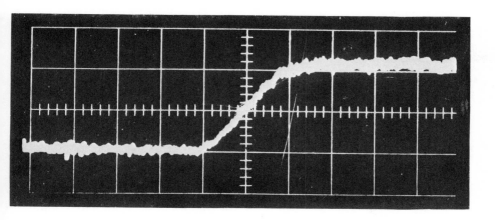

Fig. 29. Oscilloscope traces of time response of laser amplifier. Top trace shows the input signal with amplifier discharge off; bottom trace with discharge on horizontal scale 10 nsec cm. Vertical scales for the two traces are different. [After Bridges et al. (*31*).]

$\alpha_0 \simeq 4$ dB/m, but for three different T_2 values. Experimentally, for a given gas pressure, the measured result yields one net linewidth. A number of measurements were made at different amplifier gas pressures but with a constant mixing ratio of the three gases (CO_2–N_2–He). The measured gain values were fitted to the theoretical curves for a homogeneously broadened line from which one can infer a value of T_2.

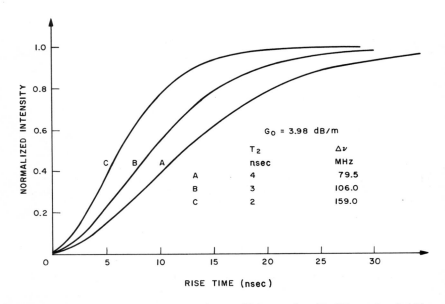

Fig. 30. The calculated step response for amplifying media with different bandwidths. [After Bridges et al. (31).]

Figure 31 shows a plot of the linewidth as a function of total gas pressure. The error bars indicate the uncertainty in the experimental values. The best straight line is drawn through the error bars and yields an intercepting value, 59 MHz at zero pressure, corresponding to the Doppler linewidth for a temperature of 372°K. The slope of this linear plot yields a collisional broadening of 4.67 MHz/Torr as compared with the absorption measurements (30) of 4.25 MHz/Torr for $T = 372°$K. The total bandwidth of a CO_2 laser under an optimum operating condition ($P \gtrsim 10$ Torr) is $\gtrsim 100$ MHz, which is about twice that of the Doppler width, therefore the CO_2 laser lines are essentially homogeneous owing to many collisions of CO_2 ($00^0 1$) with other species during its lifetime.

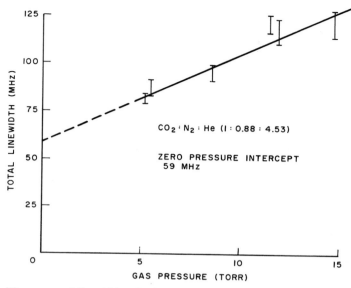

Fig. 31. The measured linewidths of a CO_2 laser amplifier as a function of gas pressure. The uncertainty in the experimental values is indicated by the error bars. [After Bridges et al. (*31*).]

Collision effects on gain saturation characteristics of the CO_2 laser have been studied(*164*) in a passive absorption cell inside the laser cavity. More recently the investigation of saturation characteristics in complex molecular systems, such as PF_5 and SF_6 gases, at CO_2 laser frequencies has led to the achievement of extremely high frequency stabilization for the laser(*165,166*). The advantage of using a passive cell instead of the usual method, frequency scanning over the linewidth of the oscillator, is that one can investigate the collisional effects over a much wider range of gas pressures. The effects of homogeneous broadening on gain are shown in Fig. 32. The dashed curve shows the gain profile of an inhomogeneously broadened line. It should be noted that collision broadening reduces the peak of the unsaturated gain but increases the linewidth, therefore the homogeneously broadened medium is a lossier system than the inhomogeneous case. When an intense field interacts with an inhomogeneously broadened line, "hole burning" occurs in the contour at the line center, reflecting the depletion of the upper laser level population by stimulated emission. In the case of a homogeneously broadened line, collisions tends to smooth out the hole-burning effect in the velocity distribution of the molecules. If the collision time $\tau_c \ll \tau_u$ which is the case

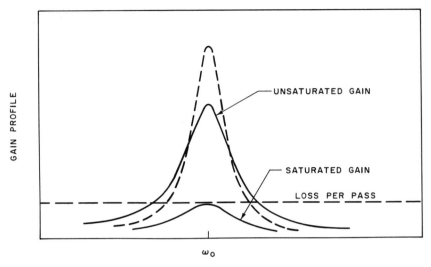

Fig. 32. Effects of homogeneous broadening on gain. Dashed curve represents the gain profile of an inhomogeneously broadened line.

in the CO_2 laser, washing out of the hole occurs, resulting in uniform reduction of the gain profile. Total saturation of the system is reached when gain is reduced below the total loss of the system.

At low gas pressure, the "Lamb dip" or hole burning has been observed(32,167) in the power tuning curve by a frequency sweep of the cavity with a piezoelectric ceramic transducer. For these experiments, the CO_2 laser must be constructed with a high degree of frequency and gain stability. Also a dispersive element such as a prism or a grating is required to eliminate the competition effects between tightly coupled rotational levels. Otherwise, these effects can cause complications in the interpretation of the results. Studies of the relative depth of the central tuning dip have been made(32) as a function of excitation current or gas pressure. Figure 33 shows the lineshape of the output of a $P(22)$ 10.6-μ CO_2 laser as a function of the cavity length. It may be noted that the relative depth of the central tuning dip increases with increasing excitation at a fixed gas pressure, consistent with the Lamb theory(168). The depth of the dip was found(32) to decrease with increasing pressure. For pressures higher than 1 Torr the profile tends to be flattened. The pressure dependence of the intensity profile can be described by the expression given by Szöke and Javan (169),

$$I \propto \frac{G_0 - L \exp(\omega - \omega_0)^2/2\Delta\omega^2}{1 + \Delta\Delta'[\Delta'^2 + (\omega - \omega_0)^2]^{-1}} \tag{88}$$

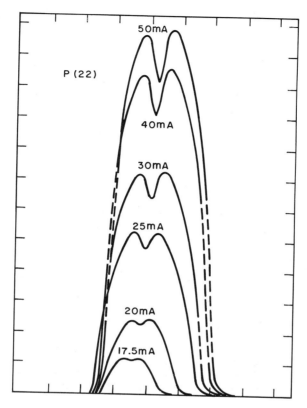

Fig. 33. Gain profile of the $P(22)$ 10.6-μ CO$_2$ laser line at different discharge currents. [After Bordé and Henry(32).]

where G_0 is the gain at the center of the Doppler line in absence of stimulated emission, L is the total loss per unit length, ω_0 is the frequency of the line center, ω is the oscillation frequency, Δ and Δ' are the "hard" and "soft collision" linewidth(170), and $\Delta\omega$ is the full width at half the maximum intensity. By fitting the measured profiles with Eq. (88), Bordé and Henry(32) obtained a plot of Δ and Δ' as a function of the total gas pressure, as shown in Fig. 34. From these results, one obtains a collision-broadening linewidth ~ 5 MHz/Torr, which is in reasonable agreement with the measurements by others(30,31) who used entirely different methods.

Other interesting aspects involving the use of the Lamb dip are: (i) the studies of rotational competition effects resulting in asymmetric gain profiles when two or more vibration–rotation lines are allowed to

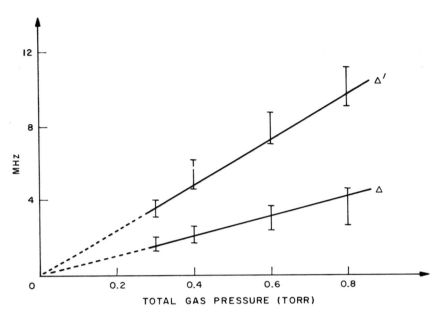

Fig. 34. The measured "hard" Δ and "soft" Δ' linewidths versus gas pressure for $P(22)$
10.6 μ CO_2 laser. [After Bordé and Henry(*32*).]

oscillate simultaneously, and (ii) the stabilization of the laser at the
center of the dip by means of standard feedback techniques.

2. Rotational Level and Mode Competition

Previously we pointed out that rotational level competition plays an
important role in the generation of high CW and Q-switched laser output in a single transition or in obtaining a high-gain saturation parameter in a CO_2 laser. Because of the strong collisional coupling responsible for the mixing of rotational levels, a small change in cavity length
of a small CO_2 laser, whose cavity mode spacing is large compared to
the linewidth, can cause the single-frequency oscillation to switch from
one rotation level to another over a certain frequency interval. Level
competition effects have been investigated in both stabilized standing-
wave(*32,33*) and traveling-wave(*33*) lasers and in laser amplifiers
(*26,27*) (see Section IV.B.5).

When two lines are oscillating simultaneously in a stabilized laser, a
push–pull effect is evident from the central tuning curves of the two
lines when their centers are swept simultaneously (Fig. 35). The profile

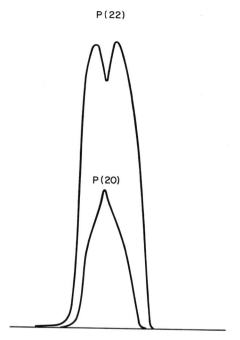

Fig. 35. Gain profile of the $P(20)$ and $P(22)$ 10.6-μ CO_2 laser lines when centers of the two lines are scanned simultaneously. [After Bordé and Henry(*32*).]

of the lower gain line, $P(20)$, shows only a sharp symmetric peak whereas the higher gain line, $P(22)$, shows a larger dip. Under high pressure, the dip in the $P(22)$ line profile normally would not be so distinct in the absence of strong level competition. Oscillation can occur on a number of lines simultaneously only at low pressure. The rotational relaxation rate increases proportionally with pressure and causes strong mixing among the levels so that eventually, at high pressures, only one line can oscillate at a time.

Quantitative investigations(*33*) were made of level competition effects both by the synchronous detection of a low-frequency variation of the heterodyne beat signal of two stabilized CO_2 lasers and by means of precise frequency control and biasing techniques of a ring laser oscillating with two opposite traveling waves. The bistable traveling-wave oscillations in a ring reveal several interesting characteristics of level competition effects and power-dependent gain profiles. In a ring laser, the Doppler shift owing to the gas flow is sufficient to cause a gain anisotropy as a function of frequency. By proper tuning of the cavity the laser can be operated as a unidirectional oscillator. However, over a narrow frequency interval, two rotational levels can be made to oscillate with two opposite traveling waves, one in a clockwise and the other in a counterclockwise (ccw) direction around the ring, with an intensity crossover between the two line centers. The measured gain profiles indicate that both the gain and intensity-dependent anomalous dispersion play a role in crossing from one rotational level to another. The power-dependent gain and index for a homogeneously broadened transition can be expressed by the equations (171),

$$\alpha(\nu) = \frac{1}{I}\frac{dI}{dx} = \frac{\alpha_0\Delta^2}{\Delta^2(1 + I/I_s) + 4\pi^2(\nu_0 - \nu)^2} \tag{89}$$

and

$$n(\nu) = 1 - \frac{c}{\nu}\frac{\alpha_0\Delta(\nu_0 - \nu)}{\Delta^2(1 + I/I_s) + 4\pi^2(\nu_0 - \nu)^2} \tag{90}$$

where α_0 is the small-signal gain coefficient, Δ the linewidth, ν_0 the center frequency of the transition, and I_s the saturation parameter. From these equations we see that both the gain and the refractive index profiles are strongly dependent on I and flattened with increasing I. If the frequency of oscillation is at line center, the dispersion effect vanishes and Eq. (89) reduces to Eq. (76). The power-dependent dispersion represents a mode-pushing term pushing the oscillation frequency away from the line center. Thus by tuning the cavity away from the line center, a point is reached where a different transition with an equal gain will compete for the total population in the laser medium.

The frequency tuning curves were obtained (33) for a ring laser which is made of a solid quartz block with a triangular shape. The perimeter of the active loop was 50 cm in length corresponding to a longitudinal mode spacing of 600 MHz. The laser medium consisted of a flowing CO_2–N_2–He gas mixture, introduced at one of the vertices, streaming down through both legs with flow rates either identical or biased in the two directions. The cavity can be tuned over a wide frequency range by means of a piezoelectric transducer attached to one of the three cavity mirrors. With an equal flow rate in the two legs, i.e., $v_{cw} = v_{ccw}$, the gain profiles for the two opposite traveling waves were identical and both were broadened as a result of the superposition of an up-shifted and a down-shifted gain profiles. The introduction of an asymmetric flow, i.e., $v_{cw} > v_{ccw}$, changed the gain profiles of the two waves significantly. The cw traveling wave had higher gain in the higher flow path and its center frequency ν_0 was up-shifted by $\Delta\nu$ and that in the lower gain path down-shifted by $\Delta\nu'$ with $\Delta\nu > \Delta\nu'$. Similar considerations applied to the ccw wave whose gain profile is a mirror image of that of the cw wave.

By tuning the cavity length from the high frequency side in the case of an asymmetric flow, the cw wave with a higher gain was brought into oscillation unidirectionally. As the cavity was being continuously tuned, a point was reached at which the ccw wave had an equal gain on a different rotational line. Over a few hundred KHz range, two oppositely directed traveling waves were made to oscillate on two different rotational transitions. Figure 36 shows the intensity of the relative amplitudes of the cw $P(16)$ and the ccw $P(18)$ lasers as a function of frequency excursion in the competition region. In addition, the frequency width of the competition region increases with increasing laser intensity. These results confirm that the competition effects are owing to intensity-dependent anomalous dispersion arising from gain saturation.

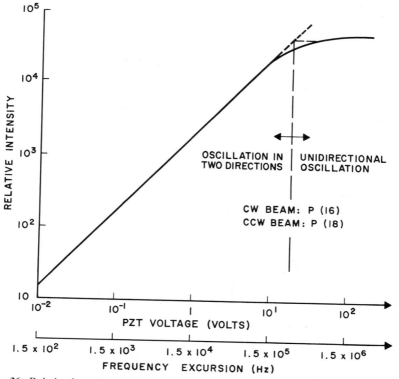

Fig. 36. Relative intensity versus frequency excursion in the competition region between clockwise (cw) $P(16)$ and counterclockwise (ccw) $P(18)$ transitions in a ring laser. [After Mocker(33).]

The interaction of modes in a CO_2 laser has been studied by Witteman(172) using a phenomenological approach rather than the usual density matrix formulation(168). In this analysis, competition between TEM_{00} and TEM_{01} modes owing to spatial overlap of the two fields E_1 and E_2 was treated. The rate equations for population inversion densities n_1 and n_2 associated with the two modes are(172)

$$\frac{dn_1}{dt} = \Gamma_1 - An_1 - \frac{c^3 A}{16\pi h\nu^3}[g(\nu_1)\rho_1 n_1 + \tfrac{1}{2}g(\nu_2)\rho_2 n_1]$$

$$\frac{dn_2}{dt} = \Gamma_2 - An_2 - \frac{3c^3 A}{64\pi h\nu^3}[g(\nu_2)\rho_2 n_2 + 2g(\nu_1)\rho_1 n_2]$$

(91)

where Γ_i is the excitation rate of the ith mode and An_i is the decay rate of the inversion density (where A is the Einstein A coefficient). The last term represents the mixing of the two modes, where $g(\nu_i)$ is the lineshape function of the ith mode, and ρ_i is the radiation flux density ($\rho_i = \tfrac{1}{2}\epsilon \mathscr{E}_i^2$). Solutions of Eqs. (91) were obtained from which estimations for the oscillation conditions could then be made. Figure 37 is a

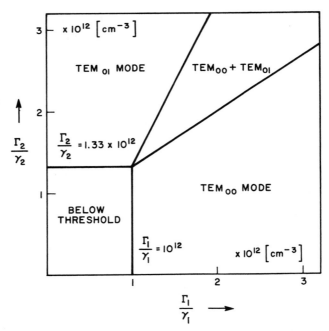

Fig. 37. The ratio of excitation parameter Γ to loss factor γ for the TEM_{00} mode versus that for the TEM_{01} mode. There are three oscillation regions, TEM_{00} or TEM_{01} separately or both modes together. [After Witteman(172).]

plot of the ratio of excitation parameter Γ to the loss factor γ for TEM_{00} mode against that for TEM_{01} mode. The threshold values for excitation are shown in Fig. 37 where ρ_i/γ_i is equal to $A/\alpha_i h\nu$ with estimated values for α_1 and α_2 taken as 5×10^5 and 3.75×10^5 erg^{-1} sec^{-1}, respectively. Above threshold, only the TEM_{00} mode oscillates if $\Gamma_1/\Gamma_2 > 1.5$, and only the TEM_{01} mode oscillates if $\Gamma_1/\Gamma_2 < 0.5$. In the region where $1.5 < \Gamma_1/\Gamma_2 < 0.5$, both modes can be made to oscillate. These results are in qualitative agreement with the observations[172] that in a high-power single-mode CO_2 laser, the oscillating mode switches from TEM_{00} to TEM_{01} when the discharge current is increased from 25 to 36 mA.

3. Frequency Fluctuation

The oscillating frequency of a CO_2 laser can be changed by varying the operating parameters such as pressure and discharge currents. A 5–8 MHz/Torr downshift in frequency owing to an increase in total pressure, and a 0.5–0.9 MHz/mA upshift in frequency owing to an increase in discharge current have been observed[174] upon heterodyning two stabilized linear CO_2 lasers (3 parts in 10^{10}), both oscillating in a single axial and transverse mode on the P branch of the 00^01–10^00 band. From two independent experiments, Mocker[174] concluded that the pressure-dependent frequency shift is caused by a change in the index of refraction of the gas discharge. The largest pressure-dependent shift, 7.4 MHz/Torr, was observed at a CO_2 partial pressure of 3.6 Torr which is near the optimum value for a small 8-mm-bore laser tube. On the other hand, the oscillating frequency increases with increasing discharge current and has a higher shift (900 KHz/mA) at lower current (5 mA/cm^2) and a lower shift (500 KHz/mA) at higher current (15 mA/cm^2). The amount of frequency shift caused by excitation was found to be independent of the location of the cavity resonance with respect to the line center. The current-dependent frequency shift is attributed also to the change in the index of refraction caused by the change in the electron density of the plasma. In a CO_2 gas discharge, the refractive index n is determined by all constituents in the plasma such as the neutral molecules and electrons:

$$n = 1 - \frac{1}{2} \frac{n_e e^2}{m_e \epsilon_0 \omega^2} + 2\pi \alpha_m n_m \tag{92}$$

where n_e and m_e are the electron density and mass, respectively; α_m and n_m are the polarizability and density of the molecular gas, respectively. For gas mixtures and pressures typically used in a CO_2 laser, the change in the refractive index contributes a shift in frequency of 10–12 MHz/Torr compared with the measured values of 5–8 MHz/ Torr. On the other hand, the calculated frequency shift owing to electron density in the plasma (assuming $n_e \simeq 10^{10}$ cm^{-3}) is a somewhat higher value than the measured frequency shift but is within a reasonable range of values for the large current densities and pressures used in the experiments(174). It should be pointed out that a significant reduction of the refractive index may arise from dissociation of CO_2 molecules, which may account for the difference in the calculated and measured results.

Optical line modulation by electron-plasma oscillations at 4 GHz has recently been observed(176) in an argon plasma tube which was placed in a microwave resonantor. By exciting the plasma oscillation in the Tonks–Dattner (TD) resonances(177) at two microwave frequencies, f_1 and f_2, such that the difference frequency Δf is much less than either f_1 or f_2 or the linewidth of the TD resonances, one can measure directly the local oscillations in the spontaneous emission intensity of the plasma. These local oscillations correspond to the localized fluctuations in the plasma density. By means of such an efficient nonlinear mixing between plasma oscillations in gases, this technique allows direct measurements(176) of the microwave-modulated CO_2 laser beam scattered from the plasma oscillation. From these experiments much information can be obtained about the plasma characteristics.

4. Nonlinear Phenomena

Coherent interaction between an intense CO_2 laser field and a resonant medium has resulted in a number of interesting phenomena including (i) passive Q switching and mode locking of the CO_2 laser, (ii) self-induced transparency, (iii) photon echo, (iv) free-induction decay and edge echo, (v) optical transient nutation, and (vi) parametric amplification. The physics of these phenomena involve coupled nonlinear differential equations describing the fields and the induced polarization by the fields. Detailed analysis of these topics is beyond the scope of this chapter; however, we shall present here some experimental results and give qualitative explanations of those which appear to have some practical interest in terms of CO_2 laser systems and devices.

Spontaneous Q switching and mode locking of CO_2 lasers by use of SF_6 gas as a saturable absorber inside the laser cavity was first observed by Wood and Schwarz(47,52) (detailed discussion will be given in Section V.E). This has subsequently led to an intensive study of the coherent interaction between the CO_2 laser and SF_6 or other media in resonance with the CO_2 lasers. Passing a rotating-mirror Q-switched CO_2 laser beam through an SF_6 cell, Patel and Slusher(99) observed self-induced transparency, a phenomenon which is similar to that observed earlier by McCall and Hahn(178) using an intense ruby laser pulse through a ruby inhomogeneous absorbing medium. The optical energy in the first portion of the input pulse, which is absorbed by the medium to create induced dipoles, returns entirely back to the field through a stimulated emission of these dipoles in the presence of the latter portion of the optical pulse. For the case of two nondegenerate quantum states with an inhomogeneously broadened line, McCall and Hahn(179) have studied theoretically the self-induced transparency problem, with emphasis on the existence of stable conditions by propagation. Results show that a stable solution exists for an input pulse of arbitrary shape with an intensity corresponding to a pulse angle $\theta = 2\pi$, where θ is defined by

$$\theta \equiv \frac{2\mu}{\hbar} \int_{-\infty}^{\infty} \mathscr{E} \, dt \tag{93}$$

In Eq. (93), μ is the dipole moment of the absorbing medium, and $\mathscr{E}$ is the field amplitude. For $\theta = 2\pi$, the input pulse will traverse the absorbing cell without suffering any loss in strength and will develop into a hyperbolic secant pulse I_t with a time delay τ_d from the peak of the input pulse, as given by

$$I_t \propto \text{sech}^2\left(\frac{t}{\tau_\omega}\right) \tag{94}$$

and

$$\tau_d = \tau_0 \frac{aL}{2} \tag{95}$$

where τ_0 and τ_ω are the pulse width of the input and the transmitted pulses, respectively, and a is the absorption coefficient of the medium.

Using a short and flat-topped CO_2 laser pulse from a laser with an electrooptic switch(56), Cheo and Wang(107) have examined these characteristics by careful study of the dynamic evolution of this pulse through an optically thin as well as thick SF_6 absorbing medium. Figure 38 shows the waveforms of both the input and the output pulses. The width of the cavity-dumped pulse is about 20 nsec, which corresponds to the round-trip transit time of light in the cavity. Trace b shows the output waveform for an input pulse intensity corresponding to a "π" pulse. Trace c shows the waveform of a "2π" pulse. It is evident from traces b and c that both the pulsewidth and pulse delay decrease as the pulse angle increases from π to 2π. As θ approaches 2π, the pulse

(A) INPUT CO$_2$
      P(20) LASER
      PULSE I$_0$

(B) OUTPUT PULSE
      FROM 2 m LONG
      SF$_6$ CELL (50m Torr)
      I$_0$ = 150 W cm^{-2}

(C) OUTPUT PULSE
      FROM 2 m LONG
      SF$_6$ CELL (50m Torr)
      I$_0$ = 800 W cm^{-2}

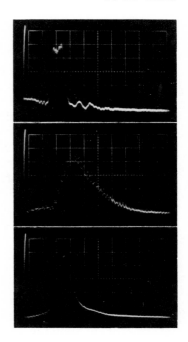

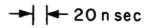

→| |← 20 n sec

Fig. 38. (a) Waveform of a cavity-dumped CO_2 laser pulse at $P(20)$ 10.6 μ. (b) Output pulse shape from a 2-m-long SF_6 cell at 50 mTorr pressure for an input pulse intensity nearly corresponding to a π pulse. (c) Output pulse shape from the same absorption cell for a 2π pulse.

shape becomes more symmetric with the exception that a long tail is always associated with each transmitted pulse. This tail is attributed (*107*) to a phenomenon which is an optical analog of the nuclear magnetic resonance (NMR) phenomenon known as free induction decay and occurs in an inhomogeneous magnetic field(*180*). It arises from the decay of the induced dipoles while undergoing free precession. The effects of propagation(*179*) could also produce such a tail on the transmitted pulse but a detailed study(*107*) of the pulse shape as a function of SF_6 gas pressure rules out this possibility. An examination (*107*) of the shape of the completely transmitted pulse reveals that the measured pulse shape is somewhat steeper than the ideal $\text{sech}^2(t/\tau_\omega)$ function. The deviation is likely caused by the participation of the absorbing lines of higher angular momentum states in SF_6(*105,181*).

The measurements(107) of pulsewidth and pulse delay time as a function of input pulse intensity I_0 indicate that local maxima in τ_ω and τ_d exist at I_0 corresponding to a π pulse. This result is consistent with the calculation(106) based on a simple model involving two nondegenerate quantum states.

Photon echo is also an optical analog of an NMR phenomena known as spin echo. When two Q-switched CO$_2$ laser pulses were passed colinearly through a 4-m-long SF$_6$ cell with a time separation $\tau_s \lesssim 1$ μsec between the two pulses, Patel and Slusher(100) observed that a small echo pulse occurred at a time about τ_s after the second excitation pulse. The optimum echo occurred when the intensities of the two excitation pulses, having identical polarization ($\Delta\phi_p = 0$) vectors, were ~ 1 W/cm^2 and 4 W/cm^2, corresponding to a $\pi/2$ and a π pulse, respectively. The echo amplitude decreased with either decreasing or increasing input pulse intensity and with increasing $\Delta\phi_p$. The echo polarization was always along the polarization of the second input pulse. Analysis(102) showed that the levels responsible for producing the observed echo (100) in a molecular gas such as SF$_6$ must arise from either $J = 1 \leftrightarrow J = 0$ or $J = 1 \leftrightarrow J = 1$ transitions. In performing this analysis one must consider the spatial degeneracy of the molecular energy levels.

Optical transient nutation(101) is another interesting phenomenon which occurs because the intense CO$_2$ laser pulse drives the SF$_6$ molecules successively from the lower state to a coherent superposition state where the induced dipoles oscillate in phase and result in a large microscopic polarization before undergoing the transition to the upper state. Similar steps are followed in the returning path. The cycle repeats but is subject to damping by the relaxation processes in the medium. This oscillation in the population between states reacts back on the field, resulting in an amplitude modulation of the incident laser signal. The frequency of the modulation Ω is proportional to the product of the dipole moment μ and the input field amplitude $\mathscr{E}$

$$\Omega = \frac{\mu\mathscr{E}}{\hbar} \tag{96}$$

Measurements(101) were made on the nutation frequency Ω as a function of $\mathscr{E}$. The dipole moment of the transition responsible for the coherent interaction with the $P(20)$ 10.6-μ CO$_2$ laser was found from the slope of the linear plot of Ω vs $\mathscr{E}$ to be 0.032 ± 0.003 D.

Parametric amplification of a CO$_2$ laser in an SF$_6$ cell has been reported by Gordon et al.(182). This phenomenon is attributed to the interaction between two optical waves in a resonant medium whose microscopic polarization is not linearly proportional to the fields.

Consequently energy of one wave (i.e., the "pump" field) is transferred to another (the "signal" field). In the experiment(*182*), the
pump for the SF_6 parametric amplifier is provided by an intense CW
CO_2 laser. The signal beam consists of a weak CO_2 laser beam with
an amplitude modulated by means of a chopper prior to amplification.
After the two beams are recombined into a single parallel beam within
1 mrad and are propagating through the SF_6 cell, the amplitude of the
sidebands is increased with decreasing modulating frequency. Analysis based on a simple two-level model with a homogeneously broadened
line shows that the gain α for the sidebands with frequency $\omega_s \ll T_2^{-1}$
is(*182*)

$$\alpha = \frac{a}{(1+I_0/I_s)} \frac{(I_0/I_s)^2 - [1+(\omega_s T_1)^2]}{(1+I_0/I_s)^2 + (\omega_s T_1)^2} \tag{97}$$

where a is the absorption coefficient of the medium, I_0 is the intensity
of the pump, I_s is the saturation parameter of the medium, and T_1 is
the collisional relaxation time of the absorber. From Eq. (97), it is
clear that a has a positive value, provided that

$$\omega_s T_1 < \left[\left(\frac{I_0}{I_s}\right)^2 - 1\right]^{1/2} \tag{98}$$

From measurements of the fluorescent decay time T_1 (2.6 msec) and
the saturation parameter of SF_6 (1.8 W/cm²), Eq. (97) yields gain
values in good agreements with the measured results. In addition to
the phenomena described, work on second harmonic generation
(*94–96*), adiabatic rapid passage(*183*) of 10.6-μ CO_2 laser pulses
through SF_6, and possibility of generating ultrashort CO_2 laser pulses
(*184*) can be found in the literature.

D. CW POWER OUTPUT

In this section experimental results are presented on the CW power
output from a conventional low-flowing CO_2–N_2–He laser and the
variation of power yield versus various experimental parameters,
such as flow rates, gas pressure of each mixture, discharge currents,
coolants, and output coupling schemes. Also included in this section
are discussions on the operating characteristics, lifetime, and stabilization in output power spectrum of small sealed-off CO_2 lasers.

1. *Flowing Laser Systems*

Early studies on CO_2 laser systems yielded power outputs varying
over a wide range (a few watts to 40 W/m discharge length), owing

primarily to inefficient output coupling. With advances in thin-film technology, low-loss ($< 1\%$) mirrors with thin dielectric layers increased the optimum output coupling to about 80 W with a multimode (transverse) spectrum from a 1-m discharge column. For a flowing CO$_2$ laser, tube construction is extremely simple, consisting of a straight pyrex tube surrounded by a water jacket. Near the ends of the tube, cold hollow-cathode electrodes of nickel, molybdenum, or simply kovar (glass-to-metal) seal can be used for a reasonable length of time without causing too much sputtering. The ends of the tube are usually sealed with windows of KCl, NaCl, or other IR transmitting materials such as Ge, GaAs, TI-1173 glass, or Irtran II set at the Brewster angle. Typically, the voltage–current requirement for a 1-m discharge tube with a 5-cm bore is about 4 kV and 100 mA, and about 10 kV and 40 mA for a 2-cm bore tube. The efficiency (ratio of output optical power to the input electrical power applied to the discharge tube) can be as high as 20%.

The investigation and optimization of CO$_2$ laser power and efficiency have been done by many workers during the past few years, and it is hardly possible to give complete references. A report prepared by Whitehouse(62) presents the state-of-the-art on mainly large conventional CO$_2$ laser systems (3–20 m in length). In a sequence of experiments, tests on mode control and output coupling were made involving the use of three mirrors of different radii of curvature ($R = 4$, 10, and 25 m). The total reflecting mirror was mounted directly on one end of the 3-m long and 2-in.-bore tube with a NaCl Brewster window on the other end. The output coupling was obtained from a NaCl flat, dielectrically coated on both surfaces to provide 35% transmission. The total cavity length L was 3.6 m corresponding to a longitudinal mode spacing $c/2L = 41.6$ MHz. The spot sizes ω_1 and ω_2 of the fundamental mode at, respectively, the curved and flat mirrors are(185)

$$\omega_1 = \left(\frac{\lambda}{\pi}\right)^{1/2}\left[\frac{R^2 L}{R-L}\right]^{1/4} \qquad \omega_2 = \left(\frac{\lambda}{\pi}\right)^{1/2}[L(R-L)]^{1/4} \qquad (99)$$

The cavity with $R = 25$ m provides a fundamental mode spot that best fills the tube volume. For the cavity with $R = 10$ m, the fundamental was sufficiently small but higher order transverse modes had the necessary gain to oscillate by filling the entire 2-in. bore. Table 19 gives the outputs for optimum gas mixtures and discharge currents and for a gas flow rate of 100 cu ft/min. At a discharge current of 120 mA, the cavity with $R = 10$ m yields an output of 80 W/m as opposed to 58 W/m with $R = 25$ m. With $R = 4$ m, the spot size is very small in the tube and is biggest at the spherical mirror. In this case the higher-

TABLE 19

The Optimum Output Power from Three CO_2–N_2–He Lasers with
Different Cavity Configuration(62)

(L = 3.6 m, bore = 2 in.; flow rate = 100 ft³/m)

R (m)	ω_1 (cm)	ω_2 (cm)	Power (W/m)	Mode
25	0.59	0.53	58	~ 1 in. fundamental
10	0.49	0.39	80	Fundamental + higher order
4	0.62	0.20	50	~ 0.5 in. fundamental

order transverse modes are not allowed to oscillate and output is limited to only 50 W/m.

Power dependence on pressure, flow rate, and discharge current have also been investigated(62), and values at which the maximum laser power can be achieved are in general very similar to those obtained in the unsaturated gain measurements(19,35) (see Section V.A). The output power depends critically on the partial pressure of CO_2, and to a lesser extent on the partial pressure of He and N_2. The optimum CO_2 partial pressure for a 2-in. bore is about 0.5 Torr. The values of 6 Torr for helium and 1.5 Torr for N_2 are reasonably close to the optimum condition for maximum power output. For a flowing gas laser operating at maximum output, a relatively weak dependence on wall temperature is observed (a 20–40% decrease in power output if the water flow is shut off). However, with an increase in tube diameter, the power is greatly decreased, indicating that water cooling is inadequate in the case ot larger-bore tubes (i.d. > 2 in.). Water cooling is extremely important for sealed-off lasers. Without water cooling, the gain of a sealed-off tube becomes so low that in most cases oscillation cannot be maintained. Other processes, such as dissociation and spontaneous radiation losses also can contribute to the reduction of gas temperature, in addition to the conduction through the tube wall and convection by gas flow, but they are negligibly small compared with the conduction and convection losses.

The power dependence on excitation voltage and current is shown in Fig. 39 for two laser tubes. In this figure the peak power is reduced by ~ 25% from the maximum power achieved with the same tube when the dielectric coating on the NaCl mirror was not deteriorated owing to continuous use at high power level. Nevertheless the V–I characteristics and the power dependence are clearly demonstrated. Results with two tube-bore sizes show that the output power is relatively independent of tube diameter D, in contrast with the small-signal gain measurements

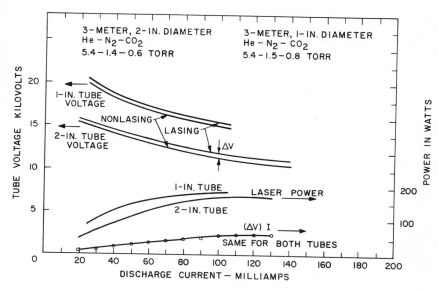

Fig. 39. Voltage, laser power, and incremental electrical power versus current for two laser tubes. Both tubes have the same length (3 m) and same pumping speed (65 ft³/m). A 1-in.-diam tube has two NaCl Brewster windows, one total reflecting mirror (50 m), and a flat mirror with $T = 65\%$. A 2-in.-diam tube has one NaCl Brewster window, one total reflecting internal mirror (10 m), and a flat mirror with $T = 35\%$. [After Whitehouse (62).]

which show a $1/D$ dependence. This result and the requirement of higher discharge currents (about a factor of two) for obtaining optimum output power constitute the principal differences between an unsaturated and a saturated CO$_2$ laser system.

It should be pointed out that the discharge tube has a negative dynamic resistance of approximately 20–200 $k\Omega$ which requires the use of a series of ballast resistances. From Fig. 39, it is clear that laser action affects the V–I characteristics by a noticeable increase in excitation power. This is caused by the radiative power density generation in the lasing medium, acting as a heat sink. The actual increase in electrical power accounts for about $\frac{1}{3}$ of the laser power extracted from the tube. Other heat loss mechanisms include conduction loss to the wall and convection by gas flow, in addition to the radiative loss. By additional measurements on the power dissipated via the cooling water, Whitehouse (62) showed (Table 20) the relative importance of the three dominant heat loss mechanisms for a 3-m-long, 2-in.-bore discharge tube at a flow rate of 65 cu ft/min. Results indicate that when a certain

TABLE 20

The Principal Heat Loss Mechanisms in a Conventional CO_2 Laser (62)

	Nonlasing		Lasing	
	Power (W)	Efficiency (%)	Power (W)	Efficiency (%)
Laser power	0	0	200	15
Power dissipated by H_2O cooling	880	70	790	60
Power dissipated by gas flow	370	30	330	25

amount of power can be coupled out off the cavity in a laser beam, other power losses, such as conduction by water coolant and convection by gas flow, are substantially reduced (> 10%). A simple calculation using the equation of heat flow shows that the gas temperature on the axis is approximately 400°–450° K and the average temperature in the discharge is about 400° K. Coolants other than water also have been attempted but they do not lead to a substantial increase in output power.

2. Large Laser Systems

The first large CO_2 laser system was constructed at the Raytheon Company in Waltham, Massachusetts in 1967 with a folded 20-m-long water-cooled discharge tube. Each leg consists of a 10-m-long 2-in.-diam Corning conical pyrex pipe with one end joined by a brass pumping port through which the flowing CO_2–N_2–He mixture is pumped out and also serves as the common cathode. Mounted on the other end is a brass flange with holes through the edge for admitting the gas. It serves also as the anode for the discharge tube. The two identical 10-m sections are placed side by side on the same physical structure and optically connected in series by means of two flat total reflecting mirrors (gold coated on sapphire or copper) oriented at 45° with respect to the incident beam. A totally reflecting mirror with radius of curvature $R = 50$ m and a freshly polished NaCl flat for the output coupling were used in this folded laser. A CW laser power of 1.2 kW was achieved with an effective pumping speed of ~ 15 cu ft/min. This corresponds to an average power of 50 W/m and an efficiency of 17%. At present, it appears that only salt flats such as NaCl and KCl are suitable for output coupling at kilowatt power levels. Several short-comings of these materials, in addition to their hygroscopic nature, were observed under high-power conditions: (i) dust particles tend to "burn in," (ii) the internal surface acquires an "orange peel" texture causing serious deterioration in optical transmission quality, (iii) the blank loses its flatness under thermal strain, and (iv) it cracks after several hours of continuous operation. Without the salt flat, an output of 100 W over an area about 1 in. diam was obtained in the form of superradiance. Spectrum analysis showed that this output consisted mainly of a single rotational line of the $P(20)$ 10.59-μ transition. The fact that only one transition oscillates is a possible indication that a cavity mode has been built up by feedback from scattering at the Brewster window.

This work was further extended by Roberts et al.(63) who constructed a laser over 170 ft in length. The entire system consisted of 18 modules, each 9 ft 8 in. long. The operating characteristics and the output of this laser as a function of length are given in Table 21. The highest output from this giant laser with 18 modules was reported to be 2.3 kW, and 1.45 kW with only one mirror (total reflecting mirror). Other large CO₂ laser systems(43,64,65), involving the combination of a small CO₂ laser source and a large laser amplifier, will be discussed in Section V.E.

3. Sealed-Off Lasers

Several advantages for operating a sealed-off laser, instead of flowing gas, are that (i) a nonflow laser is inherently more stable in its output spectrum, (ii) it can be made more compact and portable, and (iii) it conserves gases, thus reducing operating and maintenance costs, a factor especially significant in countries where He and other gas components are not abundant. The difficulties with a sealed-off laser lie, of course, in the fact that the gain is considerably reduced and that the operating life is limited. In 1966, Cheo(186) carried out a life test on a sealed-off CO₂–He discharge tube with a water-cooled wall and obtained a total life of ~450 hr under continuous operation at a constant current of 25 mA, which was the optimum current for maximum laser power (26 W/m) when the tube was freshly filled. This tube (1 m in length with a 22-mm bore), which has been extensively used for the studies of CO₂ laser amplifiers(19), consisted of a pyrex envelope with a large ballast (~10 liters) attached to the tube. Two internal mirrors, one total-reflecting and the other dielectrically coated ($T = 10\%$), were bonded (Torr-Seal) onto brass bellows at the two

TABLE 21

Operating Characteristics and Power Output as Functions of Length(63)

Number of modules	$P(CO_2)$ (Torr)	$P(N_2)$ (Torr)	$P(He)$ (Torr)	Impedence (Ω)	Power in (kW)	Efficiency (%)	Power out (W)	Current (mA)
2	0.45	0.75	4.80	1.1×10^5	1.77	13.4	237	125
3	0.45	0.75	4.85	1.75×10^5	2.9	12.3	358	130
6	0.45	0.75	4.85	4.1×10^5	6.4	10.0	640	125
8	0.46	0.74	6.7	6.1×10^5	9.3	10.5	980	125
10	0.47	0.73	5.82	7.1×10^5	11.15	10.4	1160	125
12	0.40	0.80	4.80	8.6×10^5	14.7	9.7	1440	127
14	0.45	0.75	4.85	9.1×10^5	14.4	10.6	1540	127
16	0.45	0.75	4.80	—	—	—	1760	125
18	0.45	0.75	5.8	10.3×10^5	24.3	9.4	2300	150

ends. Two cold hollow Mo electrodes were used to maintain a dc discharge. The choice of using internal mirrors is that the life and performance would not be affected by a gradual deterioration of the salt windows. No special high vacuum processing was undertaken other than that the tube was cleaned by running a discharge for a period of a few minutes, and then evacuated down to a pressure of 10^{-6} mm Hg several times prior to the final filling. Similar results were obtained also by Bridges et al.(61), Whitehouse(62), and Carbone(66).

Another test was performed by Reeves(68) who repeated the previous experiment(186) by using a CO$_2$–N$_2$–He mixture. In this case the discharge current was set at 40 mA throughout the entire experiment. The initial output power was 30 W and reached a maximum of 35 W after 200 hr of continuous operation and then gradually decayed to zero in about 580 hr. At the maximum power point, the CO$_2$–He laser(186) yielded an efficiency of about 16% and the CO$_2$–N$_2$–He laser operated at about 10% efficiency. The side-light emission spectra from the CO and N$_2$ species were monitored during the entire life test by a photomultiplier through a grating spectrometer with a 2.8 f number and a horizontal entrance slit. Almost all the identified lines in the range 3000–6000 Å were found to belong to the Angstrom and third positive bands of CO or second positive bands of N$_2$. When the discharge was initially activated, a marked decrease in current by a factor of about $\frac{1}{3}$ was observed during the first minute. A corresponding increase in CO side-light intensity was also observed, indicating that the increase in impedance which causes a decrease in current was owing to the dissociation of CO$_2$ into CO and O$_2$. The intensities of the CO and N$_2$ spectra remained relatively constant during the life of the laser, although CO showed a slight increase and N$_2$ a slight decrease. One noticeable observation was the gradual development of a dark film deposited on the cathode glass envelope, owing to sputtering of the Mo electrode. The other interesting observation was the occurrence of cataphoresis. This process becomes so severe, especially within a few hours after the laser oscillation has stopped, that the visible glow discharge changes from blue (primarily CO Angstrom) to pink (primarily N$_2$ second positive), and in less than six hours of continuous operation the discharge color changes from pink to orange (solely He line spectra). The migration of the pink–orange interface takes place from anode to cathode over a period of about 1 min. These results indicate that gas cleanup at the cathode by the sputtered film causes nonequilibrium in the CO$_2$ dissociation–recombination process and an eventual depletion of CO$_2$ and the cessation of laser output.

By keeping the cathode and its envelope at an elevated temperature
(71) (~ 300°C), the rate of absorption of CO_2 and the dissociated products at or near the cathode can be reduced, thus prolonging the life
of a sealed-off laser by over 1000 hr. A different approach suggested by
Witteman(69) is to use an inert metal such as platinum for the electrodes and to add a small amount of water vapor (or 0.2 Torr of H_2 and
0.1 Torr of O_2) to the CO_2–N_2–He mixture. Water vapor was believed
to produce OH radicals in the discharge, which combine with CO, the
dissociated product, to form CO_2 and H. Witteman showed that not
only can the life of an operating CO_2–N_2–He–H_2O (or H_2) laser be
extended beyond 2000 hr, but also the output power can be increased
by a factor of nearly two over the same laser filled without H_2O. This
large increase in output power is attributed to the effective relaxation
of the lower laser level by collisions with H_2. The laser tubes used in
this experiment were made of fused quartz and filled with 1 Torr of
CO_2, 2.5 Torr of N_2, 11 Torr of He, and 0.2 Torr of H_2O (the gas mixture ratios used by Witteman differ somewhat from those by other
investigators). A germanium flat (uncoated) was used as the output
coupling mirror and was shown to yield the best output (63 W/m) and
highest efficiency (15%) when used with a 3-m-long laser tube. Work by
Clark and Wada(72), on the other hand, showed that long life (~ 2800
hr) can be achieved also with a pyrex tube (1 cm i.d. and 50 cm long)
and cold nickel and oxidized-tantalum electrodes (instead of a quartz
tube and Pt electrodes) when a small amount of Xe (1 Torr) is added to
a CO_2–He (3.5–12 Torr) mixture instead of N_2 and H_2O. The effect
of Xe is attributed to a reduction of electron temperature, which thus
retards the CO_2 dissociation rate. The maximum output from this small
laser is about 6 W.

Some of the conflicting reports on the use of H_2O (or H_2) and electrode materials have been resolved by Deutsch and Horrigan(73) who
performed a series of life tests by using various approaches and
concluded that H_2 is essential for achieving long life of sealed-off CO_2
lasers. By using a very clean system (well baked at 300°C) filled with a
CO_2–N_2–He–Xe mixture, they found that the tube life is very short
(~ 170 hr) even though heated nickel cathode electrodes were used.
However, the initial laser power can be as high as that reported by
Witteman(69) using a CO_2–N_2–He–H_2O mixture. With an unbaked
system which has been contaminated by H_2 (by running an H_2 discharge
for several hours prior to filling), a steady laser output at a relatively
high power level was observed for almost 900 hr with no indication of
failure. This laser was made of pyrex tubing with water cooling and
with hollow cathode nickel electrodes fitted over with a glass sleeve to

minimize the sputtering owing to collisions of ions with sharp edges of the cylinder. The cathode envelope was thermally insulated to maintain a temperature of 250° to 300°C and a discharge current of 40 mA. The termination of this test was caused by a leak at one of the cemented mirrors. It was also concluded(73) that heating of a cathode alone cannot lead to long life unless the cathode has been processed under an H$_2$ atmosphere. Results of various investigators, as shown in Fig. 40,

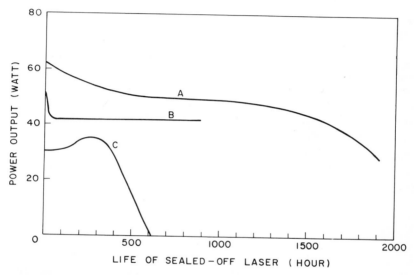

Fig. 40. The laser output versus continuous operating time of the sealed-off CO$_2$ laser. Curve A [Witteman(69)] uses a 1.5-m-long quartz tube filled with 1 Torr CO$_2$, 2.5 Torr N$_2$, 11 Torr He, 0.2 Torr H$_2$, and 0.1 Torr O$_2$. Curve B [Deutsch and Horrigan(73)] uses a 1.23-m-long 2.5-cm i.d. pyrex tube filled with 1.6 Torr CO$_2$, 2.2 Torr N$_2$, 11 Torr He, and 1.1 Torr Xe. Curve C [Reeves(68)] uses a 1-m-long 2.2-cm i.d. pyrex tube filled with 2 Torr CO$_2$, 1 Torr N$_2$, and 6 Torr He.

clearly indicate that the life of a sealed-off CO$_2$ laser can be made longer than 1000 hr under continuous operation provided that the tube is satisfactorily processed, a proper mixture of gases is selected, and the electrode material is correctly chosen. Questions in regard to quartz versus pyrex and the total life of a laser including the time that the laser is not in use still remain to be answered. Reynolds(219) reports a sealed-off CO$_2$ laser with over 10,000 hr of operation. The tube is made of quartz and has two self-heated electrodes. The initial output was about 2 W; after 10,000 hr of continuous operation the power output decreased by a factor of about $\frac{2}{3}$.

As mentioned in the previous section, water cooling is extremely important for a sealed-off laser. Results show that when wall temperature T_w is above 20°C, both the gain(19) and power output(73) decrease rapidly with increasing T_w. However for $T_w < 20°C$, the power output of a sealed-off laser increases slowly with a further decreasing T_w (as low as $-60°C$)(57), indicating that little can be gained by cooling the tube wall below 15°C.

4. Frequency and Intensity Stabilization

A number of applications in communications and in basic research require stable CO_2 lasers with extremely small fluctuation in both amplitude and frequency. An ideal laser would be one oscillating in a single cavity mode with a Gaussian profile far above threshold. Furthermore, under ideal operating conditions, the spectral width of this laser would be limited only by the phase fluctuation caused by the quantum noise(187). For a homogeneously broadened line operating above the threshold of oscillation, the linewidth $\Delta\nu_{osc}$ may be approximated by the Schawlow–Townes formula(188) as

$$\frac{\Delta\nu_{osc}}{\nu} \sim \frac{\pi h}{P_\ell} (\Delta\nu_{cav})^2 \tag{100}$$

where $\Delta\nu_{osc}$ and $\Delta\nu_{cav}$ are full widths at half the maximum intensity, and P_ℓ is the power in the beam. Under ideal operating conditions of the CO_2 laser, $\Delta\nu_{osc}$ is of the order of 10^{-3} sec^{-1}, which corresponds to a frequency stability of one part in 10^{16}. However, this limit has never been reached. For He–Ne or ion lasers, the Doppler linewidth is so wide (~ 1.5 GHz) that a large number (> 10) of axial modes would oscillate in a typical 1-m-long laser cavity even though the Fresnel number is chosen so that the higher-order transverse modes are suppressed by diffraction losses. Single-mode oscillation often is achieved by either shortening the laser tube or operating the laser near threshold. With these schemes the power output is severely limited. The situation is quite different for the CO_2 laser. It is quite easy to maintain a single-mode and single-frequency output at a high power level (> 20 W) because the linewidth of this laser is of the order of 100 MHz under typical operating conditions. Nevertheless, the cavity resonance is, in most cases, narrower than the molecular resonance and consequently the instability in the laser medium owing to fluctuation in refractive index (as discussed in Section V.C.3) as well as the fluctuation in cavity resonance have a large effect on the stability of the laser frequency. Therefore it is imperative to operate a CO_2 laser with a well regulated power supply, and closely controlled gas pressure and temperature. Changes in ambient temperature and mirror vibrations owing to environmental conditions are also a source of instability and lead to both long-term and short-term drifts in cavity resonance. At a given time, the deviation of frequency $\Delta \nu$ from the resonance frequency ν is determined by the fluctuations in the refractive index Δn and cavity length ΔL as given by

$$\frac{\Delta \nu}{\nu} = \frac{\Delta n}{n} + \frac{\Delta L}{L} \tag{101}$$

A frequency shift on the order of 0.5–1 MHz/mA and 5–10 MHz/Torr is expected as a result of the change in refractive index. On the other hand, a 1-Å displacement of a mirror in a 50-cm-long cavity owing to acoustic noise or thermal instability will cause a frequency shift $\simeq 6$ kHz. Therefore in addition to maintenance of a very stable discharge, adequate isolation from thermal and acoustic disturbances must be provided. A small change in cavity length or in refractive index can also cause the single-frequency oscillation to switch from one rotational level to another because of strong collisional coupling between rotational levels. Hence it is desirable to use a dispersive element(58) such as a diffraction grating as one of the cavity mirrors and to operate the laser in a single transverse mode.

Several review articles on laser stabilization techniques are available (189–191), therefore discussions will be limited to those which have already been applied to CO₂ laser systems. For ordinary laboratory use, the usual feedback technique by means of a piezoelectric translator (PZT) attached to one of the cavity mirrors, a lock-in stabilizer providing a small modulated signal and high voltage correction signals to the PZT, and a detector is sufficient to stabilize the cavity length. The work on frequency stabilization of sealed-off CO₂ lasers by stabilization of the cavity structure has been done extensively by Freed(75). Other techniques, either by utilization of competition effects between the cw and ccw waves in a ring laser(33) or by means of laser-saturated molecular absorption in SF₆(165,166), also have been used to stabilize the CO₂ laser.

The most widely used methods of determining the relative stability of lasers is to examine the time variation of the beat frequency of two independently free-running and stabilized lasers by recombining the two beams onto a photodetector(192), usually liquid-nitrogen-cooled gold-doped germanium ($\tau \gtrsim 30$ nsec) or liquid-helium-cooled Cu- or Hg-doped germanium ($1 \leqslant \tau \leqslant 20$ nsec) with a 50-Ω load. The spectral analysis of the beat signal yields information concerning both the long-

and short-term stability of the lasers. By this technique Freed(75) showed that sealed-off CO_2 lasers, without any feedback stabilization and subjected to the normal laboratory environment, have a short-term stability of about 5 parts in 10^{12} for an observation time of 0.05 sec, and about 5 parts in 10^{13} if the 60-Hz disturbance owing to the power supply ripple is disregarded. Typical results of the beat frequency spectrum are shown in Fig. 41. The total frequency excursion

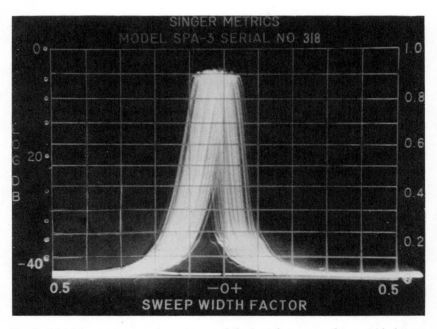

Fig. 41. Multiple-spectrum analyzer traces of the beat frequency of two stable lasers. Film exposure time, 4 sec; horizontal scale, 2 kHz/div; scanning rate, 60 sec. [After Freed(75).]

in this figure covers less than 2 kHz for 4 sec exposure time. The lasers used in these measurements were dc-excited discharge tubes with internal mirrors supported by four superinvar bars. At room temperature, the coefficient of thermal expansion of this alloy is claimed to be at least one order of magnitude lower than fused silica and passes through zero near room temperature. These rods are covered with acoustic and thermal insulating materials. The total length of the lasers is about 50 cm.

In the case of a ring laser, the intensity ratio of the two oppositely directed traveling waves oscillating on two adjacent rotational levels (see Section V.C) is a function of frequency excursion. Therefore a discriminant can be derived without modulation of any of the laser parameters. This was achieved(33) by amplitude modulation of both beams alternatively external to the cavity before recombining. A phase-lock amplifier receives either a positive or negative signal, depending on whether the cw or the ccw beam is larger in intensity, and closes a feedback loop on a piezoelectric transducer on the ring laser. The operation can be understood with reference to Fig. 36. A slight increase in frequency from the equal-intensity point, 150 kHz, can cause the ring laser to change from dual-beam oscillation to unidirectional oscillation. Since a difference in intensity of 10^{-3} can be easily detected with a lock-in amplifier, a stabilization to ~ 5 parts in 10^{12} can be achieved. Other unique features of this operation are that this laser yields two different rotational lines locked in at the inter-rotational level competition region, and that changes in excitation will affect both transitions to an equal extent if the competition region lies exactly half-way between the two line centers, thus providing a first-order cancellation of the instabilities produced by fluctuations in the discharge.

Research on the stabilization of the CO_2 laser by means of locking the laser to the Lamb dip of a closely matched resonant line in SF_6 gas is presently in progress. This technique is similar to that used by Barger and Hall(193) for the 3.39-μ He–Ne laser with CH_4 as the resonant medium. However, exact measurements of the frequency stability of the CO_2 laser by this technique are not available at this time.

E. Q SWITCHING

The Q switching technique by which very intense laser pulses can be generated from a lasing medium is usually applicable when the upper level lifetime τ_u is long compared with the round-trip transit time τ_R of light in the cavity. The process in general involves a time-varying Q (quality factor) of the resonant cavity by means of an intracavity switch. The energy stored in the medium during the quiescent phase will build up far beyond the steady-state power level and then decay to the CW value when a high Q condition is suddenly established. This technique was first proposed by Hellwarth(194) for use with solid-state lasers and recently has been successfully applied to the CO_2 laser.

To understand the transient behavior for the rise and decay of the inversion process, Garrett(195) has presented a simplified analysis similar to that of Hellwarth(194), adaptable to a gas laser. The rate of change of the field $\mathscr{E}$ in a given mode and the rate of change of the upper state population n_u are given by

$$\frac{d\mathscr{E}}{dt} = \left(An_u - \frac{1}{2}\frac{\omega}{Q}\right)\mathscr{E} \tag{102}$$

and

$$\frac{dn_u}{dt} = \Gamma_u - \left(\frac{1}{\tau_u} + \frac{\epsilon A\mathscr{E}^2}{4\pi h\nu}\right)n_u \tag{103}$$

where the assumption is made that the lower-level population relaxes very rapidly so that terms containing $n_\ell(\ll n_u)$ have been neglected. The coefficient A is given by

$$A = \frac{c\lambda^2}{8\pi\tau_r}g(\omega) \tag{104}$$

where $g(\omega)$ is the lineshape function as given by either Eq. (64) or Eq. (65) depending upon the line broadening process. The term Γ_u in Eq. (103) is the pumping rate, expressed as molecules/cm³/sec. Equation (102) is closely related to Lamb's equation of the form,

$$\frac{d\mathscr{E}}{dt} = \alpha\mathscr{E} - \beta\mathscr{E}^3 \tag{105}$$

where the cubic term in $\mathscr{E}$ is contained in the first term on the right-hand side of Eq. (102) through the first-order saturation effect. The last term in Eq. (103) represents the loss term owing to stimulated emission.

By solving these coupled equations, Garrett (195) showed that, as the feedback is suddenly introduced, the $\mathscr{E}$ field in a cavity of length L will build up exponentially in time to a peak intensity I_{max}, given by

$$I_{max} \simeq h\nu n_u L\left(\frac{\alpha_0}{\tau_R}\right) \tag{106}$$

which is the product of energy per photon $h\nu$, the number of available molecules in the upper state per cm² of cross-sectional area, and the single-pass gain per round-trip transit time. The field will then decay exponentially to the CW level as given by

$$I_{CW} \simeq \frac{h\nu n_u L}{\tau_u} \tag{107}$$

where τ_u is the effective lifetime of the upper level. By combining Eqs. (106) and (107), one obtains

$$I_{max} = \frac{\alpha_0 \tau_u}{\tau_R} I_{CW} \tag{108}$$

Equation (108) shows that I_{max} is increased from I_{CW} by a factor $\alpha_0\tau_u/\tau_R$ which is the gain and lifetime product divided by the round-trip transit time.

For a 1-m-long 20-mm-bore laser tube, $\alpha_0 \simeq 3$, $\tau_R \simeq 6$ nsec, $I_{CW} \simeq$ 20 W/cm^2. With the assumption $\tau_u = 10^{-3}$ sec in a typical CO$_2$–N$_2$–He gas mixture, $I_{max} \simeq 10$ MW/cm^2 if Q switching is performed with an ideal switch under an optimum coupling condition. This value, however, has never been achieved in practice. The maximum peak power of a Q-switched CO$_2$ laser achieved to date is about two orders of magnitude below the above theoretical limit and the average power of the Q-switched laser is always less than the available CW power from the same laser.

It must be pointed out that the analysis by Garrett(195) is based on a simple two-level model, and was carried out by making a number of approximations. One important consideration which has been overlooked is that since the CO$_2$ laser is a multilevel system, the energy stored in the laser medium must be distributed among all the rotational sublevels. This is in a way completely opposite to the problem encountered in the analysis of the gain saturation of the CO$_2$ laser. As discussed in Section V.B, the two-level model cannot explain the gain saturation for the case of CW oscillation, and the expression derived from this model must be modified by a weighted average of the relaxation rate owing to the cross-coupling effects among all the sublevels. In the case of very fast Q switching, thermal equilibrium in all the rotational sublevels simply does not exist as in the continuous case so that Q switching can occur on any one of the rotational transitions independent of the others in the band. Therefore the energy stored in the vibrational level of CO$_2$ (00^{0}1) cannot be extracted completely in a short duration. The effects of cross relaxation become important only if the switching time and pulse width are relatively long. In spite of these complications, a modest amount of peak power on the order of a few kilowatts, corresponding to a few millijoules of energy per pulse, can be obtained from a 1-m-long CO$_2$ laser. The repetition rate can be made as fast as the cross-relaxation rate ($\sim 10^7$ Torr^{-1} sec^{-1}) as in the case of spontaneous pulsing (or mode locking), or at a lower rate ($\sim 10^3$ sec^{-1}) for a longer (high average power) Q-switched pulse. In the following we shall present several Q-switching techniques and discuss in detail the characteristics of Q-switched CO$_2$ laser output in terms of pulse shape, width, energy and repetition rates, and the physical processes associated with these pulsing phenomena.

1. *Pulsed Discharge*

The CO$_2$ laser action was first observed(1,2) in a pulsed discharge with a few-microsecond-long excitation pulse. It is therefore reasonable to assume that the gain of a CO$_2$ laser transition in a pulsed lasing medium may be somewhat higher than that in the CW condition. This has in fact been confirmed by Cheo(24) and subsequently by others (64,196). A large enhancement in gain (about one order of magnitude) is observed in the afterglow period owing to a rapid relaxation of the lower laser level before the relaxation of the upper laser becomes significant. By pulsing the discharge with a relatively long current pulse (~ 100 μsec), laser pulses with peak power on the order of 1 kW and 150 μsec in duration have been obtained by Frapard(197). Using very high excitation voltage in the range of 200 kV–1 MV and long pulse-width (6–50 μsec), Hill(88) obtained a large increase in laser peak power output on the order of 200 kW, corresponding to about 5 J per pulse at repetition rates of about 50 pps from a 2.5-m-long, 3-in.-bore tube filled with a flowing CO_2–N_2–He mixture at high concentration. The typical gas pressure used in these experiments exceeds 50 Torr. Efficiency varies from 4 to 10% depending on tube bore and gas mixtures. Figure 42 shows the waveforms of the excitation currents and laser pulse. The laser-pulse waveforms, Figs. 42b and 42c, can vary substantially upon variation of the gas-mixture ratio. In high-pressure gas discharges, it is expected that plasma instabilities can occur, causing a nonuniformity in the laser output across the tube diameter.

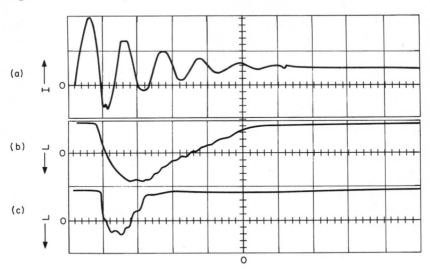

Fig. 42. (a) The excitation-current waveform I in a 2-in. bore, 6-ft-long laser tube filled with 5 Torr CO_2, 13 Torr N_2, and 30 Torr He at a flow rate 40 ft³/hr. The initial break-down voltage is ~ 500 kV and $I_m \simeq 40$ A. (b) The laser output L caused by the excitation (a). (c) Variation of the laser pulsewidth by change of CO_2–N_2 mixing ratios. Horizontal scale, 10 μsec/division. [After Hill(88).]

Recently, Beaulieu(*220*) has reported the development of an electrically pulsed, atmospheric-pressure CO_2 laser with 20-MW peak power
in a submicrosecond pulse. The efficiency is 17%.

2. *Mechanical Q Switching*

The most straightforward method of spoiling the cavity Q is either
by rotating one of the cavity mirrors with a synchronous motor or by
chopping the beam in the cavity with a mechanical shutter. The
rotating-mirror technique has been commonly used in solid-state lasers,
and was first applied to obtain Q-switched CO_2 laser pulses by Kovacs
et al.(*44*). From a 2-m-long laser, they reported laser pulses with a
peak power in excess of 10 kW corresponding to a few mJ of energy
per pulse up to a maximum rotation rate of 500 Hz. A typical output
pulse from a rotating-mirror Q-switched CO_2 laser cavity is shown in
Fig. 43. With a single transverse mode operation, the optimum rotating

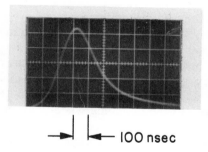

⟶| |⟵ IOO nsec

Fig. 43. Output waveform of a rotating-mirror Q-switched CO_2 laser pulse. The laser
cavity consists of a 1-m-long flowing CO_2–N_2–He discharge tube terminated by two
KCl Brewster-angle windows. The rotating mirror is coated with gold film and has
$R = 10$ m. Power is coupled out from the diffraction grating mirror at zeroth order.
Horizontal scale, 100 nsec/division; peak power, $\simeq 1$ kW.

speed occurs at 200 Hz. As the rotation rate is increased from the optimum value, both the pulse width and the pulse energy decrease sharply
and reduce to zero at a rate of around 400 Hz. The output pulse shape
is asymmetric with a full width at half power of about 300 nsec. Let
a_1 and a_2 be the radii of the laser beam on the rotating and the stationary
mirrors, respectively. The rotating mirror reflects an image of the
stationary mirror back on itself in a time $t \simeq 2a_2/\omega L$ where ω and L are
the angular velocity and length of the cavity, respectively. For a 10-mm
spot size on the stationary mirror, $\omega = 1800$ rad/sec and $L = 2$ m, one
obtains an average time t_Q of about 300 nsec during which the rotating

mirror is aligned with the stationary mirror, consistent with the observed pulsewidth. The buildup time τ_b is typically of the order of 200 nsec, as shown in Fig. 43, and depends on the net gain and the turn-on time of the switch.

One complication with the rotating-mirror technique is the occurrence of time-varying frequency or chirping in the output pulse spectrum. During the pulse buildup and decay, the cavity mirror continuously rotates and the cavity length as well as the modes change as a function of time. Workers(198,199) have observed this effect in a rotating-mirror Q-switched CO_2 laser pulse by heterodyning the Q-switched laser output with a stable CW local oscillator. The beat frequency ν is found to be a linear function of time. The measured(199) value of $d\nu/dt$ is 70.4 MHz/μsec for a mirror rotation rate of 120 Hz. The result is consistent with a simple estimate of $d\nu/dt = -\nu r\omega/L$ where $r \simeq 3$ mm is the offset of the mirror from tube axis, ω is the angular velocity of the mirror, and L is the cavity length.

Using a mechanical shutter in the cavity near the flat mirror at which the laser beam is focused by two lenses to a very small spot, Meyerhofer(46) was able to produce pulses of about the same energy as those obtained with a rotating-mirror technique but at much higher repetition rates up to 5000 pps. The beam size at the focus between two antireflecting coated germanium lenses is about 95 μm so that the shutter-opening time can be minimized to about 700 nsec. By means of appropriate notches cut into the chopping wheel, any pulse repetition rate can be obtained; however, pulse energy will decrease in amplitude (Fig. 44) with decreasing pulse separation.

3. Reactive and Passive Q Switching

An approach that differs from the conventional Q-switching technique, and is known as the "reactive" process for obtaining high repetition rate Q switching in a CO_2 laser, has been described by Bridges(45). By moving one of the cavity mirrors along the laser axis at velocities between 16 and 30 cm/sec, he could obtain Q-switched pulses at repetition rates between 30 and 60 kHz in a single transition [$P(20)$ at 10.6 μ] with a width in the order of 1 μsec and an average power about the same as the CW output. To achieve a series of uniform single-transition pulses, the mirror speed must be maintained above 16 cm/sec, as shown in Fig. 45a. At higher speed (>30 cm/sec) the pulses begin to alternate in size, a large pulse occurring for every complete wavelength of motion, as shown in Fig. 45b. The recovery time between pulses is attributed to the relaxation time of

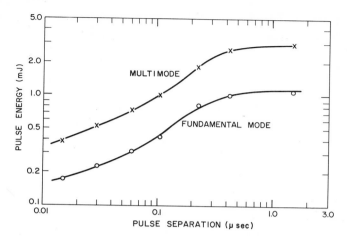

Fig. 44. Measured values of pulse energy as a function of pulse separation. The linear velocity of the mechanical shutter is maintained at a constant value $(0.8 \times 10^4 \text{ cm/sec})$ in all cases. [After Meyerhofer(46).]

the lower laser level. The measured time $(\sim 30\ \mu\text{sec})$ between pulses at the onset of a single-line oscillation is consistent with pulse-gain measurements on the lower-laser lifetime (at $P \sim 7$ Torr) made by Cheo(24).

At lower speed $(< 16 \text{ cm/sec})$, the situation becomes rather complex. The output pulse spectrum consists of a number of lines during a half-wavelength movement, as shown in Fig. 45c. The pattern repeats itself but the process is subject to hysteresis effects. At slow speeds, the time required to travel a half-wavelength is much longer than the relaxation time of the lower laser level. A quasi-CW operation is maintained therefore, and switching occurs over several lines, owing to conditions more favorable for that particular transition in coincidence with a longitudinal mode at that instant.

Passive Q switching is another technique which can be used to obtain short CO_2 laser pulses by means of a saturable absorber. The first successful operation of passive Q switching in a CO_2 laser, using SF_6 gas as the saturable absorber, was reported by Wood and Schwarz (47). Subsequently similar phenomena were observed by various investigators using other gases, such as formic acid vapor, propylene, heated $CO_2(48)$, vinyl chloride(49), $BCl_3(50)$, CH_3F, $PF_5(51)$, CF_2Cl_2, and $C_2F_3Cl(200)$. Undoubtedly many more can be found to produce Q-switched CO_2 laser pulses. The requirements for passive Q switching are that the medium must possess a strong absorption

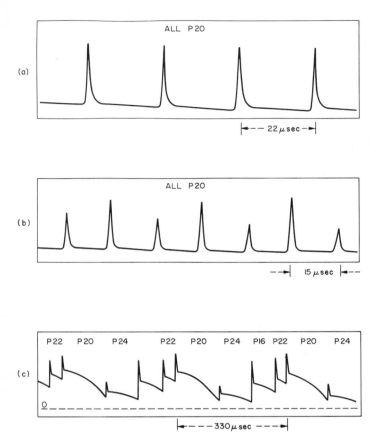

Fig. 45. Typical output of a CO_2 laser with one end mirror moving along the laser axis at the following speeds: (a) 24 cm/sec, (b) 35 cm/sec, (c) 1.6 cm/sec. The time taken for the mirror to travel a half wavelength is indicated. [After Bridges(45).]

cross section σ at CO_2 laser wavelengths and a long relaxation time T_1 for the upper state of the absorber so that $I/I_s \gg 1$, where I is the laser intensity and $I_s = h\nu/\sigma T_1$ the saturation parameter. In the case of SF_6, the I_s has a value of a few watts per cm² at a pressure of about 100 mTorr. When the absorbing gas is placed inside the laser cavity where I is in the order of 10^2 W/cm², bleaching will occur repetitively at a time period ($\sim 500~\mu$sec) corresponding to the collisional relaxation time T_1 of the upper vibrational state of SF_6. The repetition rate can be increased to some extent by the addition of buffer gases such as helium. With the help of a prism or a diffraction grating, the bleaching, similar to that observed in other laser systems(201), can permit a burst of laser radiation at a wavelength in resonance with the absorber. The typical pulse is about 0.5–2 μsec in duration and has a peak power 20–200 times that of the CW output, depending on the gain of the laser medium and the concentration of the absorbing gas. Table 22 gives the performance of passive Q switching by various absorbers. Upon extension of the cavity length to 16.25 m, corresponding to an axial mode spacing of 9 MHz, self locking of a number of longitudinal modes has been observed(52) in a CO$_2$–N$_2$–He laser with an SF$_6$ absorber inside the cavity. However, mode locking of the CO$_2$ laser had been reported(53) earlier through the use of a GaAs intracavity acoustooptic switch as will be discussed in the following section. With the bleachable gas SF$_6$ inside the cavity, laser pulses as short as 20 nsec, as shown in Fig. 46a, with peak power in excess of 10^4 W have been obtained. The duration of the mode-locked pulses increases with increasing cavity length and decreasing absorber concentration. The fast rise and slower fall of these mode-locked pulse trains cover a period of approximately 1 μsec or longer (Fig. 46b), which is comparable to that of a passive Q-switched pulse. The situation repeats

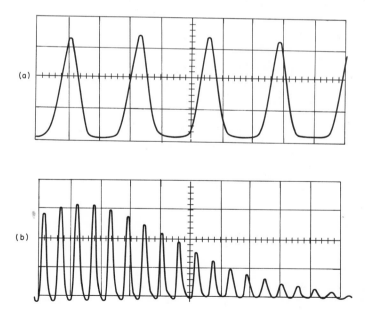

Fig. 46. Oscilloscope traces (reproduced) of mode-locked operation of a CO$_2$ laser with a cavity length $L = 16.25$ m. Horizontal scale: (a) 50 nsec/division, (b) 200 nsec/division. [After Wood and Schwarz(52).]

TABLE 22

Performance and Operating Characteristics of Passive Q Switching in a CO_2 Laser by Various Saturable Absorbers

Absorbing gas	Absorber transition	Abs. coef. ($Torr^{-1} cm^{-1}$)	Cell length (m)	Pressure (mTorr)	Temp (°C)	Laser tube length (m)	Laser transition	Pulsewidth (μsec)	Pulse separation (μsec)	Peak power	Ref.
CH_3F	$Q_{Q(12)}$	0.018	1	15–20	300	1	$P(20)$, 9.6 μ	1–2	40–100	$12P_{CW}$	51
	$Q_{R(4)}$	0.0033	1	40–160	300	1	$P(32)$, 9.6 μ	1–2	40–100	$12P_{CW}$	51
CO_2-propylene			0.75	500–500	450	1	$P(18)$, $P(20)$, 10.6 μ	1–2	18		48
Formic acid			0.2	~500	300	1	~9.2 μ			$\sim P_{CW}$	48
Vinyl chloride		0.01	0.1	2000	300	3.4	$P(18)$, $P(20)$, 10.6 μ	2	20	$4P_{CW}$	49
BCl_3 (He, NH_3, or air)		0.01–0.02	0.2	~1000	300	0.8	P and R branches 10.4 μ		40–100		50
PF_5		0.02–0.12	1	5–70	300	1	$P(2)$–$P(38)$, 10.4 μ	0.6–1	150–500	$\sim 50P_{CW}$	51
SF_6		0.1–0.5	0.1	20–50	300	3	$P(18)$–$P(26)$, 10.4 μ	0.4–2	250–450	$\sim 20P_{CW}$	47,49
C_2F_3Cl		0.019	0.25	500	300	2.5	$P(6)$–$P(20)$ $R(22)$–$R(26)$, 9.4 μ	1.8–5	15–40	$\sim 10P_{CW}$	200
CF_2Cl_2		0.076	0.25	20–50	300	2.5	$P(16)$–$P(46)$, 10.4 μ	0.8–3	15–100	$\sim 20P_{CW}$	200

every millisecond. Cavity dumping of a CO$_2$ laser by means of a bistable Fabry–Perot resonator filled with a saturable absorber (SF$_6$ gas) has been attempted(202). Preliminary results show that this technique can be used to achieve dumping but requires considerable improvement. Furthermore, passive Q switching and mode locking by use of saturable absorbers coupled with the present cavity dumping techniques do not allow simple control over the operating parameters and consequently the laser output. The most desirable methods are either by acoustooptic or by electrooptic means, which, as discussed below, provide a convenient way to control the shape and repetition rate of Q-switched CO$_2$ lasers.

4. Acoustooptic and Electrooptic Q Switching

Mode locking of a small (4 m long) CO$_2$ laser corresponding to an axial mode of 36 MHz was first reported by Caddes et al.(53), using a GaAs crystal as an intracavity acoustooptic loss modulator. These authors claim that mode locking of as many as five axial modes can be obtained by transferring energy from the fundamental oscillating mode to nearby below-threshold modes by sweeping the frequency back and forth across the oscillating linewidth (FM)–instead of the usual mode-locking operation (AM) by driving the cavity resonance at half the fundamental frequency $c/2L$ as first introduced by Hargrove et al.(203). In the AM case, the cavity loss is periodically modulated by the standing acoustic waves at a frequency $\nu_m = c/4L$, causing all excited cavity modes within the atomic linewidth to be phase locked. The observation by Caddes et al.(53) has been analyzed by extending the work on FM of McDuff and Harris(204) to include the effects of saturation. For homogeneous saturation, the ratio of the peak intensity I_p of the mode-locked laser to that of the CW laser I_{cw} is given by(53)

$$\frac{I_p}{I_{cw}} = \frac{\left[\left(\frac{\alpha_0}{\ell}\right) - \left(1 + \frac{2d}{\ell}\right)^{1/2}\right](1+\gamma)^2}{\left[\left(\frac{\alpha_0}{\ell}\right) - 1\right]\left(1 + \frac{2d}{\ell}\right)^{1/2}(1-\gamma)^2} \tag{109}$$

where α_0, ℓ, and d are the small-signal gain coefficient, the unmodulated loss parameter, and the depth of modulation. The parameter γ describing the below-threshold modes is given by(204)

$$\gamma = 1 - \left[\left(1 + \frac{2d}{\ell}\right)^{1/2} - 1\right]\frac{\ell}{d} \tag{110}$$

Equation (109) is plotted in Fig. 47 as a function of α_0/ℓ. Notice that

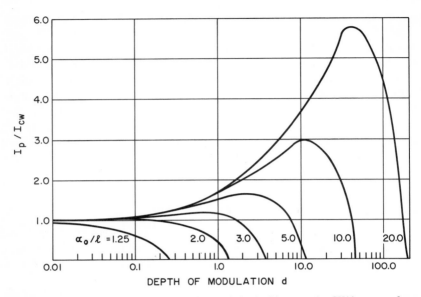

Fig. 47. The calculated intensity ratios of a mode-locked laser to the CW laser as a function of the depth of modulation d for various values of the gain α_0 over loss ℓ parameters (α_0/ℓ). [After Caddes et al. (53).]

an enhancement of 5.7 in peak power from that of CW power is expected for a homogeneously broadened medium with $\alpha_0/\ell = 20$. On the other hand, no peak power enhancement is expected for an inhomogeneously saturated medium (results are not included here). Using a GaAs acoustooptic modulator driven by a quartz transducer at 18 MHz in a 4-m-long CO_2 laser cavity, an attempt was made (53) to produce short mode-locked laser pulses; unfortunately the photoconductive detector used to monitor the laser output did not have the bandwidth to detect the mode-locked pulses directly. However, Caddes et al. (53) were able to obtain beat signals at frequencies corresponding to $c/2L, c/L, \ldots, 5c/2L$. From the measured intensities of the beat signals, the ratio of d/ℓ is found to be 2.5, corresponding to an estimated pulsewidth of 5 nsec. It should be pointed out that the frequency response of a similar detector (SBRC 9145-Sb compensated Ge:Cu) has been shown (192) to have a 3-dB rolloff at 150 MHz, therefore the poor frequency response of the particular detector used in the above experiment is owing strictly to a unique situation of that detector.

In another experiment Bridges and Cheo (56) observed spontaneous self-pulsing in a CO_2 laser without a saturable absorber, using an intracavity GaAs crystal as an electrooptic Q switch. Each Q-switched pulse contains a train of 10 or more 20-nsec pulses with a total duration of ~ 400 nsec in a single $P(20)$ transition in the 10.4-μ band. The switch is capable of dumping any one of the short pulses from the train out of the cavity through an output coupling polarizer, to produce a single output of ~ 10 kW peak power and ~ 20 nsec duration. Figure 48

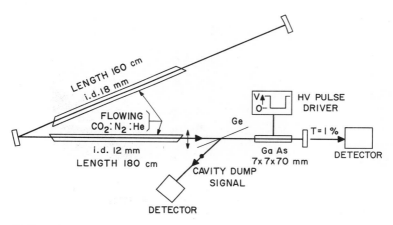

Fig. 48. Experimental arrangement for self-pulsing and dumping in a CO₂ laser with a GaAs electrooptic Q switch.

shows the experimental arrangement. The laser cavity consists of two discharge tubes filled with a flowing gas mixture of CO₂–N₂–He, a Brewster-angle Ge plate polarizer, a GaAs electrooptic crystal, and two end mirrors (one a totally reflecting mirror with 10-m radius of curvature and the other a plane mirror with 1% transmittance). Two cavity lengths, 5.5 and 8.2 m, were used in this study. The total available single-pass small-signal gain was 15 dB at 10.6 μ. Because of optical inhomogeneities in the GaAs, the net gain was considerably lower than this value, but the exact amount was not measured. The crystal, when suitably pulsed in conjunction with the polarizer, provided Q switching and cavity dumping. Similar arrangements have been used for solid-state lasers (205).

A spark gap pulser produced the switching voltages. During the quiescent phase, 5.5 kV dc was applied to the GaAs ($7 \times 7 \times 70$ mm) converting the normally isotropic crystal into a quarter-wave plate. The combination of biased crystal and polarizer prevented oscillation by introducing a large coupling loss to the resonator. The Q switching was achieved by pulsing the voltage to zero in less than 10 nsec

(10–90%). The coupling loss was then reduced to the 1% of the plane mirror. Output signals were detected with a GE:Cu:Sb photoconductor and a Tektronix-454 oscilloscope, with a combined time response of ≈ 2.7 nsec, which is limited primarily by the rise time of the scope. A typical signal observed at the slightly transparent end mirror is shown in Fig. 49(a). Following the Q switch, after a delay of ≈ 100 nsec, a

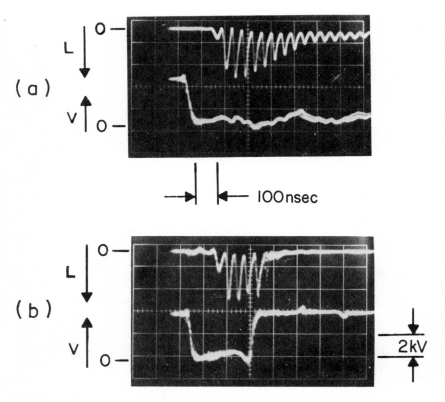

Fig. 49. Oscilloscope traces of the self-pulsing waveforms in a CO_2 laser as observed from the partially transmitting plane mirror ($t = 1\%$). (a) The cavity Q is suddenly switched on by a GaAs electrooptic intracavity modulator. (b) The cavity Q is switched on and then switched off in a time approximately 250 nsec. Cavity length = 5.5 m; horizontal scale: 100 nsec/cm.

train of 10 or more short 20-nsec pulses is seen, rising to a peak intensity after ≈ 250 nsec and falling gradually to zero in about 500 nsec. This envelope is comparable to the single long (> 200 nsec) pulse typically obtained with other types of Q switching. The period between the short pulses was the round-trip transit time in the resonator, about 37 nsec for the 5.5-m and 55 nsec for the 8.5-m resonator. The pulses were locked to the voltage pulse and were stable and repetitive from one Q-switch even to the next. Tests with a spectrometer showed that the output under these conditions consisted of only the strongest, $P(20)$, vibration–rotation line.

It was noticed that the pulse train appeared only when the oscillation built up to its peak in less than ≈ 300 nsec. The buildup time t_b was minimized by the use of small external coupling during the Q-switching phase, and by careful adjustment of the resonator. A sharp minimum in t_b was observed as the mirrors were brought into correct alignment. Cavity length tuning also varied t_b periodically, the behavior repeating with each half-wavelength displacement of the mirror. Close to the optimum points, $P(20)$ pulse trains were noted, while at intermediate points the oscillation built up in a much longer time (> 500 nsec) and consisted of several lines with no rapid pulsing. Related effects were noticed with length tuning in CW CO_2 oscillators. The optimum adjustment corresponds to the best coincidence between resonator modes and the strongest line, $P(20)$. The oscillation on $P(20)$ then builds up rapidly and, because of strong collisional coupling$(26,27)$, takes precedence over other lines. The decay of the pulses after the peak is attributed to depletion of the vibrational inversion.

During the period preceding pulsing, a dc bias voltage is applied to the GaAs intracavity electrooptic Q switch which spoils the cavity Q. When the bias voltage is switched off, radiation builds up to its peak from the noise after about six round-trip passes (see Fig. 49). By pulsing the dc voltage on again in a time within ~ 10 nsec of the peak of oscillation, the resonator energy could be dumped via the germanium polarizer. After dumping, the Q is again spoiled and oscillation ceases. Without spontaneous self-pulsing, the dumped pulsewidth would equal the round-trip transit time of light in the resonator with a square-topped pulse shape as shown in Fig. 38a. When spontaneous pulsing was present and reapplication of the dc voltage occurred at a null between pulsations, a single short laser pulse of about 25-nsec duration and about 10-kW peak power was obtained as shown in Fig. 50a. When voltage was reapplied near a peak of the pulsation rather than near a null, a split pulse output was obtained as shown in Fig. 50b. The output power, coupled from the 1% transmitting mirror when dumping occurred, is shown in Fig. 49b. Optical inhomogeneities in the GaAs reduced the power output considerably from that potentially available. By replacing the 1% transmitting mirror with a diffraction grating, a short cavity-dumped CO_2 laser pulse from any one of the transitions

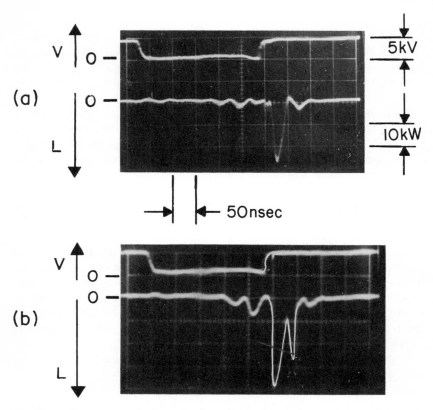

Fig. 50. Oscilloscope traces of a cavity-dumped CO_2 laser pulse coupled out of the cavity from the germanium polarizer. (a) Dumping occurs at a null of pulsations (cavity length = 5.5 m), (b) Dumping occurs near the peak of pulsations, resulting a split pulse (cavity length = 8.2 m). Horizontal scale: 50 nsec/division.

$P(14)$ to $P(30)$ of the 10-μ and 9-μ bands, with a 180-cm-long CO_2–N_2–He flowing-gas discharge tube, has been obtained as shown in Fig. 38a. Pulses from the R branch of both bands also can be obtained, but the number of rotational lines is reduced.

Since no saturable absorber was present during observation of self-pulsing, it was concluded that the effect results from nonlinear satura-tion phenomena in the laser medium itself. Spontaneous self-pulsing of this type has been previously reported for other gas lasers(206), and has been explained in terms of phase locking of a number of longitudinal modes, or in terms of a π pulse circulating in the resonator(207,208). These may be considered equivalent interpretations of the same phenomenon in the frequency and time domains, respectively. The observations will be compared with the circulating pulse picture as follows.

It was previously assumed(207) that the circulating pulse duration (τ) had to be less than T_2 (the average molecular dephasing time). This is believed to be an unnecessary restriction and the short pulse train, where $\tau = 8T_2$ [Ref.(31)], also can be described as π pulses. In the closely related problem concerning the buildup of a steady-state π pulse in a homogeneously broadened amplifier, theory(208) shows that the pulse duration is unrestricted and depends not only on T_2, but also on the gain and loss parameters of the system. The definition of a π pulse (for linear polarization), independent of pulse length is

$$\pi = \frac{\mu}{\hbar} \int_{-\infty}^{+\infty} \mathscr{E} \, dt \tag{111}$$

where μ is the dipole moment and $\mathscr{E}$ is the electric field amplitude of the pulse. Assuming that similar considerations apply to the oscillator, we can now compare the experimental results(56) quantitatively with the circulating π-pulse theory. From Eq. (111) the electric field strength required for a π pulse can be calculated using a graphical integration of the observed pulse shape. The value of $\mu/\hbar$ is derived from the measured (30) value (5 sec) of the 001–100 transition lifetime. The calculated peak field strength is 1.4×10^5 V/m corresponding to a power density of 2.4 kW/cm². Estimating the average beam diameter (multimode) as 15 mm and the peak power of the strongest pulse as 10 kW gives an actual peak field strength 1.5 times the calculated values for the π pulse. This is close enough to be significant in view of the approximate nature of the theory, which neglects diffraction, variation of field strength across the beam, and cross-relaxation effects.

As mentioned previously, the observed laser pulses were synchronized with the switching pulse. Recent computer simulation studies showed how spontaneous pulses can build up from noise(209) or from a small Gaussian disturbance(210). In the experiment with an electrooptic switch, the pulse builds up in a similar way from a "noise step" produced when the GaAs switch is turned on. About six round trips are required for this initial disturbance to build up to peak oscillation. The major difference between this experiment(56) and previous studies with a rotating mirror is that the electrooptic switch has a switch-on time less than the round-trip transit time. The recovery (in ≈ 37 nsec) of population inversion (pumping) between the short pulses can be accounted for by the rapid transfer(26,27) of the rotational energy from other rotational states to the oscillating rotation–vibration level.

5. High-Power Pulsed Radar Systems

As discussed previously, high-power pulsed CO_2 laser output can be obtained from a high-pressure gas ($P > 50$ Torr) breakdown by using very high voltage at E/P values ranging from 12 to 70 V/cm Torr. Laser outputs under these conditions have very high energy (on the order of several joules per pulse) from a relatively compact system, but are very unstable. Therefore, it is not suitable for certain applications, such as in communication and pulsed Doppler radar systems. Work toward the development of a high average power (> 1 kW) IR radar system to be installed at the optical radar facility(92) of Lincoln Laboratory in Westford, Massachusetts was begun in 1967 almost simultaneously at Raytheon(64) and at Hughes Research Laboratories (65) in Malibu. The coherent laser source at 10 μ offers several advantages over conventional microwave radar systems in terms of beam divergence, large backscattered cross sections at optical frequency (211), jamming, and other interference effects. The transmitter was designed for signal processing at low signal-to-noise ratios backscattered from a moving target over periods of between 10 to 100 μsec, depending on the nature of the target. The Doppler frequency-shifted signal is detected by heterodyning with that of the transmitted pulse, which must be maintained at a constant and stable frequency with a well-defined wavefront at the transmitting plane. The most straightforward way to achieve these requirements is to use a low-power frequency-stabilized oscillator whose output pulse train is subsequently amplified through a sequence of laser amplifiers.

Figure 51 shows the design of the 1-kW transmitter. A 25-W, single-mode output from a small CW oscillator is collimated by an array of reflectors and first directed through two stages of amplification to yield a maximum output of 400 W. This output is then modulated by a mechanical chopper at high repetition rates (up to 12 kHz) using a focusing technique similar to that described in Section V.E.2. These rectangular pulses of $\gtrsim 0.2$ mJ energy are then directed into the 50-m-long power amplifier with a small-signal gain of ~ 60 dB, a value chosen to avoid self-oscillation. At each end of the four-power amplifier sections, irises are used to minimize off-axis scattering. The folded amplifier-tube design is similar to those described in Section V.D.2. The performance of this transmitter is shown in Fig. 52. The average output power reaches 1 kW at a pulse repetition rate above 7 kHz, and continuously increases toward the equivalent CW level near 1400 W if the mechanical modulator is removed from the system. When operated in multimode oscillation, this system is capable of delivering an output

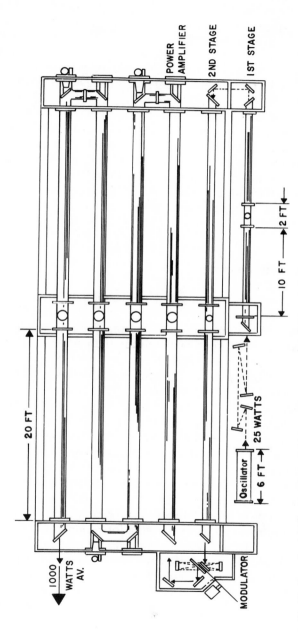

Fig. 51. Design of a 1-kW transmitter, which consists of a single-mode 30-W (CW) laser oscillator, first- and second-stage simplifiers, the modulator section, and the power amplifier section. [After Miles and Lotus (64).]

CW power in excess of 2.5 kW. The "self-oscillation background" line
(in Fig. 52) refers to the spurious oscillations in the power amplifier.
This is caused mainly by diffuse scattering from the mechanical
chopper blade during the quiescent phase of operation. However,
these spurious oscillations can be suppressed at pulse-repetition rates
above 5 kHz. A test of beam divergence in the focal plane showed
that the angular spread is less than 3×10^{-4}, 6×10^{-4}, and 10^{-3} rad at
respective CW power levels of 100, 200, and 500 W. In this system,
there is a considerable amount of backscattered signals from various
optical components within the first two amplifying stages and in
particular from the chopper blade; therefore, for heterodyne detection
of weak return signals this spurious noise in the output channel must
be reduced by either an efficient isolator between the oscillator and
the amplifier array or replacement of the mechanical shutter with an
electrooptic switch.

VI. Novel Laser Systems

Previous sections dealt mainly with conventional CO_2 laser systems,
employing the usual electrical discharges in either a nonflowing or a
low-speed longitudinal flowing gas mixture. For these systems, considerable advances have been made during the past few years both in
the understanding of the physical processes involved in various lasing
materials and in their optimum performance. Continuous-wave power
as high as 8.8 kW(212) with efficiency of $\sim 25\%$, Q-switched pulses of
$\gtrsim 0.2$ MW peak power(213,214), and a single-pass unsaturated gain of
7.8 dB/m(35) have been achieved. On the other hand, concentrated
effort has been placed on the development of various novel laser
systems with the aim of producing high average power ($\gtrsim 1$ kW) from a
small and compact laser system by means of nonconventional approaches, for example, systems employing either a continuous transverse flow at supersonic speed(82,87), or a sudden expansion of hot
gas through a nozzle in a shock tube(215). The concept of utilizing thermal pumping and rapid adiabatic cooling for production of population
inversion in CO_2 molecules was first introduced by Basov et al.(78) as
early as 1966. Experimental results on these systems are in general not
available, and information which recently appeared in literature(87,214)
are of a preliminary nature. However, experiments with transverse
flow at subsonic speeds have shown that high power(86) and gain(159)
can be obtained from a relatively compact laser system. Other ap-

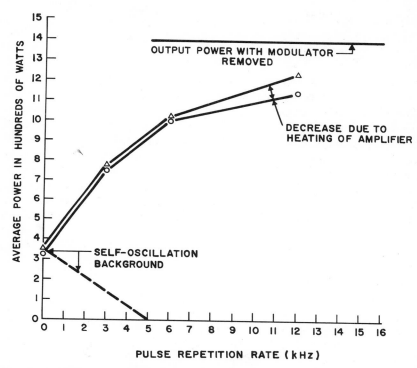

Fig. 52. Measured average power output of the transmitter shown in Fig. 51 versus repetition rate of the 10-μsec CO₂ laser pulse train. [After Miles and Lotus(64).]

proaches involving chemical(83,216) and thermal(77–82) pumping also have led to some significant advances.

The basic principles of a high-speed transverse-flowing-gasdynamic laser are based on differences in relaxation times for energy levels in the subsystems and the exchange rates of internal energies to translational temperature in the lasing medium. The rapid cooling of the translational temperature of the gas by sudden expansion or high flow rate is particularly important for thermally pumped lasers. Relatively high speed is also essential for chemically pumped lasers because reactions which are suitable for laser action are time dependent, and the chemical species are controlled by the gas flow. In the following a brief discussion of the physical mechanisms involved in each of these novel systems is given, and results on their performance under various experimental conditions are presented.

A. High-Speed Subsonic Gas Flow

1. *Transverse Flow*

An analysis of power output and gain saturation characteristics of an idealized high-speed transverse-flow laser has been made by Cool(85). The assumption is made that the pumping process (electric discharge, flame, or vibrational energy transfer from various excited metastables of diatomic molecules such as N_2, DF, HCl, HBr, and HI) which establishes an upper laser level population occurs in the region located upstream before the gas enters the laser cavity. Laser action takes place within an optical cavity downstream where the vibrational relaxation occurs but excitation ceases with the exception of a resonant transfer from the metastables. Figure 53 describes the coordinates used

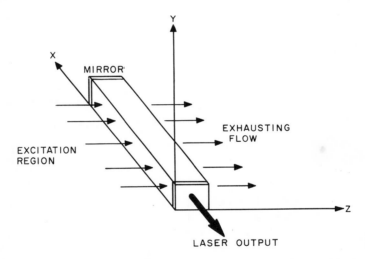

Fig. 53. Schematic diagram of a high-speed flow laser with the optical axis transverse to the gas flow. [After Cool(85).]

in the analysis. The gas flow is in the positive Z direction between two cavity mirrors which are aligned perpendicular to the X axis, and in this region ($Z \geq 0$) excitation is completed and the vibrational populations decay primarily by collisions and by stimulated emission within the laser cavity. A simple kinetic model (Fig. 54) as proposed by Moore et al.(25) is used by lumping a group of strongly coupled asymmetric vibrational states 00^0n, where $n = 1, 2, \ldots$, as the common upper laser level (designated by group 2), and similarly by combining a group

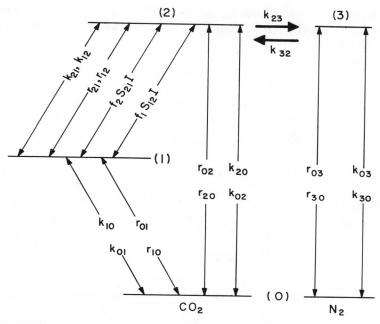

Fig. 54. Rate-controlling processes in a CO_2 laser. Groups 1 and 2 represent the collective lower and upper laser levels. Group 3 represents the excited metastables such as N_2, DF, HCl, etc.

of closely coupled states, i.e., 10^00, 02^00, 02^20, and 01^10 as the lower laser level (denoted by group 1). Group 3 represents the metastables of diatomic molecules with vibrationally excited states up to $v = 4$ that are strongly coupled to the $v = 1$ state through a rapid intramolecular $V \rightarrow V$ process. The energy stored in group 3 is subsequently transferred to group 2 via the resonant exchange process, as described by Eq. (33) for the case of a CO_2–N_2 mixture. Therefore the number densities n_2 and n_3 of the respective groups 2 and 3 quickly reach an equilibrium through the coupled equations

$$\frac{dn_2}{dt} = -\frac{dn_3}{dt} = k_{32}n_3 - k_{23}n_2 \qquad (112)$$

where k_{32} and k_{23} are related by

$$k_{32} = k_{23} \exp\left(\frac{E_{N_2} - E_{CO_2}}{kT}\right) = 0.914\, k_{23} \qquad (113)$$

The measured value(25) for k_{32} is $1.9 \times 10^4\,\text{Torr}^{-1}\,\text{sec}^{-1}$. The rate equations describing the intramolecular coupling between groups 0, 1, and 2 are given by

$$\frac{dn_0}{dt} = -(r_{02} + r_{01})n_0 + r_{20}n_2 - k_{10}n_1$$

$$\frac{dn_1}{dt} = r_{21}n_2 + r_{01}n_0 - k_{10}n_1 + k_{21}n_2 + \text{SI} \qquad (114)$$

$$\frac{dn_2}{dt} = r_{02}n_0 - r_{20}n_2 - (r_{21} + k_{21})n_2 - \text{SI}$$

where the rate coefficients are defined in accordance with those indicated in Fig. 54; $r_{ij} \equiv$ excitation or deexcitation rates from ith to jth level; k_{ij} are the effective or over-all relaxation rates of these groups, and SI is the stimulated emission rate which connects the upper and lower laser levels as

$$\text{SI} = (x_2 S_{21} n_2 - x_1 S_{12} n_1) I \qquad (115)$$

where S_{12} is the line strength of that transition as given by Eq. (69), x_1 and x_2 are mole fractions of molecules in group 1 and 2, respectively, and I is the stimulated photon energy flux within the cavity, expressed in W/cm^2. These equations provide the initial population inversion in the upstream region. As the fluid enters the cavity, only vibrational relaxation processes occur ($Z \geqslant 0$) and the relationships determining the performance of the high-speed transverse-flow laser can be derived from the continuity equations for the number densities of these groups of vibrational states. For a one-dimensional isothermal flow with a uniform constant velocity U, Cool(85) obtained a set of equations describing the relaxation behavior within the laser cavity as

$$U\left(\frac{\partial n_3}{\partial Z}\right) = -k_{30}n_3 - k_{32}n_3 + k_{23}n_2$$

$$U\left(\frac{\partial n_2}{\partial Z}\right) = -k_{21}n_2 - k_{23}n_2 + k_{12}n_1 + k_{32}n_3 - \text{SI}$$

$$U\left(\frac{\partial n_1}{\partial Z}\right) = k_{21}n_2 - k_{12}n_1 - k_{10}n_1 + k_{01}n_0 + \text{SI} \qquad (116)$$

$$U\left(\frac{\partial n_0}{\partial Z}\right) = k_{10}n_1 - k_{01}n_0 - k_{32}n_3 + k_{23}n_2$$

where the rate constants are the same as those defined in accordance with Fig. 54. In Eqs. (116), the relaxation rate k_{03} and spontaneous emission processes have been neglected during the relaxation period.

Also the population densities n_i and stimulated emission term SI are the total average values within a cavity of length L along the X axis. From Eqs. (116) one obtains an oscillation condition relating to the relaxation processes in a steady flow as given by (85)

$$SI + U\lambda\left(\frac{dS}{dZ}\right) = d_2(k_{32}n_3 - k_{23}n_2) + d_1(k_{10}n_1 - k_{01}n_0)$$
$$- (k_{21}n_2 - k_{12}n_1) \tag{117}$$

where

$$\lambda = (x_1 S_{12} - x_2 S_{21}), \qquad d_1 = \lambda x_1 S_{12}, \qquad d_2 = \lambda x_2 S_{21} \tag{118}$$

From Eq. (117), we see that contributions to the population inversion are provided by the first (resonant transfer to group 2) and second (collisional relaxation of group 1) terms on the right-hand side. The last term represents the dissipative effect owing to a direct collisional coupling between groups 1 and 2. The solution of Eqs. (117), for a special case of constant mirror reflectances, yields the following relationship:

$$S = \frac{S_0}{(1 + I/I_S)} \tag{119}$$

where

$$S_0 I_S = C_1 \exp\left[\frac{-\alpha(1+\beta)Z}{U}\right] + C_2 \exp\left[\frac{-\alpha(1-\beta)Z}{U}\right] + C_3 \tag{120}$$

and

$$I_s = C_4 \exp\left[\frac{-\alpha(1+\beta)Z}{U}\right] + C_5 \exp\left[\frac{-\alpha(1-\beta)Z}{U}\right] + C_6 \tag{121}$$

with constants C_i, α, and β expressed in terms of various rate constants. The quantity I_s is the local saturation parameter at Z, and $h\nu S_0$ is the unsaturated gain at the optical frequency ν. Since the inversion density is a function of Z, it is also practical to use a nonuniform reflecting mirror $R(Z)$. The total laser power output P_ℓ per effective cavity volume V can be obtained by integrating Eqs. (120) and (121) over the cavity width W along the flow as

$$\frac{P_\ell}{V} = \frac{h\nu U}{W}\left(\frac{t}{a+t}\right)\left[\int_0^{W/U} S_0 I_s d\xi + \frac{\ln(1-a-t)}{h\nu L}\int_0^{W/U} I_s d\xi\right] \tag{122}$$

where $\xi \equiv Z/U$, a, and t are dissipation loss and transmission coefficients of the laser cavity. Figure 55 shows the dependence of the local saturation parameter I_s along the flow direction Z as calculated

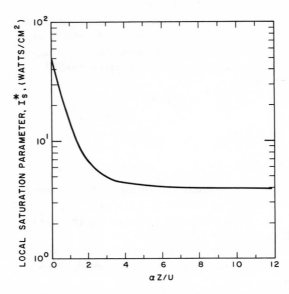

Fig. 55. Calculated local saturation parameter I_s as a function of $\alpha Z/U$, where $\alpha = 1.36$ $\times 10^4 \sec^{-1}$, Z is the distance along the direction of gas flow, and U is the fluid velocity. [After Cool(85).]

from Eqs. (120) and (121). The initial value $I_s(0)$ is determined mainly by the transfer rate constant k_{32}; further downstream, I_s is primarily limited by the $V \to T$ relaxation rate of level 1 characterized by the rate constant k_{10}.

The optimized laser output coupling t_m can be obtained from Eq. (122) by setting $\partial P_\ell /\partial t = 0$. The optimum coupling parameter S_m is given by

$$S_m = -\frac{1}{h\nu L} \ln (1 - a - t_m) \tag{123}$$

Equation (123) yields the maximum power outputs for given values of L/a. The total radiation energy per optimum cavity volume $HW_m L$ is given by

$$E = h\nu \int_0^{W_m/U} SI \, d\xi \tag{124}$$

and the total laser output per optimum cavity cross-sectional area HW_m is shown in Fig. 56 for $a = 0.04$ as a function of the cavity length L. The results of Figs. 55 and 56 were obtained by using rate constant measurements, most of which were made in an absorption cell(25)

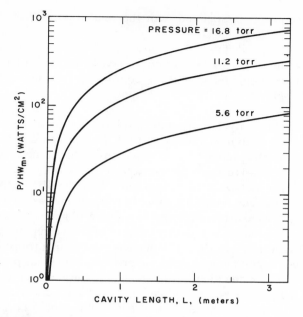

Fig. 56. Calculated laser output per effective cavity cross-sectional area (P/HW_m) as a function of the cavity length L. [After Cool(85).]

near room temperature and employing a laser system(85) with relatively low gain coefficient $\alpha = (1/I)(\partial I/\partial X) = 5.5$ m^{-1}. These values can vary substantially from one laser system to another depending upon the experimental conditions, such as gas flow rate and temperature. In addition. the treatment of gain saturation and power output, as discussed in Section V.B, must take into account the cross relaxation and diffusion effects. Despite these uncertainties the calculated results of Figs. 55 and 56 indicate some interesting characteristics of the high-speed flowing CO$_2$ gas laser, namely:

(a) The power output is directly proportional to the flow velocity U.

(b) The ratio of optimum laser output to cavity volume varies as the square of the gas pressure, provided that the gas temperature and mixture are fixed and a constant percentage of excited molecules is maintained.

(c) The localized power output and saturation parameter decay exponentially in two stages with increasing Z. The initial decay is characterized by a time constant associated with the rate of energy transfer to the upper CO$_2$ laser level, and the second stage overlaps the first and has a time constant associated with the relaxation rate of the lower laser level.

(d) For high speed flow, where flow transit time is short compared with the buildup time for the boundary layer and molecular diffusion time, the gain is independent of the Y coordinate and the power output is directly proportional to the cavity height.

Almost concurrently Tiffany et al. (86) have obtained a very impressive CW power output in excess of 1 kW from a self-contained CO_2 laser with 1-m active length, by using a recirculated forced-convection flow of a CO_2–N_2–He mixture transverse to the optical cavity. The major difference between this experiment and the above theoretical model is that the electrical discharge is used in the laser cavity. The flow velocity U used in this experiment is above 30 m/sec corresponding to a transit time across a few centimeters of active region within the lifetime of the upper laser level. Since gas cooling in a transverse-flow laser does not depend upon a diffusion process, higher gas pressure can be used, thus providing higher gain and power output. Figure 57 shows the power output for several gas mixtures as a function of electrical input power to the discharge. For a gas mixture of 2 Torr CO_2, 11 Torr N_2, and 5 Torr He, an output in excess of 1 kW/m in a beam of 40 mm diam can be coupled out of the cavity through a 35% transmitting

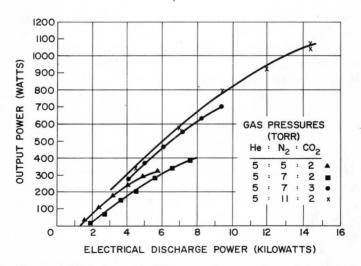

Fig. 57. Measured CO_2 laser output (CW) power for several gas mixtures as a function of electrical input power in a high-speed ($\sim$ 30 m/sec) transverse-flow laser. [After Tiffany et al. (86).]

Ge mirror at an input electrical power of 14 kW. This corresponds to an efficiency of about 7%. The large amount of working gas (~ 4000 ft³/m) is recirculated in a closed cycle by a blower and a heat exchanger at a total power consumption of about 200 W.

Subsequently, measurements on gain and saturation parameter have been made(*159*) in a 1-m-long subsonic transverse-flow laser amplifier using a stable low-power laser as the input signal source, with a beam diameter of 6 mm. Figure 58 shows the small-signal gain of the transverse-flow laser amplifier as a function of distance downstream (in the

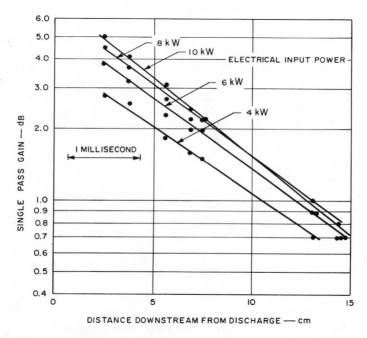

Fig. 58. Measured small-signal gain (dB) as a function of Z, the distance (cm) downstream from the electrical discharge at various input power levels in a high-speed (~ 35 m/sec) transverse-flow laser. [After Targ and Tiffany(*159*).]

flow direction) from the discharge region for a number of input electrical power levels. The gas-flow velocity in this case is 35 m/sec and gas mixture consists of 2 Torr CO₂, 11 Torr N₂, and 5 Torr He. Results show that the gain decreases exponentially with increasing distance from the excitation region and drops to its $1/e$ value at a length of about 7 cm which corresponds to a time constant of about 2 msec, consistent with the collisional relaxation time of the upper CO₂ laser level. The measured gain(*159*) and power output(*86*) dependence on gas pressure, flow velocity, and distance Z downstream are only in qualitative agreement with those characteristics as predicted by the theory(*85*). From the measurements of the small-signal gain (7.6 dB/m) in the discharge region and the corresponding power output (1100 kW) that can be coupled out through a 35% transmitting mirror under the same experimental conditions, a gain saturation value of 246 W/cm² has been estimated(*86*) for the transverse-flow laser. Clearly this laser system represents one of the most significant advances in the CO_2 laser in recent years, and undoubtedly more information and an improved theoretical model about this system will soon become available.

Recently Buczek et al.(*221*) have reported a transverse flowing, magnetically stabilized CO_2 laser. Premixed gas flows through a rectangular channel transverse to the axis of a cylindrical discharge. If the discharge is unstabilized, the perpendicular gas flow forces the discharge to assume a curved path between the electrodes. However, by application of a transverse magnetic field the discharge can be forced back into a straight line coincident with the optical axis of the oscillator.

2. *Longitudinal Flow*

High CW power (140 W at about 10% efficiency from a 10-cm-long discharge tube) has been obtained from high pressure (10–120 Torr) and high-speed (several hundred meters/sec) flowing CO_2–N_2–He lasers, although the direction of gas flow is along the optical axis. In a similar experiment, Brown and Davis(*222*), obtained a power output of 2300 W at 13% efficiency in a 1-m by 3.8-cm discharge. The laser configuration is shown in Fig. 59. A water-cooled copper drift section about 30 cm in length is attached to the discharge tube and provides a continuous transfer of energy from N_2 ($v = 1$) to CO_2 molecules. Actually, for this purpose the drift section is much longer than needed, because, at the pressures used in this experiment, the vibrational transfer time is of the order of a few microseconds, corresponding to about 1-cm path length at the maximum pumping speed of the system. The same laser device was used also in another experiment by injecting CO_2 about 10 cm downstream from a N_2–He discharge section. This scheme produced a CW output of 98 W at 11.7% efficiency, but required a higher flow rate (by a factor of about four) for CO_2 gas than for CO_2 injected directly into the discharge region. The measured values for saturation parameter varied from about 40 W/cm² at low pressure (~10 Torr) to about 600 W/cm² at high pressure (~100 Torr). The increase in gain saturation at higher gas pressure

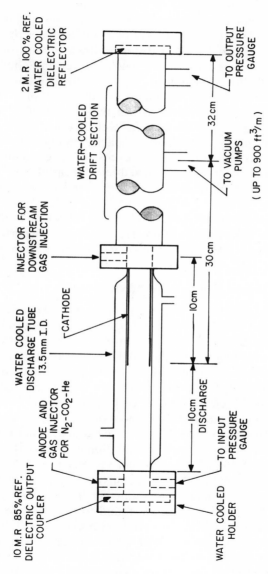

Fig. 59. Design of a high-pressure longitudinal-flowing laser tube. [After Deutsch et al. (89).]

is owing primarily to an increase in the relaxation rate of the upper laser level. The corresponding small-signal gain values were found to be 3.7 dB/m (high pressure) and 10.6 dB/m (low pressure). The fact that the laser output is relatively insensitive to the discharge length and the drift tube length (from 30 cm to 82 cm) indicates that the laser action occurs only within a short distance downstream from the discharge section, and suggests that a much shorter and compact laser can be made with the configuration shown in Fig. 59 but with much reduced lengths both for the discharge and drift sections. It must be realized that both schemes discussed in this section, i.e., transverse and longitudinal high-speed gas flow, as well as those in the following sections, require enormous pumping capacity and consequently consume a tremendous amount of gas (in excess of several hundred liters per minute). For this reason a closed-loop recirculating system is extremely desirable and should be incorporated as an integral part of such a laser.

In another investigation in which CO_2 was introduced downstream from a N_2–He discharge section DeMaria(223) reported a CW power output of 11.2 kW at 13% efficiency in a 1-m by 2-cm channel.

B. THERMAL PUMPING AND SUPERSONIC FLOW

Since the first proposal of thermal pumping and adiabatic cooling of CO_2 gas mixtures by Basov et al.(78), several attempts have been made both in the United States and in the Soviet Union to achieve population inversion between the $00^0 1$ and $10^0 0$ vibrational states of the CO_2 molecules, using various techniques. Fundamental processes in these gasdynamic lasers involve initial thermal pumping of CO_2 molecules by either burning or heating the gas or mixture of gases followed by a sudden expansion of the heated gas mixture through a nozzle. The rapid change in the translational temperature of the gas mixture causes the excited vibrational states of the molecules to relax toward an equilibrium through various $V \rightarrow V$ processes at different rates (see Section IV for details), resulting in a temporary population inversion immediately following the expansion. In these laser systems, N_2 and He are often found to be useful for the same reasons as in conventional lasers. To date, both CW and pulsed laser actions at relatively high power level have been achieved by expansion of the heated gas through a supersonic nozzle. Gerry(224) has recently reported 30 kW of CW power from a gasdynamic laser operating in a limited beam close to diffraction. In a mode-divergent beam he obtained 60 kW of CW power.

1. *Some Early Experiments*

In 1967, Makhov and Wieder(*79*) demonstrated that the upper CO$_2$ laser level 00^{0}1 can be pumped by heated nitrogen. In the meantime Wieder(*77*) also reported the observation of CW oscillation ($\sim$ 1 mW) at the 10.6-μ by exciting the CO$_2$ molecules with a CO–air diffusing flame at a temperature of $\sim$ 3000°K. The laser linewidth in this system was deduced from the absorption measurements to have a value roughly one to two orders of magnitude higher than that in a conventional discharge tube. The apparatus used in this experiment consisted of a 4-m-long quartz tube with a 24-mm bore. Two burners were placed on either side of the tube in the parallel direction. By heating N$_2$ in an oven initially to a temperature $\sim$ 1150°C at pressures ranging from 0.1 to 1.5 atm, Wisniewski et al.(*80*) have attained population inversion in the 00^{0}1–10^{0}0 CO$_2$ vibration–rotation transitions upon mixing CO$_2$ with heated N$_2$ molecules. With some modification of the experimental arrangement (Fig. 60), Fein et al.(*81*) have obtained pulsed laser action with a peak power output of 20 mW and a peak small-signal gain of 11%. As shown in Fig. 60 the N$_2$ flow velocity at the inlet is about 0.25 m/sec at a pressure of 400 Torr and reaches a value of $\sim$ 1.25 m/sec in the oven before passing through the nozzle. At these low flow rates, there is sufficient time to allow the

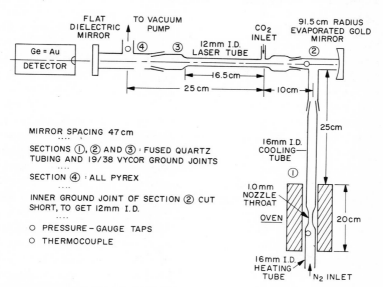

Fig. 60. Thermally-pumped CO$_2$–N$_2$ laser at a subsonic flow velocity $\sim$ 30 m/sec. [After Fein et al.(*81*).]

vibrational temperature to come to equilibrium with the translational temperature T_t. After expansion of N_2 through the nozzle, T_t reduces to near room temperature through the diffusion process in the cooling tube. Therefore, more than half of the excited N_2 molecules are lost before reaching the laser cavity. Note that this experiment does not utilize rapid cooling near the nozzle at which the fluid is accelerated to a supersonic speed. We shall see later that laser action with considerable increase in output power has been achieved in the region immediately following a sudden expansion of a heated gas mixture.

As the excited N_2 enters the laser tube at a flow velocity of about 30 m/sec, its pressure reduces to about 6 Torr, and it is mixed with 1 Torr of CO_2 at room temperature. Resonant transfer of vibrational energy from N_2 to CO_2 occurs in a time in the order of 50 μsec. In this system, both R and P branches at 10.6 μ oscillate, suggesting that a "total inversion" has been achieved(3) by this scheme. The laser power increases almost linearly with increasing flow velocity in the range from 20 to 35 m/sec for an optimum gas mixture. The maximum gain as shown in Fig. 61 occurs at a distance $5 < Z < 15$ cm from the inlet port of CO_2 gas and depends somewhat on the mixture ratio used. Since this thermal laser is relatively simple to construct and involves no electrical discharge, it can be a useful tool for studies of gas kinetics and molecular-energy transfer processes.

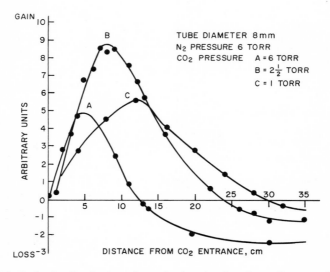

Fig. 61. Optical-gain distribution along the length of a thermally-pumped CO_2-N_2 laser tube from the CO_2 inlet. [After Fein et al.(81).]

2. Supersonic Flow

Two independent experiments on sudden expansion of a heated gas mixture through a supersonic nozzle have been reported recently. Kuehn and Monson(215) obtained pulse-laser energies up to ~ 10 J/cm² extracted from a 100-cm³ cavity with a gain path length of 14 cm; Bronfin and Hall(87) reported a CW laser power up to 60 W per 6 cm³ active volume.

The apparatus of the shock tube and the expansion chamber used for the pulsed-laser experiment is shown in Fig. 62. The heated gas

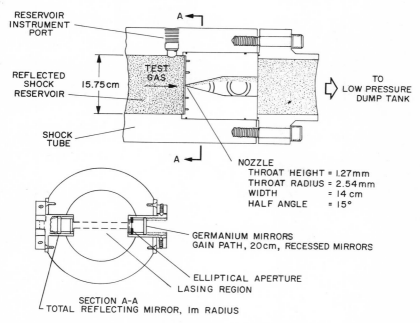

Fig. 62. Cross-sectional diagram of a shock-wave-generated supersonic flowing laser. [After Kuehn and Monson(215).]

mixture in a conventional shock tunnel is expanded through a two-dimensional supersonic nozzle with a slit width of 14 cm. The laser cavity is located behind the nozzle with an optical axis transverse to the direction of gas flow. The cavity mirrors are spaced 20 cm apart and recessed from the nozzle wall to avoid physical damage to the mirror surfaces by the solid particles generated by the shock waves. In this type of laser system, instabilities in the laser beam and extremely large beam divergence are generated as a result of gas turbulences but improvements on the beam shape and instability to some extent can be made by using apertures. Two elliptical apertures (one 3 cm long and 1.27 cm diam; the other 3 cm long and 2.54 cm diam) were used in this experiment. Figure 63 shows the laser output as a function of the initial

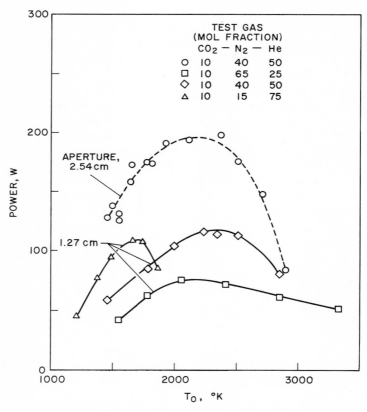

Fig. 63. Average power of the pulsed supersonic flowing laser as a function of the initial test-gas temperature before expansion through the nozzle. The output coupling mirror has a transmission of 7%. [After Kuehn and Monson(215).]

gas temperature before expansion. Data were obtained by using a 7% transmitting mirror in the laser cavity and for fixed total gas pressure $P_0 = 15.5 \pm 2.5$ atm before expansion. With optimum output coupling ($\sim 12\%$), a maximum average output power of ~ 250 W (with the aperture 2.54 cm diam) has been obtained from a heated CO_2–N_2–He mixture in the ratio $1:4:5$.

Figure 64 shows the apparatus used to achieve a CW supersonic flow ($\sim 10^3$ m/sec). Arc-heated N$_2$ at a temperature of 1000–3000°K and a pressure of 1 atm is rapidly expanded through a two-dimensional supersonic nozzle 5 cm in length with 1 mm throat height, providing a Mach number of 3.4. Mixing of CO$_2$–He with the vibrationally excited N$_2$ near room temperature is accomplished by injection through two narrow slots positioned from both sides of the tunnel normal to the supersonic flow. The injection velocities are varied continuously up to sonic values. The typical total gas flow rates were about 10 liters per sec (STP) with a total gas pressure, measured downstream in the test chamber, of about 50 Torr at a mixing ratio of 1 : 2 : 2 (CO$_2$: N$_2$: He). The advantages of this arrangement over the previous experiments (Fig. 60) are that (1) very little loss of the vibrationally excited CO$_2$ or N$_2$ occurs through diffusion, (2) the present arrangement, by placing the laser cavity very near the nozzle slit, allows the use of high pressure, thus yielding much higher laser power, and (3) the high convection flow rate removes the bottleneck formation at the 01^{1}0 level owing to thermal excitation. Under the experimental conditions, the transit time of fluid through the laser cavity is about 20 μsec, compared with ~ 3 μsec resonant transfer time between CO$_2$ and N$_2$ and ~ 75 μsec relaxation time of the upper laser level. Relaxation time of the 01^{1}0 on the other hand is in the order of 10 μsec.

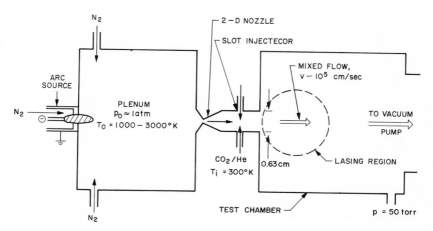

Fig. 64. Cross-sectional diagram of the thermal mixing laser. Optical gain, transverse to flow, is 5 cm; 2-dimensional nozzle throat height is 1 mm; and exit height is 6.3 mm (at 15° expansion angle). Injector slots for CO$_2$–He are 5 cm × 1 mm perpendicular to the expanded N$_2$ flow. The lasing region is indicated by the dotted circle. [After Bronfin et al. (82).]

Small-signal gain measurements as a function of flow rates R are shown in Fig. 65. The optical gain increases with increasing R_{CO_2} and R_{He}, and the highest gain value (about 0.8%/cm optical path length) is obtained at volumetric flow rate ratios $R_{He}/R_{N_2} = 0.1$ and $R_{CO_2}/R_{N_2} = 0.3$. These relatively low gain values for a supersonic flowing laser compared with those obtained from the conventional laser systems (35) indicate that gain saturation in a supersonic flowing laser must occur at much higher values. As mentioned before, both the stability and beam divergence in supersonic flowing laser are much more inferior than in

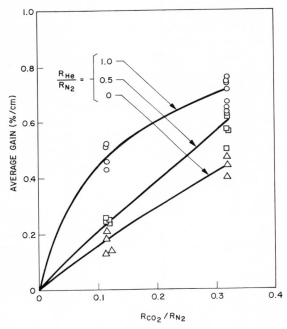

Fig. 65. Effect of composition on 10.6-μ gain. The N_2 flow rate is 3.63 liters/sec (STP), expanding from 2000° K, 1 atm. [After Bronfin et al. (82).]

the conventional low-speed CO_2 laser systems, owing primarily to the aerodynamic turbulences in the fluid. These turbulences which are properties of supersonic fluids limit the application of this laser.

C. CHEMICAL LASERS

Cool et al. (83) have obtained CW laser oscillation at 10.6 μ by means of chemical reactions forming vibrationally excited states of DF, HF, and HCl molecules which subsequently transfer their vibrational energies to excite CO$_2$ molecules to the 00^{0}1 level(*217,218*). In the early experiments(*83*), rf power was applied to the system through a side-arm tube to produce dissociated chlorine, fluorine deuterium, or hydrogen reactants before the mixture flowed into a coaxial Teflon reaction tube of 9-mm bore and 21-cm length, located in the upstream portion of the laser cavity. Recently it has been found that this external rf energy source is unnecessary and it has been removed from the system(*216*). As shown in Fig. 66, the NO, F$_2$, and He gases are

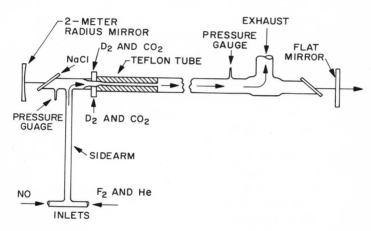

Fig. 66. A CW chemically pumped CO$_2$ laser with no external energy source. Laser action occurs mainly in the upstream portion of the tube. The cavity length is 1.5 m and the average flow velocity is $\sim$ 600 m/sec(*83*).

injected into an 11-mm-bore quartz tube in which the reaction

$$F_2 + NO \longrightarrow NOF + F \tag{125}$$

takes place. The F and F$_2$ subsequently are mixed with D$_2$ in a 9-mm-bore Teflon tube to produce vibrationally excited DF molecules via the reactions

$$F + D_2 \longrightarrow (DF)^* + D \tag{126}$$

and

$$D + F_2 \longrightarrow (DF)^* + F \tag{127}$$

These excited DF molecules then transfer their energy to CO$_2$ molecules and produce a very impressive CW laser power of 8 W in a drift tube of less than 1 m in length. However, a supersonic speed ($\sim$ 600 m/sec) is required for the fluid flow in this CW chemical laser; therefore, this laser consumes tremendous amounts of gases in order to sustain a continuous operation. Further studies on the chemistry and on the possibility of utilizing a recirculating loop are definitely required before a practical chemical laser can be realized.

LIST OF SYMBOLS

A	Einstein A coefficient for spontaneous emission
A_{mn}	transition probability between states m and n
a_0	Bohr radius
a_{ij}	internuclear force constants
a	absorption coefficient
a_1, a_2	mirror radius
B, B_v	rotational constant
b	distance of closest approach
C_{ij}	coefficients of quadratic terms of the kinetic energy
c	velocity of light
D, D_v	rotational constant (fine structure)
D_{ij}	coefficients of quadratic terms of the potential energy
$D_{\infty h}$	a point group
d	modulation depth
$\mathscr{D}$	diffusion coefficient
E_i, E_v, E_ν	vibrational energy of ith level or of ν mode
E_r	rotational energy
$\mathscr{E}$	electric field
$F(J)$	term values for rotational energy level
f_i	f number of ith emission line
f_P, f_R	beat frequencies in the P and R branch, respectively
$G(v)$	term values for vibrational energy level
$G(t, z)$	step response
g_l, g_u	statistical weight of the lower and upper level, respectively
$g(\nu), g(\omega)$	line-shape function
g_{ii}	ℓ-type doubling constants
$g(J)$	statistical weight of the Jth rotational energy level
H	width of a square laser cavity
H_v	Hermite polynomials
$\hbar = h/2\pi$	Plank constant times $1/2\pi$
I, I_e	moments of inertia
I	laser intensity
I_0, U_t	initial and transmitted laser intensities
I_s	gain saturation parameter

J	rotational quantum number
K_{iJ}^{u}	cross-relaxation rate constant between ith and Jth rotational levels in the upper laser level
K_{iJ}^{ℓ}	cross-relaxation rate constant between ith and Jth rotational levels in the lower laser level
K_i, K_δ	valence-force constant
k	Boltzmann constant
k_e, k_e'	resonant exchange-rate constant
k_{vi}	vibrational relaxation-rate constant
k_{A-B}	collisional relaxation-rate constant between species A and B
k_{rot}, k_{A-B}^{rot}	rotational relaxation-rate constant
k_{ij}	vibrational relaxation-rate constant between states i and j
L	laser cavity length; also, patch length
ℓ	equilibrium internuclear distance between C and O
ℓ_i	angular momentum quantum number
M	collisional partner M; also, magnetic quantum number
M_O, M_C	mass of oxygen and carbon atoms
m	reduced mass
$m = -J$	P-branch rotational quantum number
$m = J+1$	R-branch rotational quantum number
N_e	average electron density
N_0	ground state population per unit volume
N_v	normalizing constant
n	index of refraction
n_e	electron density
n_i	population density of ith level
n_v	vibrational population density of Vth level
n_J	population density of Jth rotational level
n_J^{u}, n_J^{ℓ}	population density of the upper and lower rotational levels
n_T	population density of Tth rotational level
P	gas pressure
P_ℓ	power output of laser beam
$P(J)$	P-branch vibrational–rotational transition
p	dipole moment
$\mathscr{P}$	probability of excitation or deexcitation processes
Q	quality factor of resonant cavity; also, quadrupole moment
R	mirror radius of curvature
$R(J)$	R-branch vibrational–rotational transition
R_x	flow rate of x specie
R_y	Rydberg constant

r_0	laser beam radius at $1/e$ intensity
r_{ij}	excitation or deexcitation rates between states i and j
$r(t)$	classical trajectory
S_i	symmetric coordinates
S_{mn}, S	line strength between states m and n
STP	standard temperature and pressure; i.e., $273°$ K and 1 atm
T, T_t	translational temperature of gas
T_1	collisional or dissipative relaxation time
T_2	homogeneous relaxation (atomic or molecular dephasing) time
T_v	vibrational temperature of a given mode of oscillation
t, t_m	transmission coefficient of a mirror
t	spectral transmittance
U	fluid velocity
U_0	impulse response
V'	interaction potential
$\bar{v}$	average thermal velocity
v_i	vibrational quantum number
W, W_m	width of a square laser cavity
$\mathscr{W}$	induced emission rate
X	Cartesian coordinate
X_T	electronic excitation probability
x	mole fraction of a component in a mixture
x_{ij}	anharmonic force constants
Y	Cartesian coordinate
Z	average number of collisions; also Cartesian coordinate along the direction of gas flow
α_0	small-signal gain coefficient
α_m	polarizability of molecule
β	wall reflection coefficient
Γ_j	pumping rate of population from ground state to the jth level
γ_{ij}	excitation and deexcitation rates between states i and j
Δ_J	collisionally broadened linewidth
Δ, Δ'	hard and soft collisional linewidths
δ_+, δ_-	sum and difference, respectively, of energy between two states
ϵ	dielectric constant; also average electron energy
θ_{JM}	Jacobi polynomials
λ	wavelength

λ_i	roots of the secular determinant
μ	micron; also reduced mass of CO$_2$ molecule
μ_{mn}	dipole matrix elements between states m and n
ν	optical frequency
ν_0	center frequency of an optical line
ν_i	frequency of the three fundamental modes of vibration
ν_c	collision frequency
$\Delta\nu$	width of a spectral line at one-half maximum intensity
ξ_i	normal coordinates
ρ_i	radiative flux density of the ith mode
$\Sigma_g, \Sigma_u, \pi_u$	species associated with ν_1, ν_2, and ν_3 modes
$\Sigma, \pi, \Delta, \ldots$	spectroscopic notations to designate the vibrational levels
σ	cross section
τ_0	initial pulse width
τ_R	round-trip transit time of light in a cavity
τ_u, τ	collisional lifetimes of the upper and lower vibrational laser levels
τ_r	radiative lifetime of vibrational level
τ_ν	effective vibrational lifetime
τ_c	time between successive collisions
τ_w	width of a laser pulse
τ_d	delay time of a laser pulse
ϕ	azimuth angle about a spatially fixed axis
Ψ_e	wave function of an electronic state
Ψ_v	wave function of a vibrational state
Ψ_r	wave function of a rotational state
Ω	nutational frequency of induced dipoles
ω	angular frequency
ω_1, ω_2	mode spot size at flat and curved mirrors, respectively

ACKNOWLEDGMENTS

I wish to express my appreciation to the Bell Telephone Laboratories (Whippany) for encouragement and cooperation during the preparation of the manuscript. I am particularly grateful to T. J. Bridges, H. G. Cooper, and other members of the Laboratories for valuable discussions and comments, and to T. J. Bridges and W. B. Gandrud for proofreading the manuscript.

REFERENCES

1. C. K. N. Patel, W. L. Faust, and R. A. McFarlane, *Bull. Am. Phys. Soc.*, **9**, 500 (1964).
2. C. K. N. Patel, *Phys. Rev.*, **136A**, 1187 (1964).
3. C. K. N. Patel, *Phys. Rev. Letters*, **12**, 588 (1964).
4. N. Legay-Sommaire, L. Henry, and F. Legay, *Compt. Rend.*, **260**, 339 (1965).
5. P. Barchewitz, L. Dorbec, R. Farrenq, A. Truffert, and P. Vautier, *Compt. Rend.*, **260**, 3581 (1965).
6. P. Barchewitz, L. Dorbec, A. Truffert, and P. Vautier, *Compt. Rend.*, **260**, 5491 (1965).
7. F. Legay and N. Legay-Sommaire, *Compt. Rend.*, **260**, 3339 (1964).
8. C. K. N. Patel, *Phys. Rev. Letters*, **13**, 617 (1964).
9. C. K. N. Patel, *Appl. Phys. Letters*, **7**, 15 (1965).
10. G. Moeller and J. D. Rigden, *Appl. Phys. Letters*, **7**, 274 (1965).
11. C. K. N. Patel, P. K. Tien, and J. H. McFee, *Appl. Phys. Letters*, **7**, 290 (1965).
11a. A. J. DeMaria, to be published.
12. W. J. Witteman, *IEEE J. Quant. Electron.*, **2**, 375 (1966).
13. M. J. Weber and T. E. Deutsch, *IEEE J. Quant. Electron.*, **2**, 369 (1966).
14. P. K. Cheo, *Appl. Phys. Letters*, **11**, 38 (1967); *IEEE J. Quant. Electron.*, **4**, 587 (1968).
15. R. L. Taylor and S. Bitterman, *Rev. Mod. Phys.*, **41**, 26 (1969).
16. P. O. Clark and M. R. Smith, *Appl. Phys. Letters*, **9**, 367 (1966); for more recent measurements see D. C. Tyte and R. W. Sage, *Proc. IERE, Conf. On Lasers and Opto-Electronics, March, 1969, Southampton, England.*
17. R. D. Hake and A. V. Phelps, *Phys. Rev.*, **158**, 70 (1967).
18. M. J. W. Boness and G. J. Schulz, *Phys. Rev. Letters*, **21**, 1031 (1968).
19. P. K. Cheo and H. G. Cooper, *IEEE J. Quant. Electron.*, **3**, 79 (1967).
20. W. J. Witteman, *Phys. Letters*, **18**, 125 (1965); *IEEE J. Quant. Electron.*, **QE-2**, 375 (1966).
21. N. N. Sobolev and V. V. Sokovikov, *JETP Letters*, **4**, 204 (1966); *ibid.*, **5**, 99 (1967).
22. B. F. Gordietz, N. N. Sobolev, and L. A. Shelepin, *Soviet Phys. JETP*, **26**, 1039 (1968).
23. L. O. Hocker, M. A. Kovacs, C. K. R. Rhodes, G. W. Flynn, and A. Javan, *Phys. Rev. Letters*, **17**, 233 (1966).
24. P. K. Cheo, *J. Appl. Phys.*, **38**, 3563 (1967).
25. C. B. Moore, R. E. Wood, B. Hu, and T. J. Yardley, *J. Chem. Phys.*, **46**, 4222 (1967).
26. P. K. Cheo and R. L. Abrams, *Appl. Phys. Letters*, **14**, 47 (1969).
27. R. L. Abrams and P. K. Cheo, *Appl. Phys. Letters*, **15**, 177 (1969).
28. C. K. Rhodes, M. J. Kelly, and A. Javan, *J. Chem. Phys.*, **48**, 5730 (1968).
29. T. K. McCubbin, R. Darone, and J. Sorrell, *Appl. Phys. Letters*, **8**, 118 (1966).
30. E. T. Gerry and D. A. Leonard, *Appl. Phys. Letters*, **8**, 227 (1966).
31. T. J. Bridges, H. A. Haus, and P. W. Hoff, *IEEE J. Quant. Electron.*, **4**, 777 (1968).
32. C. Bordé and L. Henry, *IEEE J. Quant. Electron.*, **4**, 874 (1968).
33. H. W. Mocker, *IEEE J. Quant. Electron.*, **4**, 769 (1968).
34. T. Deutsch, *IEEE J. Quant. Electron.*, **3**, 151 (1967).
35. P. K. Cheo, *IEEE J. Quant. Electron.*, **3**, 683 (1967).

36. G. J. Dezenberg and J. A. Merritt, *Appl. Optics*, **6**, 1541 (1967).
37. C. Rossetti and P. Barchewitz, *Compt. Rend.*, **262**, 1199 (1966).
38. N. Djeu, T. Kan, and G. J. Wolga, *IEEE J. Quant. Electron.*, **4**, 256 (1968).
39. T. A. Cool and J. A. Shirley, *Appl. Phys. Letters*, **14**, 70 (1969).
40. H. Kogelnik and T. J. Bridges, *IEEE J. Quant. Electron.*, **3**, 95 (1967).
41. D. F. Hotz and J. W. Austin, *Appl. Phys. Letters*, **11**, 60 (1967).
42. C. P. Christensen, C. Freed, and H. A. Haus, *IEEE J. Quant. Electron.*, **5**, 276 (1969).
43. R. C. Crafer, A. F. Gibson and M. F. Kimmitt, *Brit. J. Appl. Phys.*, **2**, 1135 (1969).
44. M. A. Kovacs, G. W. Flynn, and A. Javan, *Appl. Phys. Letters*, **8**, 62 (1966).
45. T. J. Bridges, *Appl. Phys. Letters*, **9**, 174 (1966).
46. Dietrich Meyerhofer, *IEEE J. Quant. Electron.*, **4**, 7621 (1968).
47. O. R. Wood and S. E. Schwarz, *Appl. Phys. Letters*, **11**, 88 (1967).
48. P. L. Hanst, J. A. Morreal, and W. J. Henson, *Appl. Phys. Letters*, **12**, 58 (1968).
49. J. T. Yardley, *Appl. Phys. Letters*, **12**, 120 (1968).
50. N. V. Karlov, G. P. Kuzmin, Yu N. Petrov, and A. M. Prokhorov, *JETP Letters*, **7**, 134 (1968).
51. T. Y. Chang, C. H. Wang, and P. K. Cheo, *Appl. Phys. Letters*, **15**, 157 (1969).
52. O. R. Wood and S. E. Schwarz, *Appl. Phys. Letters*, **12**, 263 (1968).
53. D. E. Caddes, L. M. Osterink, and Russel Targ, *Appl. Phys. Letters*, **12**, 74 (1968).
54. T. Walsh, *RCA Rev.*, **27**, 323 (1966).
55. J. E. Kiefer and A. Yariv, *Appl. Phys. Letters*, **15**, 26 (1969).
56. T. J. Bridges and P. K. Cheo, *Appl. Phys. Letters*, **14**, 262 (1969).
57. T. J. Bridges and C. K. N. Patel, *Appl. Phys. Letters*, **7**, 244 (1965).
58. J. D. Rigden and G. Moeller, *IEEE J. Quant. Electron.*, **2**, 365 (1966).
59. C. Frapard, *Phys. Letters*, **20**, 384 (1966).
60. J. Fantasia, Honeywell Inc., Syst. and Res. Div., Final Rept., HRC66-34, November (1966).
61. W. B. Bridges, P. O. Clarke, and A. S. Halstead, *CFSTI*, Doc. No. AD807363, January (1967).
62. D. R. Whitehouse, *CFSTI*, Doc. No. AD65303, May (1967).
63. T. G. Roberts, G. J. Hutcheson, J. J. Ehrlich, W. L. Hales, and T. A. Barr, *IEEE J. Quant. Electron.*, **3**, 605 (1967).
64. P. A. Miles and J. W. Lotus, *IEEE J. Quant. Electron.*, **4**, 811 (1968).
65. M. R. Smith and D. C. Forster, 1968 International Quantum Electronics Conference, Miami, Florida, May 14–17, 1968, Paper 10J-2.
66. R. J. Carbone, *IEEE J. Quant. Electron.*, **3**, 373 (1967).
67. W. J. Witteman and H. W. Werner, *Phys. Letters*, **26A**, 454 (1968).
68. R. P. Reeves, unpublished work, 1967.
69. W. J. Witteman, *Appl. Phys. Letters*, **11**, 337 (1967).
70. J. Fantasia, *CFSTI*, Doc. No. N68-13277, October (1967).
71. R. J. Carbone, *IEEE J. Quant. Electron.*, **4**, 102 (1968).
72. P. O. Clark and J. Y. Wada, *IEEE J. Quant. Electron.*, **4**, 263 (1969).
73. T. F. Deutsch and F. A. Horrigan, *IEEE J. Quant. Electron.*, **4**, 972 (1968).
74. R. S. Reynolds, NASA Contract No. NAS5-10309, January (1968).
75. C. Freed, *IEEE J. Quant. Electron.*, **3**, 203 (1967); *ibid.*, **4**, 404 (1968).
76. F. A. Horrigen, C. A. Klein, R. I. Rudko, and D. T. Wilson, Army Missile Command Contract No. DAAHO1-67-1589, September (1968); Laser Technology, Laser Optics Inc., Jan. 1969.

77. I. Wieder, *Phys. Letters*, **24A**, 759 (1967).

78. N. G. Basov, A. N. Ordevskill, and V. A. Scheglov, *Soviet Phys. Tech. Phys.*, **12**, 243 (1967).

79. G. Makhov and I. Wieder, *IEEE J. Quant. Electron.*, **3**, 378 (1967).

80. E. E. Wisniewski, M. E. Fein, J. T. Verdeyen, and B. E. Cherrington, *Appl. Phys. Letters*, **12**, 257 (1968).

81. M. E. Fein, J. T. Verdeyen, and B. E. Cherrington, *Appl. Phys. Letters*, **14**, 337 (1969).

82. B. R. Bronfin, L. R. Boedeker, and J. P. Cheyer, *Bull Am. Phys. Soc.*, **14**, 857 (1969), Paper FE-10; *Appl. Phys. Letters*, **16**, 214 (1970).

83. T. A. Cool, T. J. Falk, and R. R. Stephens, *Appl. Phys. Letters*, **15**, 318 (1969); T. A. Cool and R. R. Stephens, *J. Chem. Phys.*, **51**, 5175 (1969).

84. V. K. Konyukhov and A. M. Prokhorov, *JETP Letters*, **3**, 436 (1966).

85. T. A. Cool, *J. Appl. Phys.*, **40**, 3563 (1969).

86. W. B. Tiffany, R. Targ, and J. D. Foster, *Appl. Phys. Letters*, **15**, 91 (1969).

87. B. R. Bronfin and R. J. Hall, *Bull. Am. Phys. Soc.*, **14**, 857 (1969), Paper FE-11.

88. A. E. Hill, *Appl. Phys. Letters*, **12**, 324 (1968).

89. T. F. Deutsch, F. A. Horrigan, and R. I. Rudko, *Appl. Phys. Letters*, **15**, 88 (1969).

90. N. McAvoy, H. L. Richard, J. A. McElroy, and W. E. Richards, "A 10.6 Micron Laser Communication System Experiment for ATS-F and ATS-G", Goddard Space Flight Center, Greenbelt, Maryland, *NASA TMX-524-68-206*, May (1968).

91. For a comprehensive evaluation and analysis, see *Deep Space Communication and Navigation Study*, Vols. I, II, and III, prepared by Bell Tel. Labs. for NASA, Contract No. NAS5-10293, May (1968).

92. For example, see *Lincoln Laboratory Optics Research Report*, Air Force Contract No. AF19(628)-5167, July (1969).

93. F. P. Gagliano, R. M. Lumley, and L. S. Watkins, *Proc. IEEE*, **57**, 114 (1969).

94. C. K. N. Patel, *Phys. Rev. Letters*, **20**, 1027 (1968); *ibid.*, **16**, 613 (1966).

95. C. K. N. Patel, P. A. Fleury, R. E. Slusher, and H. L. Frisch, *Phys. Rev. Letters*, *971* (1966).

96. G. D. Boyd, T. J. Bridges, and E. G. Burkhart, *IEEE J. Quant. Electron.*, **4**, 515 (1968).

97. J. Warner, *Appl. Phys. Letters*, **12**, 222 (1968).

98. W. B. Gandrud, G. D. Boyd, J. H. McFee, and F. H. Wehmeier, *Appl. Phys. Letters*, **16**, 59 (1970).

99. C. K. N. Patel and R. E. Slusher, *Phys. Rev. Letters*, **19**, 1019 (1967).

100. C. K. N. Patel and R. E. Slusher, *Phys. Rev. Letters*, **20**, 1087 (1968).

101. G. B. Hocker and C. L. Tang, *Phys. Rev. Letters*, **21**, 591 (1968).

102. J. P. Gordon, C. H. Wang, C. K. N. Patel, R. E. Slusher, and W. J. Tomlinson, *Phys. Rev.*, **179**, 294 (1969).

103. C. K. Rhodes, A. Szöke, and A. Javan, *Phys. Rev. Letters*, **16**, 1151 (1968).

104. E. B. Treacy and A. J. DeMaria, *Phys. Letters*, **29A**, 369 (1969).

105. C. K. Rhodes and A. Szöke, *Phys. Rev.* **184**, 25 (1969).

106. F. A. Hopf and M. O. Scully, *Phys. Rev.*, **179**, 399 (1969).

107. P. K. Cheo and C. H. Wang, *Phys. Rev.*, A1, 225 (1970).

108. See for example, C. B. Moore, *Acc. Chem. Rev.*, **2**, 103 (1969).

109. I. Burak, A. V. Nowak, J. I. Steinfeld, and D. G. Sutton, *J. Chem. Phys.*, **51**, 2275 (1969).

110. A. M. Ronn and D. R. Lide, Jr., *J. Chem. Phys.*, **47**, 3669 (1967).

111. C. K. N. Patel, *J. Chimie. Phys.*, **64**, 82 (1967).

112. V. P. Tychinskii, *Soviet Phys. — Usp.*, **10**, 131 (1967).

113. N. N. Sobolev and V. V. Sokovikov, *Soviet Phys. — Usp.*, **10**, 153 (1967).

114. H. Herzberg, *Molecular Spectra and Molecular Structure — II. Infrared and Raman Spectra of Polyatomic Molecules*, Van Nostrand, Princeton, N.J., 1962.

115. J. B. Howard and E. B. Wilson, Jr., *J. Chem. Phys.*, **2**, 620 (1934).

116. W. H. Shaffer and R. R. Newton, *J. Chem. Phys.*, **10**, 405 (1942).

117. P. E. Martin and E. F. Barker, *Phys. Rev.*, **41**, 291 (1932). [Ref. for ν_2.]

118. E. F. Barker and T. Y. Wu, *Phys. Rev.*, **45**, 1 (1934). [Ref. for ν_3.]

119. I. Hanson, *Phys. Rev.*, **46**, 122 (1931).

120. I. Wieder and G. G. McCurdy, *Phys. Rev. Letters*, **16**, 565 (1966).

121. E. Fermi, *Z. Physik*, **71**, 250 (1931).

122. C. Frapard, P. Laures, M. Roulot, X. Ziegler, and N. Legay–Sommaire, *Compt. Rend.*, **262**, 1340 (1966).

123. T. J. Bridges and A. R. Strnad, *IEEE J. Quant. Electron.*, **3**, 335 (1967).

124. J. A. Howe, *Appl. Phys. Letters*, **7**, 21 (1965).

125. J. A. Howe and R. A. McFarlane, *J. Mol. Spectry.*, **19**, 224 (1966).

126. B. Hartman and B. Kleman, *Can. J. Phys.*, **44**, 1609 (1966).

127. N. Djeu, T. Kan, C. R. Miller, and G. J. Wolga, *J. Appl. Phys.*, **39**, 2157 (1968).

128. H. Statz, C. L. Tang, and G. F. Koster, *J. Appl. Phys.*, **37**, 4278 (1966).

129. U. P. Oppenheim and A. D. Devir, *J. Opt. Soc. Am.*, **58**, 585 (1968).

130. C. Rossetti, F. Bourbonneux, and P. Barchewitz, *Compt. Rend.*, **262**, 1684 (1966).

131. T. J. Bridges and T. Y. Chang, *Phys. Rev. Letters*, **22**, 811 (1969).

132. A. V. Phelps, *Rev. Mod. Phys.*, **40**, 399 (1968).

133. A. Stamatovic and G. J. Schulz, to be published.

134. G. J. Schulz, *Phys. Rev.*, **116**, 1141 (1959); *ibid.*, **125**, 229 (1962); *ibid.*, **135**, A988 (1964).

135. K. Takayanagi, *Progr. Theoret. Phys. (Kyoto)*, **40**, 216 (1967).

136. S. S. Penner, *Quantitative Molecular Spectroscopy and Gas Emmissivities*, Addison-Wesley, Reading, Mass., 1959.

137. S. C. Brown, *Basic Data of Plasma Physics*, MIT Press, 1959.

138. A. Javan, W. R. Bennett, and D. R. Herriott, *Phys. Rev. Letters*, **6**, 106 (1961).

139. L. Landau and E. Teller, *Physik. Z. Sowjetunion*, **10**, 34 (1936).

140. R. N. Schwartz, Z. I. Slawsky, and K. Herzfeld, *J. Chem. Phys.*, **20**, 1591 (1952).

141. R. D. Sharma and C. A. Brau, *Phys. Rev. Letters*, **19**, 1273 (1967).

142. W. A. Rosser, A. D. Wood, and E. T. Gerry, 1968 International Quantum Electronics Conference, Miami, Florida, May 14–17, 1968, Paper 7G-4.

143. R. L. Taylor, M. Camac, and R. M. Feinberg, *Proceedings Eleventh International Symposium on Combustion, Pittsburgh, Pa., 14–20 August* (1966).

144. C. B. Moore, *Fluorescence* (G. G. Guilbault, ed.), Dekker, New York, 1967, Chap. 3.

145. G. Gould, *Appl. Opt., Suppl.* **2**, 59 (1965).

146. R. C. Millikan and D. R. White, *J. Chem. Phys.*, **39**, 3209 (1963).

147. K. F. Herzfeld and T. A. Litovitz, *Absorption and Dispersion of Ultrasonic Waves*, Academic Press, New York, 1959.

148. J. O. Hirchfelder, C. F. Curtis, and R. B. Bird, *Molecular Theory of Gasses and Liquids*, Wiley, New York, 1954.

149. V. N. Kondrat'ev, *Chemical Kinetics of Gas Reactions*, Addison-Wesley, Reading, Mass., 1964.

150. J. E. Lennard–Jones, *Proc. Roy. Soc. (London)*, **A106**, 441 (1924).

151. J. T. Yardley and C. B. Moore, *J. Chem. Phys.*, **46**, 4491 (1967).

152. M. Kovacs, D. R. Rao, and A. Javan, *J. Chem. Phys.*, **48**, 3339 (1968).
153. K. F. Herzfeld, *J. Chem. Phys.*, **47**, 743 (1967).
154. B. F. Gordietz, N. N. Sobelev, V. V. Solovikov, and L. A. Shelepin, *IEEE J. Quant. Electron.*, **4**, 796 (1968).
155. R. D. Sharma, *J. Chem. Phys.*, **49**, 5195 (1968).
156. R. Holmes, G. R. Jones, and R. Lawrence, *J. Chem. Phys.*, **41**, 2995 (1964).
157. R. R. Patty, E. R. Manring, and J. A. Gardner, *Appl. Opt.*, **7**, 2241 (1968).
158. A. Yariv, *Quantum Electronics*, Wiley, New York, 1967.
159. R. Targ and W. B. Tiffany, *Appl. Phys. Letters*, **15**, 302 (1969).
160. J. Tulip, *IEEE J. Quant. Electron.*, **6**, 206 (1970).
161. W. J. Witteman, *IEEE J. Quant. Electron.*, **2**, 375 (1966); *Philips Res. Rept.*, **21**, 73 (1966).
162. W. W. Rigrod, *J. Appl. Phys.*, **36**, 2487 (1965).
163. W. B. Bridges, *IEEE J. Quant. Electron.*, **4**, 820 (1968).
164. Y. V. Brzhazovsky, V. P. Chebotayev, and L. S. Vasilenko, *IEEE J. Quant. Electron.*, **5**, 146 (1969).
165. P. Rabinowitz, R. Keller, and J. T. LaTourrette, *Appl. Phys. Letters*, **14**, 376 (1969); *Bull. Am. Phys. Soc.*, **14**, 940 (1969).
166. Y. H. Pao, private communication, 1969.
167. T. Kan, H. T. Powell, and G. J. Wolga, *IEEE J. Quant. Electron.*, **5**, 299 (1969).
168. W. E. Lamb, Jr., *Phys. Rev.*, **134**, A1429 (1964).
169. A. Szöke and A. Javan, *Phys. Rev.*, **145**, 137 (1966).
170. C. Bordé and L. Henry, *Compt. Rend.*, **265**, 1251 (1967).
171. D. H. Close, *Phys. Rev.*, **153**, 360 (1967).
172. W. J. Witteman, *IEEE J. Quant. Electron.*, **5**, 92 (1969); *ibid.*, **4**, 786 (1968).
173. H. Kogelnik and T. Li, *Proc. IEEE*, **54**, 1312 (1966).
174. H. W. Mocker, *Appl. Phys. Letters*, **12**, 20 (1968).
175. J. B. Gerardo and J. T. Verdegen, *Proc. IEEE*, **52**, 690 (1964).
176. R. A. Stern and P. K. Cheo, *Phys. Rev. Letters*, **23**, 1426 (1969); 1970 Annual Joint Meeting of Am. Phys. Soc. and Am. Assoc. of Phys. Teachers, Chicago, Illinois (invited paper).
177. See, for example, P. Vandenplas, *Electron Waves and Resonances in Bounded Plasmas*, Wiley, New York, 1968.
178. S. L. McCall and E. L. Hahn, *Phys. Rev. Letters*, **18**, 908 (1967).
179. S. L. McCall and E. L. Hahn, *Phys. Rev.*, **183**, 457 (1969).
180. A. Bloom, *Phys. Rev.*, **98**, 1105 (1955).
181. R. L. Abrams and A. Dienes, *Appl. Phys. Letters*, **14**. 237 (1969).
182. P. L. Gordon, S. E. Schwarz, C. V. Shank, and O. R. Wood, *Appl. Phys. Letters*, **14**, 235 (1969).
183. E. B. Treacy and A. J. DeMaria, *Phys. Letters*, **29A**, 369 (1969).
184. W. G. Wagner, H. A. Haus, and K. T. Gustafson, *IEEE J. Quant. Electron.*, **4**, 267 (1968).
185. G. D. Boyd and H. Kogelnik, *Bell System Tech. J.*, **41**, 1346 (1962).
186. P. K. Cheo, unpublished work, 1966.
187. A. L. Schawlow and C. H. Townes, *Phys. Rev.*, **112**, 1940 (1958).
188. A. E. Siegman, B. Daino, and K. R. Manes, *IEEE J. Quant. Electron.*, **3**, 180 (1967).
189. A. D. White, *IEEE J. Quant. Electron.*, **1**, 349 (1965).
190. G. Birnbaum, *Proc. IEEE*, **55**, 1015 (1967).

191. T. G. Polanyi and I. Tobias, Lasers—A Series of Advances, Vol. II (A. K. Levine, ed.), Dekker, New York, 1968.

192. T. J. Bridges, T. Y. Chang, and P. K. Cheo, *Appl. Phys. Letters*, **12**, 297 (1968).

193. R. L. Barger and J. L. Hall, *Phys. Rev. Letters*, **22**, 4 (1969).

194. R. W. Hellwarth, *Advances in Quantum Electronics* (J. Singer, ed.), Columbia Univ. Press, New York, 1961.

195. C. G. B. Garrett, *Gas Lasers*, McGraw-Hill, New York, 1967.

196. R. C. Crafer, A. F. Gibson, M. J. Kent, and M. F. Kimmitt, *Brit. J. Appl. Phys.*, **2**, 183 (1969).

197. C. Frapard, *IEEE J. Quant. Electron.*, **2**, 225 (1966).

198. E. B. Treacy, *Proc. IEEE*, **46**, 2053 (1968).

199. R. L. Abrams, *IEEE J. Quant. Electron.*, **5**, 522 (1969).

200. S. Marcus, *Appl. Phys. Letters*, **15**, 217 (1969).

201. B. H. Soffer, *J. Appl. Phys.*, **35**, 2551 (1964); P. Kafalas, J. I. Masters, and E. M. E. Murray, *ibid.*, 2349 (1964).

202. A. Szöke, V. Daneu, J. Goldhar, and N. A. Kurnit, *Appl. Phys. Letters*, **15**, 376 (1969).

203. L. E. Hargrove, R. L. Fork, and M. A. Pollack, *Appl. Phys. Letters*, **5**, 4 (1964).

204. O. P. McDuff and S. E. Harris, *IEEE J. Quant. Electron.*, **3**, 101 (1967).

205. W. R. Hook, R. P. Hilberg, and R. H. Dishington, *Proc. IEEE*, **54**, 1954 (1966); R. N. Zitter, W. H. Steier, and R. Rosenberg, *IEEE J. Quant. Electron.*, **3**, 614 (1967).

206. P. W. Smith, *IEEE J. Quant. Electron.*, **3**, 627 (1967).

207. A. G. Fox and P. W. Smith, *Phys. Rev. Letters*, **18**, 826 (1967).

208. J. A. Armstrong and E. Courtens, *IEEE J. Quant. Electron.*, **4**, 411 (1968); F. T. Arecchi and R. Bonifacio, *ibid.*, **1**, 169 (1965).

209. J. A. Fleck, *Phys. Rev. Letters*, **21**, 131 (1968).

210. H. Risken and K. Nummedal, *J. Appl. Phys.*, **39**, 4662 (1968).

211. P. K. Cheo and J. Renau, *J. Opt. Soc. Am.*, **59**, 821 (1969).

212. 8.8 Kilowatt CW Output Produced by Experimental CO₂ System, *Laser Focus*, **4**, 3 (1968).

213. D. C. Smith, *IEEE J. Quant. Electron.*, **5**, 291 (1969).

214. A. F. Gibson, private communication.

215. D. M. Kuehn and D. J. Monson, *Appl. Phys. Letters*, **16**, 48 (1970).

216. CW Chemical Laser Requires No External Energy Source, Editor, *Phys. Today*, Search and Discovery, **22**, 55, December (1969).

217. H. L. Chen, J. C. Stephensen, and C. B. Moore, *Chem. Phys. Letters*, **2**, 593 (1968).

218. R. W. F. Gross, *J. Chem. Phys.*, **50**, 1889 (1969).

219. R. S. Reynolds, International Electron Device Meeting, Washington, D.C., Oct. 28–30, 1970.

220. J. A. Beaulieu, *Bull. Am. Phys. Soc.*, **15**, 808 (1970).

221. C. J. Buczek, R. J. Wayne, P. Chenausky, and R. J. Freiberg, *Appl. Phys. Letters*, **16**, 321 (1970).

222. C. O. Brown and J. W. Davis, private communication, 1970.

223. A. J. DeMaria, "Electrically Excited CO₂ Lasers," Meeting American Physics Society, Washington, D.C., April 1970, invited paper.

224. E. T. Gerry, "Gas-Dynamic CO₂ Lasers," Meeting American Physics Society, Washington, D.C., April 1970, invited paper.

Chapter 3 · DYE LASERS

M. BASS, T. F. DEUTSCH, AND M. J. WEBER

RESEARCH DIVISION, RAYTHEON COMPANY
WALTHAM, MASSACHUSETTS

I. Introduction

In March 1966, Sorokin and Lankard(*1*) reported laser action from a solution containing a fluorescent organic dye. The discussions, experiments, and speculations of workers who had previously considered using organic molecules to obtain stimulated emission are reviewed in detail in recent articles by Sorokin et al.(*2*), Snavely(*3*), and Stepanov and Rubinov(*4*). [For the general reader, an excellent survey of organic lasers appears in a recent article by Sorokin(*5*).] In the years since

1966, both the understanding of the physics of dye lasers and the development of dye laser technology have progressed very rapidly. The present review surveys the progress in each of these areas and, by relating the basic physics to the operation of optically pumped dye lasers, describes the current state-of-the-art. In order to offer a comprehensive survey, both flashlamp-pumped and laser-pumped dye lasers are included.

To provide an overview of present dye laser systems, the properties and performance of dyes reported in the literature are summarized in Table 1. This compilation is provided in order that the reader may easily scan the data, and compare dye laser systems with other optically pumped lasers. Detailed discussions of the reported results may be found in Sections IV and V.

One significant advantage of dye lasers is the possibility, through proper dye selection, of obtaining intense, coherent light at almost any wavelength from 3400 to 11,750 Å. In addition, the lasing wavelength of a single dye laser is continuously tunable over a few hundred angstroms, either by varying the solution or cavity parameters, or by using an intracavity wavelength selector such as a diffraction grating. Dye lasers are therefore an attractive choice for a continuously tunable source of coherent light: they offer a simple alternative to the optical parametric oscillator.

In their present stage of development, dye lasers are pulsed devices; peak powers are typically on the order of 1 MW. By comparison, gigawatt power levels can be obtained from sophisticated ruby or Nd: glass laser systems. Although the technology necessary to achieve such high peak powers from dye lasers remains to be developed, the apparatus required to achieve megawatt power levels is, for the most part, no more complicated than that used for a pulsed Nd:YAG laser.

However, since a typical dye has a fluorescent lifetime $\sim 5 \times 10^{-9}$ sec, the optical pump source for dye lasers must be capable of high pumping rates in order to overcome spontaneous emission losses. The required pump rates have been obtained by one of two methods: (1) the intense output of other lasers has been used, or (2) a variety of flashlamps, ranging from sophisticated laboratory lamps to conventional commercial units, have been used. While it may seem inefficient and inelegant to use one laser to pump another, the laser-pumped dye laser provides a simple and inexpensive optical frequency converter which can increase the applications of other lasers. Laser-pumped lasers have been shown also to have unusual capabilities as broadband optical frequency amplifiers [see Ref.(19)].

In contrast to the difficulties involved in preparing effective solid-

TABLE 1

Summary of Dye Laser Properties

Property	Typical values	Conditions	Comments
Wavelength	3400–11,750 Å	Flashlamp and/ or laser pumped	A variety of dye-solvent combinations are available to span the entire wavelength range nearly continuously
Tuning	Up to 400 Å	Prism, filter, grating, Q switch in cavity	Length, concentration, and temperature of active medium also provide tuning
Spectral width	15–150 Å	Broadband mirrors	
	~ 0.5 Å	Grating in cavity	
	~ 0.01 Å	Grating plus etalon	
Beam divergence	2–5 mrad	Flashlamp and/or laser pumped	Dependent on uniformity of pumping
	0.5 mrad	Etalon in cavity	
Efficiency	Up to 25%	Laser pumped	Measured optical efficiency
	$\sim 0.4\%$	Flashlamp pumped	Electrical energy input to laser energy output
Output			
Energy	2 J (high)– 0.1 J (typical)	Flashlamp pumped	
Power	~ 2 MW	20 MW pump	Rhodamine 6G
	0.75–2.0 MW	Flashlamp pumped	
Repetition rate	Up to 200 pps	Laser pumped	Pump laser rate limits
	20–50 pps	Linear flashlamp	System cooling and component failure are limits
	1 pps	Annular flashlamp	
Temporal			
Pulse duration	~ 20 nsec	Laser pumped	Follows pump pulse
	0.5 μsec typical; up to 140 μsec achieved	Flashlamp pumped	Shorter than pump duration
Mode locked		Mode-locked pump	Pump cavity length integral multiple of dye cavity
	Pulses $< 10^{-9}$ sec	Flashlamp pumped with intracavity saturable absorber	
	Pulses $< 10^{-11}$ sec	Mode-locked pump	Observed by two-photon fluorescence

state laser hosts, a vast number of relatively inexpensive dyes and solvents are useful as dye lasers. Furthermore, the technology required for handling dyes and the optical pumping techniques have improved considerably during the past three years, and it is now clear that it is relatively easy to assemble and operate a dye laser.

In the search for better performance from dye lasers, dye chemistry has become very important, not only because it determines lasing properties, but also because it affects dye stability. Better under-standing of the chemical aspects of dye lasers will be needed to over-come decomposition problems and to discover ways of making dye lasers more useful and practical devices.

II. Optical Properties of Organic Dyes in Solution

In the late nineteenth century a dye was defined as an organic com-pound having an intense color (a strong, broad visible absorption) which could be imparted more or less permanently to other materials. A material which might be chemically similar but uncolored was not called a dye. Today, however, the requirement that the substance have color has been dropped, and the word dye is used to refer to organic compounds with certain chemical constituents and with certain spec-troscopic properties. The first attempt to define a dye on the basis of chemical composition was made in 1876 by Witt(6), who stated that all colored organic compounds (called chromogens) contain certain unsaturated chromophoric groups (e.g., $-NO_2$, $-N{=}N-$, ${=}CO$) which are responsible for the color; should these compounds also contain certain auxochromic groups (e.g., $-NH_2$ and $-OH$), then they possess dyeing properties. In the intervening years this statement has been refined and generalized but the Witt definition is still a simple, practical, and satisfactory frame of reference for all but the dye specialist.

Materials which are called dyes today can absorb and emit in the ultraviolet and near infrared as well as in the visible(1–11). The visible absorbing dyes are responsible for the many gaily colored materials seen every day. Strongly fluorescent dyes are used to make the glowing paints and tapes used for safety markers. The UV-absorbing, blue-emitting dyes are placed in many cloths and papers to enhance their whiteness. The IR-absorbing dyes are essential to passively Q-switched Nd lasers.

Although thousands of dyes are known, only a fraction fluoresce in solution and it is among these that lasing dyes are found. The spectro-scopic properties of these materials are determined by the dye's physical and chemical structure and the manner in which the dye inter-acts with the solvent. In this section the spectral properties of dyes in solution are discussed and explained by use of a schematic energy level diagram. Dye chemistry also is described and related to the spectral properties.

A. SPECTROSCOPY

The optical absorption and emission processes of organic dyes in solution have been studied for over a century. During this time the spectral properties of many thousands of dyes and dye solutions from the UV to the IR have been cataloged (*7–11*). The wavelengths, width, structure, and strength of the spectral processes vary for different dyes or for a particular dye in different solvents. However, most dye solutions have the following spectral properties which are similar in many respects to those of rhodamine B given in Fig. 1.

1. The widths of the principal absorption and emission bands are generally on the order of $1000\ cm^{-1}$. One or more absorption bands may be found toward the UV from the principal absorption.

2. The fluorescence peak occurs at longer wavelengths than the principal absorption peak. This displacement is called the Stokes shift of fluorescence from absorption. The extent of this Stokes shift and the width of the fluorescence and absorption spectra may be such that the short wavelength tail of the fluorescence substantially overlaps the long wavelength tail of the absorption.

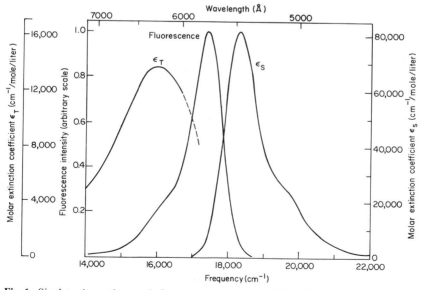

Fig. 1. Singlet absorption and fluorescence spectra of $5 \times 10^{-5}\ M$ rhodamine B in methanol, and triplet absorption spectrum for rhodamine B in polymethylmethacrylate (*33*). The subscripts S and T denote singlet and triplet plates, respectively.

3. The fluorescence spectrum is generally a mirror image of that of the principal absorption band.

4. The fluorescent lifetime is typically $\sim 5 \times 10^{-9}$ sec.

5. Excited-state absorption bands may be observed on a transient basis by monitoring the solution transmission following exposure to light in the principal absorption band. A triplet–triplet absorption band may overlap the fluorescence band and may persist for as long as 10^{-3} sec, depending on the solution treatment.

6. After optical irradiation, new long-lived absorption bands due to new chemical species formed by photochemical processes may sometimes be observed. These may be transient if the new species reform into the original dye molecules or permanent if the decomposition is irreversible.

B. ENERGY LEVELS

An energy level diagram is useful for understanding dye laser operation. However, in contrast to atomic or simple molecular gas lasers and solid-state lasers, it is presently not possible to draw a detailed, accurate picture of the energy levels of a dye in solution. A dye molecule is composed of many atoms and hence it is extremely difficult to describe the wave functions which represent the various configurations of the molecule and which are needed to calculate the energies. In addition, since we are dealing with complex molecules composed of many atoms, there is a vast number of possible states involving the different permitted combinations of electronic, vibrational, and rotational motions. Therefore, even if the energies could be found, the complete level diagram would be extremely complicated.

Various perturbation techniques can yield reasonably good approximations for the wave functions and energy levels of small molecules (*12,13*). These procedures, however, are not applicable here since most dye molecules are not small. Instead, the energy levels and absorption and emission processes of dyes in solution can be envisaged schematically by constructing an energy level diagram by analogy with a simple diatomic molecule. The diatomic molecule's energy is dependent upon a single configurational coordinate (the interatomic distance). The resulting energy level diagram is similar to that obtained for the familiar, harmonic oscillator problem. On the other hand, the configuration, and therefore the potential energy, of a particular electronic–vibrational–rotational state of a dye molecule is a function of many coordinates. However, for the purpose of a schematic diagram, it is assumed that the configuration of the state can be described by only one coordinate which is depicted by the horizontal distance in Fig. 2. For clarity the origin of the configuration coordinate for the states labeled T_i has been shifted to the position of the dashed line. Note that the coordinate corresponding to minimum energy is different for each group of states, indicating that the equilibrium configuration depends on the electronic state of the particular group. The use of such a diagram to describe a complex organic dye molecule was first proposed by Jablonski(*14*) in 1935. In Fig. 2 the radiative transitions relevant to

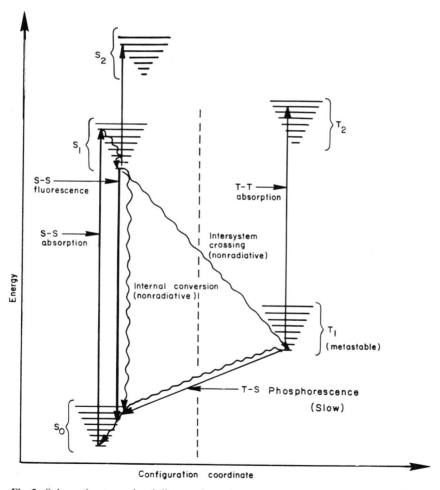

Fig. 2. Schematic energy level diagram for a dye molecule. Transitions relevant for dye laser action are shown(*33*). See text for details.

absorption and fluorescence processes are indicated by solid lines; nonradiative transitions are indicated by wavy lines.

The vibrational–rotational states shown in Fig. 2 are grouped according to the electronic state of the molecule. The separation of different electronic states is typically about 10,000 to 20,000 cm^{-1}; the separation of different vibrational states belonging to a given electronic state is of the order of 1000 cm^{-1}; the separation of different rotational states for a given vibrational state is of the order of 1–10 cm^{-1}(7,12). It follows, therefore, that for a given electronic state the vibrational and rotational energies appear as fine and hyperfine structures, respectively. The states of the dye are further grouped according to the electronic spin into singlets, labeled S, or triplets, labeled T. In reality, however, there is usually a small admixture of pure singlet and triplet states due to spin–orbit interactions.

The study of the events involved in the absorption and reemission of light begins and ends with the molecule in equilibrium in the ground state, S_0. Since this is the optical excitation-stimulated emission cycle of a dye laser, this procedure is useful for understanding the various processes involved in and competing with lasing. The relevant transitions, their rates, and other properties will be discussed at the appropriate points in this cycle. At room temperature the thermal equilibrium distribution has very few molecules that are more than 200 cm^{-1} away from the lowest ground state level. Optical absorption, an electronic process, occurs much more rapidly than the rate at which the molecule can change its internuclear configuration. This process is indicated in Fig. 2 by a vertical arrow from the thermally populated low-lying levels of S_0 to excited levels of S_1 which have the same configuration coordinate. This conforms with the Franck–Condon principle which states that the molecular configuration cannot change during an electronic process(15,16).

After absorbing light the molecule is energy rich. At this point it may reemit the absorbed photon, but it is more likely to relax nonradiatively into the low lying states of S_1 by giving up energy to the solvent. Estimates of the time required for this relaxation vary between 10^{-11} and 10^{-12} sec. The molecules in the S_1 state will, therefore, thermalize before making other transitions.*

* Until recently, this relaxation time was just an estimate based on the rates of thermalization given for other systems. However, Rentzepis(17) measured this time in azulene using picosecond pulse excitation and found a value of 7×10^{-12} sec. Evidence that excited state thermalization takes place in less than 10^{-10} sec in several other dyes is obtained from the fact that the fluorescent spectrum of dyes excited by mode-locked laser pulses is identical with that obtained when they are excited continuously(18). Dye amplifier experiments(19) and efficient spectral narrowing of dye laser emission(20) also support the notion that thermalization in S_1 is very fast.

A molecule in the lower levels of S_1 can return to S_0 by emitting a photon of light. This radiative relaxation is called fluorescence. The energy of the emitted light is less than that of the absorbed light; the fluorescence is Stokes shifted to long wavelengths from the absorption.* Some variation of the radiative lifetime for a particular dye (typically $\sim 5 \times 10^{-9}$ sec) is observed on changing the solvent, the concentration, or the temperature of the solution. The ratio of the number of emitted photons to the number of absorbed pump photons is called the fluorescent quantum efficiency and can be as high as unity for some dye solutions and as low as 0.01 for others.

A molecule in S_1 can make three other transitions that compete with fluorescence and hence with lasing. Transitions between S_1 and other excited singlet states can result in absorption at the same frequencies as the fluorescence and so cause a pump-dependent loss in a dye laser.† Nonradiative transitions between states of the same multiplicity or internal conversion can occur between S_1 and S_0 and reduce the fluorescent quantum efficiency. Nonradiative transitions between states of different multiplicity, S_1 and T_1 for example, can also occur. These are referred to as intersystem crossings. Solvent molecules, dissolved paramagnetic impurities such as O_2, or heavy atoms with large spin–orbit interactions may admix some of both spin states into each state of the dye.‡ This enhances the coupling of singlet and triplet states and increases the rate of intersystem crossing. The lifetime of a molecule in state T_1 can be as long as 10^{-3} sec in a carefully deoxygenated solution or as short as 10^{-7} sec in an oxygen-enriched solution.

Intersystem crossing reduces the population of S_1 available for radiative transitions to S_0 and so reduces the possible fluorescent quantum efficiency. Also, by populating the metastable state T_1, intersystem crossing may give rise to a triplet absorption loss which may increase

---

*Since the same Stokes shift is found in both vapor and solution phases (21), an additional mechanism for relaxation of the upper state, not involving the solvent, must exist. This involves rapid intramolecular redistribution of excess vibrational energy via anharmonic coupling. If the equilibrium configurations of the electronic states differ, the Franck–Condon principle ensures that reradiation will terminate on highly excited levels of the ground state, thereby producing the Stokes shift.

†The Q-switched lasers have recently been used both to raise a molecule into the first excited singlet state and simultaneously to produce a gas breakdown which provides a synchronized continuum source for time-resolved spectroscopy. Both excited singlet–singlet and triplet–triplet absorptions have been detected with this technique (22).

‡The heavy halogen ion present in some ionic dyes (See Table 2) has a large spin–orbit interaction which can facilitate intersystem crossing. However, when these ions are present in the solution due only to the dissociation of the dye molecule, they are too few and too widely separated from the dye molecules to be very effective for the desired purpose.

with the dye's excitation. If this loss is at the same frequencies as the fluorescent emission, it can possibly prevent or quench laser action.

The molecule which has reached T_1 may, in addition to intersystem crossing, make a weak radiative transition to S_0. The light emitted is called phosphorescence; it has a radiative decay rate many times slower than that of fluorescence light.

The process of relaxation within S_0 may be quite different from one dye to the next. Certain cyanine dyes in room temperature solutions, for example cryptocyanine in methanol, exhibit spectral "hole burning" under intense Q-switched ruby laser irradiation (23–25). This is evidence for an inhomogeneously broadened ground state on the 3×10^{-8} sec time scale of the laser pulse. Such a dye does not thermalize in less than $\sim 10^{-6}$ sec. Other dyes, for example aluminumphthalocyanine, do not show "hole burning" and can be considered to thermalize very quickly (25). Although the rates of thermalization of dyes in the ground electronic state may vary widely, it is unlikely that any dye in a room temperature solution takes much longer than 10^{-6} sec to reach thermal equilibrium.

Dye solution lasing involves stimulated emission between the low-lying levels of S_1 and the high-lying levels of S_0. The energy level diagram shows that the absorption–emission process involves four states of the dye molecule, which suggests that the inversion required for lasing should be small. In Section III we show that the fractional population inversion required to lase a dye in a typical cavity is much less than 0.5, confirming the notion of a four-level laser scheme. This lasing inversion is not easy to achieve, however, because high optical pumping rates are required to overcome depletion of the population in S_1 by spontaneous emission.

C. Dye Chemistry

Some useful guidelines to the spectral properties of dye molecules are obtained from their chemical composition. In Table 2, the chemical formulas and structural diagrams of several dyes representing important families of dyes are given. A family of dyes is a group of compounds having similar chemical structures and thus certain common spectral properties. By surveying the published literature on dye fluorescence, one finds that the most strongly fluorescent dyes belong to the oxazole, xanthene, anthracene, coumarin, acridine, azine, phthalocyanine, and polymethine families. Since fluorescence is an essential prerequisite to lasing, the properties of these families of dyes are of great interest.

The names of dye compounds describe the structural formula of their molecules. In general the last element in the name is the family of the dye while the preceding elements are the various substitutional groups attached to the basic molecule. The numbers indicate the position of the groups with respect to the basic molecule. Consider, for example, 7-diethylamino-4-methylcoumarin as shown in Fig. 3. In Fig. 3a, the coumarin molecule is sketched and the numbering of the vertices indicated. In Fig. 3b, a methyl group, CH_3, replaces the hydrogen atom in the position numbered 4 so that 4-methylcoumarin results. Finally, as shown in Fig. 3c, a diethylamine group $(C_2H_5)_2NH$, is placed in the 7 position to form the desired molecule.

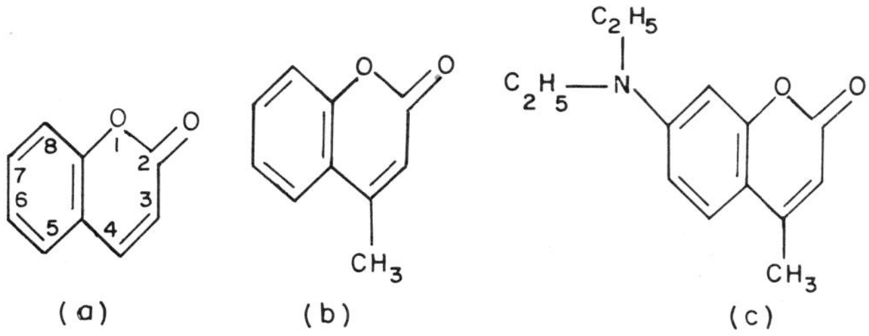

Fig. 3. Step-by-step schematic construction and labeling of a coumarin derivative. (a) Coumarin with vertex numbers indicated; (b) 4-methylcoumarin; (c) 7-diethylamino-4-methylcoumarin.

The identification of dye compounds is sometimes confused by the use of several names for the same molecule. As an example, umbelliferone and 7-hydroxycoumarin are different names for the same molecule. More descriptive names are often given to groups of dyes within one family. For example, the dye 3,3'-diethylthiatricarbocyanine iodide is a polymethine dye but since it contains sulfur atoms it is sometimes also called a thiacarbocyanine dye.

Although one must usually investigate the absorption and emission spectra of the particular dye-solvent combination of interest, the literature contains some useful general guidelines to the spectral properties of dyes in certain families. Brooker(26) in 1942 found that the absorption and emission spectra of dyes in the polymethine family shift to longer wavelengths as the number of —CH=CH— or cyanine units in the polymethine chain is increased. Miyazoe and Maeda(27) have recently studied both the conventional spectroscopic properties and lasing characteristics of polymethine dyes in great detail. Table 3 shows the broad spectral range over which polymethine dyes may be used, the shift to long wavelengths with increasing length

TABLE 2

Chemical Structure Diagrams and Data for Molecules Representative of Six Important Families of Dyes

Chemical formula	Name	Molecular weight	Family
$C_{24}H_{16}N_2O_2$	or p-Bis[2-(5-phenyloxazolyl)]benzene	364.40	Oxazole
$C_{26}H_{18}$	9,10-Diphenylanthracene	330.43	Anthracene

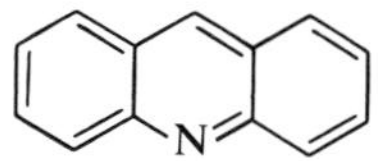

Acridine

179.22

Acridine

$C_{13}H_9N$

Xanthene

479.02

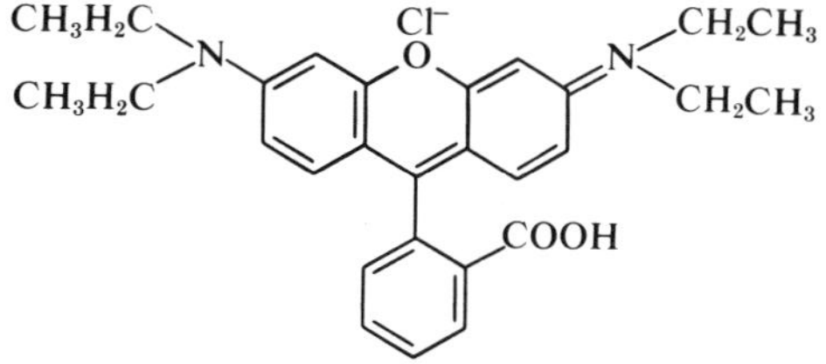

Rhodamine B

$C_{28}H_{31}ClN_2O_3$

Polymethine

523.33

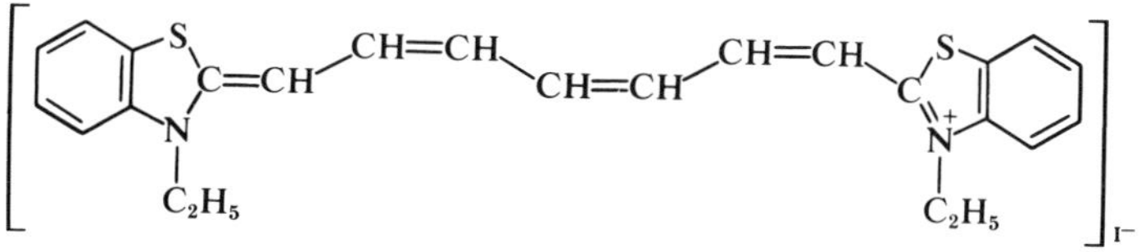

3,3'-Diethylthiatricarbocyanine iodide

$C_{23}H_{29}IN_2S_2$

TABLE 3

Spectral and Lasing Properties of Related Polymethine Dyes

Dye	Property and Value (mμ) for n =					Concentration	Solvent
	n	0	1	2	3		
	λ_{abs}^{max}	428	561	659	764		
	λ_{fluor}^{max}	a	586	686	800		
	η_{fluor} [b]	—	—	0.66	0.61	$10^{-5}M$	Ethylene glycol
	η_{lasing} [c]	—	—	0.116	0.135	$10^{-4}M$	
	λ_{abs}^{max}	377	487	586	689		
	λ_{fluor}^{max}	—	507	614	720		
	η_{fluor}	—	—	—	0.93	$10^{-5}M$	Ethylene glycol
	η_{lasing}	—	—	—	0.279	$10^{-4}M$	

				Ethylene Glycol
λ_{abs}^{max}	439	548	646	749
λ_{fluor}^{max}	—	569	680	782
η_{fluor}	—	—	—	0.50	$10^{-5}M$
η_{lasing}	—	—	—	0.087	$10^{-5}M$

[a] When $n = 0$ no fluorescence was detectable.

[b] Measured relative to a $10^{-5}M$ solution of 3,3'-diethylthiatricarbocyanine iodide in dimethyl sulfoxide. η_f (DTTC in DMSO) = 1.0.

[c] Fraction of ruby laser pump energy converted to dye laser light. Lasing was not studied in dyes which did not absorb at 694.3 mμ.

of the polymethine chain, and some lasing characteristics of three of the more than twenty dyes they studied. It was also found for those dyes studied that changing either the alkyl radical or the halogen ion had little effect on the dyes' spectral properties.

In general the xanthene dyes absorb and emit light in the visible part of the spectrum. Two dyes in which laser action is most easily obtained, rhodamine 6G and rhodamine B, are xanthene dyes which emit between ~ 5500 and 5900 Å and 5800 and 6300 Å, respectively. The anthracene, oxazole, and coumarin dyes absorb in the near UV and emit in the violet or blue-violet regions of the spectrum.

Chemical or photochemical instability is a major problem with some dyes which otherwise appear to have promising laser properties. Many polymethine dyes tend to be photochemically unstable in solution. Miyazoe and Maeda(27) have found that when the number of —CH =CH— units in the polymethine chain exceeds 4, the dyes are also chemically unstable.* Xanthene dyes are for the most part much more stable than polymethine dyes under visible and ultraviolet irradiation. They can be pumped with very intense flashlamps many times and show no noticeable decomposition. However, after ≈ 100 hr use at a pulse repetition rate of ~ 1 pps, dye solutions of rhodamine 6G must be changed in order to restore lasing performance. Dye lasers using 7-diethylamino-4-methylcoumarin achieve laser action only once when flashlamp pumped, suggesting that this dye is photochemically unstable. The 7-hydroxycoumarin in water buffered to a pH of 9 is chemically unstable and has a useful life of less than one day.

The photochemical instability can be even more of a problem for laser action if the decomposition products absorb at wavelengths where the dye emits. Recently Ferrar(28) found that flashlamp-pumped fluorescein lasers stopped lasing after three or four excitations. Spectroscopic evidence was given indicating that photochemical decomposition occurred after each exposure to the pump lamp and that the resulting molecules absorbed the light emitted by the fluorescein.

D. Solvent Effects

The position and structure of the absorption and emission spectra of molecules in solutions are dependent on the solvent. For example, the spectra of molecules dissolved in cyclohexane are generally much more structured than when dissolved in ethanol. This effect is quite evident

*In our laboratory nearly all the molecules of 3,3′-diethylthiatricarbocyanine iodide in a dimethyl sulfoxide solution were found to have decomposed under the action of fluorescent room lamps after ~ 5 hr exposure.

in the different spectral widths of the lasing bands observed for various oxazole dyes in cyclohexane and ethanol solution(29). Cherkasov(30) and Bakhshiev(31) have shown that the wavelength displacement can often be correlated with a change in solvent dielectric constant and in the index of refraction.

The radiative lifetime and fluorescent quantum efficiency of a dye solution are also functions of the solvent used. Miyazoe and Maeda(27) found a strong correlation between quantum efficiency and lasing efficiency for polymethine dyes in different solvents. On the other hand, laser action is easily obtained with solutions of xanthene dyes, even though the lifetime and quantum efficiency vary widely. As an example, rhodamine B in a water solution has a quantum efficiency of 25% and a lifetime of 0.94×10^{-9} sec(32); in a methanol solution these quantities are 62% and 2.0×10^{-9} sec, respectively(33). Laser action is easily obtained with both solutions, although their lasing efficiencies may differ.

The spectral properties of many dye solutions are determined by the pH of the solution, which controls the extent of dissociation of the dye molecules. For example, the absorptivity of fluorescein in water at 4910 Å increases by nearly an order of magnitude when the pH is raised from 1 to 9(34). In laser experiments 7-hydroxycoumarin performs best when it is in a water solution buffered to a pH of 9.

When the concentration of dye molecules is too high, there can be large absorptive losses due to the overlap of the long wavelength tail of the absorption with the fluorescence. It is thus a good rule in working with dye lasers to use solutions with concentrations less than $10^{-2} M$. An additional benefit obtained by using low concentration is that there is less chance for the dye molecules to come together in pairs, forming dimers. These molecular pairs reduce the number of molecules available for lasing, effectively lowering the concentration, and may absorb light where the dye might lase. Ethanolic solutions of 7-diethyl-amino-4 methylcoumarin at concentrations greater than $10^{-3} M$ have a distinct yellow cast. This absorption, which occurs between 4300 and 5000 Å, is very likely due to dimer formation. Since stimulated emission from this dye occurs at ~ 4600 Å, high concentrations should be avoided.

III. Analysis of Dye Laser Action

Laser action in organic dye molecules involves stimulated emission between levels of the excited singlet state S_1 and the ground state S_0.

The spontaneous emission spectrum of organic molecules in solution is broad since it arises from transitions from a number of thermally populated vibrational–rotational levels of S_1 to the manifold of vibrational–rotational levels of S_0. Optically pumped dye media may therefore exhibit gain over a broad spectral region. The specific frequency at which the medium first has sufficient gain to overcome the optical resonator losses is dependent on both the relative level populations and the frequency profiles of the stimulated absorption and emission cross sections. The analysis of dye laser action is further complicated by the fact that, in addition to ground-state absorption, excited-state absorption from S_1 and, more frequently, from the metastable T_1 state may be significant. In general, the gain of dye lasers is a function of (1) the level populations, which change with time during optical pumping, (2) the frequency, as determined by the emission and absorption spectra, and (3) other properties such as temperature, concentration, and length of the active medium.

The theory of dye laser action and the analysis of spectral and temporal properties and pumping rates have been considered by several authors(2–4,33,35–44). These have included treatment of level populations before and after the onset of stimulated emission, the photon field, the frequency dependence of the gain, and the effects of triplet losses on the gain profile. A general treatment of the characteristics of an optically pumped dye medium, beginning with the frequency dependence of the gain and proceeding to its evolution during the pumping pulse, is given below. The approach used is an extension of that developed by McCumber(45) to treat the broad absorption and emission spectra associated with phonon-terminated solid-state lasers. The results explicitly show the manner in which the gain throughout the broad fluorescence spectrum of the dye varies with frequency and with time. Detailed experimental verification of the predictions is given later in Sections IV and V.

A. FREQUENCY DEPENDENCE

The energy levels and transitions essential to understanding laser action in dyes are shown in Fig. 2. Excitation is achieved by optical pumping into S_1 or higher-lying singlet states followed by rapid decay to S_1. The molecular population within the set of vibrational–rotational levels of a given electronic state is assumed to attain a Boltzmann equilibrium distribution characteristic of the host temperature in a time that is short compared to the rates of optical pumping and decay from S_1 (although, as noted elsewhere, this may not always be satisfied for S_0). The gain per unit length for photons of angular frequency ω is

$$G(\omega) = e(\omega) - a(\omega) \qquad (1)$$

where $a(\omega)$ is the absorption coefficient of the dye medium and $e(\omega)$ is the corresponding stimulated emission coefficient. The total absorption coefficient includes transitions from the ground state and spin-allowed transitions from excited states S_1 and T_1 to higher-lying states which may also occur at the frequency ω. In terms of the absorption and emission cross sections σ_i^a and σ_i^e for electronic state i, Eq. (1) becomes

$$G(\omega) = N_{S_1}\sigma_{S_1}^e(\omega) - N_{S_0}\sigma_{S_0}^a(\omega) - N_{S_1}\sigma_{S_1}^a(\omega) - N_{T_1}\sigma_{T_1}^a(\omega) \qquad (2)$$

where N_i is the number density of molecules in the ith set of states. Note that if the excited-singlet absorption cross section $\sigma_{S_1}^a$ is greater than $\sigma_{S_1}^e$ at all frequencies, laser action is not possible.*

The stimulated emission cross section can be derived from the spontaneous emission spectrum. If the latter is described by the fluorescence function $f(\omega)$, then it can be shown(46) from the Einstein relationships that

$$\sigma_{S_1}^e(\omega) = \left[\frac{2\pi c}{\omega n(\omega)}\right]^2 f(\omega) \qquad (3)$$

where $n(\omega)$ is the refractive index at frequency ω and c is the velocity of light in vacuum. The function $f(\omega)$ in Eq. (3) is normalized such that

$$\int_0^\infty f(\omega) = \frac{\phi}{8\tau_s} \qquad (4)$$

where ϕ is the quantum yield and τ_s is the lifetime of the excited singlet state. Combining Eqs. (2) and (3), it is seen that the gain of the dye medium at any frequency can be obtained from readily measurable absorption cross sections per molecule (molar extinction coefficients ϵ) and the fluorescence spectrum; normalization of the latter requires, in addition, measurements of the lifetime and spontaneous emission probability or radiative quantum efficiency.

For broadband transitions, the absorption and emission cross sections are related by(47)

$$\frac{\sigma^a(\omega)}{\sigma^e(\omega)} = \exp\left[\frac{\hbar(\omega - \omega_0)}{kT}\right] \qquad (5)$$

---

*Excited-singlet-state absorption is thought to be the reason why certain dyes which have high fluorescent quantum yields, such as rubrene, do not achieve laser action.

where $h\omega_0$ is a temperature-dependent excitation potential given by the net free energy required to excite one molecule while maintaining the surrounding temperature T. This relationship can be used to simplify the form of $G(\omega)$. For dyes which have approximately mirror-image absorption and emission spectra, $h\omega_0$ in Eq. (5) is given by the energy of a transition between the lowest vibrational–rotational levels of S_0 and S_1 in Fig. 2. Because the ground-state absorption cross section is usually small in the region where stimulated emission occurs, Eqs. (4) and (5) can be used to express the gain in terms of the fluorescence function. Thus the gain becomes

$$G(\omega) = \left\{ N_{S_1} - N_{S_0} \exp\left[\frac{-h(\omega_0 - \omega)}{kT}\right] \right\} \left(\frac{2\pi c}{\omega n}\right)^2 f(\omega)$$
$$- N_{S_1} \sigma^a_{S_1}(\omega) - N_{T_1} \sigma^a_{T_1}(\omega) \tag{6}$$

In this form, one sees that as $(\omega_0 - \omega)$ increases, the S_0 absorptive loss term becomes smaller. This corresponds to lower frequency transitions terminating on higher-lying levels of S_0 which have smaller thermal populations.

If the excited singlet and triplet losses are small in the spectral region of interest and no significant population buildup in T_1 occurs, Eq. (6) becomes simply

$$G(\omega) = \left[\left(\frac{2\pi c}{\omega n}\right)^2 f(\omega) + \sigma^a_{S_0}(\omega)\right] N_{S_1} - \sigma^a_{S_0}(\omega) N \tag{7}$$

where the total number of dye molecules $N \approx N_{S_0} + N_{S_1}$. In this case the gain increases and its frequency profile changes with increasing excited state population N_{S_1}. To illustrate this, consider the spectra of rhodamine B in Fig. 1 and assume $k_{ST} = 0$ so that triplet effects can be neglected(41). The gain as a function of frequency for a $5 \times 10^{-5} M$ methanol solution, predicted by Eq. (7), is shown in Fig. 4 for various N_{S_1} values. Note that even for a small fractional population in S_1 $(N_{S_1}/N \ll 1)$, the medium exhibits a positive gain coefficient over a broad spectral region. The critical population inversion required to achieve stimulated emission in a typical optical resonator, as we shall see, is indeed small, $\sim 10^{14}$ molecules/cm^3. This corresponds to a fractional population of $< 1.0\%$. Therefore, dye lasers truly typify a four-level laser scheme.

The peak of the gain curves in Fig. 4 shifts to higher frequency as N_{S_1} increases and the ground-state absorptive loss decreases. Threshold for laser action occurs when, at some frequency, $G(\omega)$ becomes equal to the loss coefficient $L(\omega)$ of the optical resonator. The latter

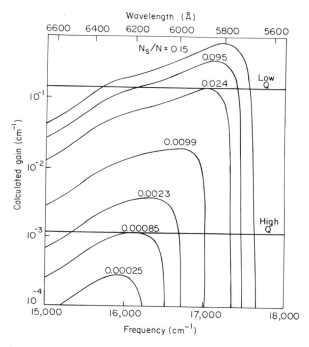

Fig. 4. Calculated gain vs. frequency curves for rhodamine B in methanol (neglecting triplet effects) for various excited singlet state population densities N_{S_1} (33).

includes losses per pass due to the mirror transmission, diffraction and scattering, and absorption due to any regions of the dye solution not exposed to the pump radiation. The initial frequencies of stimulated emission will be those for which the above condition is first satisfied. As indicated in Fig. 4, for a high-Q resonator having no dispersive losses ($L(\omega) =$ constant), the required threshold gain is small and the frequency ω_L at which $G(\omega) = L(\omega)$ is low; a cavity having greater losses requires higher gain to achieve threshold and the laser frequency will be higher. Hence, by varying the resonator Q, the laser frequency can be tuned from a low frequency, corresponding to the lowest loss possible, to a frequency corresponding to the peak of the fluorescence curve. If a dye has a large Stokes shift, the S_0 absorptive loss term will always be small and laser action will occur near the peak of the fluorescence curve. (An example of this behavior is considered in Ref. (33) for 7-hydroxycoumarin.)

The spectral emission and absorption functions in Eqs. (2) or (7) are dependent upon the population distribution in the manifold of vibrational–rotational levels of the initial electronic state. Therefore the gain has a temperature dependence which has not been indicated explicitly in Eq. (2). As the temperature is decreased, the absorption and fluorescence spectra narrow and the amount of overlap decreases, thereby causing the lasing frequency to move toward the peak of the fluorescence curve. The temperature variation of the lasing frequency has been investigated by Schappert et al.(48) using a DTTC in ethanol laser. The fluorescence spectrum at two different temperatures is shown in Fig. 5; circles indicate the lasing wavelengths at different concentrations. The frequency shift is particularly pronounced at the higher concentrations when the overlapping ground-state absorption becomes significant. Since the rate of nonradiative loss from S_1 by internal conversion also may change with temperature, an additional temperature dependence may arise through changes in ϕ and the magnitude of $f(\omega)$ in Eq. (3).

The gain of a dye medium and the oscillation frequency are also dependent on the length of the active region and the dye concentration.

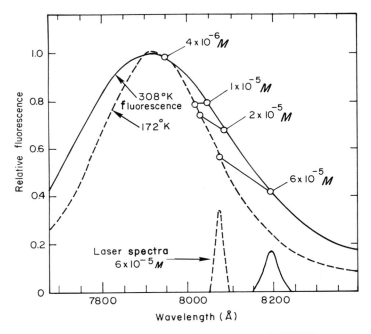

Fig. 5. Comparison of the fluorescence and laser spectra of DTTC in ethanol at two different temperatures. The circles indicate lasing wavelengths at different molar concentrations (48).

If the region of excited dye is lengthened while the resonator losses are held constant, the gain per pass will be larger. Therefore $G(\omega) = L$ will be satisfied for a smaller value of N_{S_1} and the laser frequency, determined by the peak of the corresponding gain curve (see, for example, Fig. 4*), will be lower. As the dye concentration is increased, there are initially more molecules contributing to the absorptive loss. This also shifts the peak of the gain curve to lower frequencies and thereby lowers the laser frequency. An upper limit on the concentration may be imposed by the onset of concentration quenching of the fluorescence and/or by nonuniform pumping if the absorption coefficient becomes too large. The shifts to lower frequencies with increased length and concentration of active dye solution have been discussed by several groups(42,44) and are well documented experimentally(42,49,50).

The triplet-state system has been neglected in the above simplified analysis. It can and usually does have important consequences for the gain characteristics and laser performance. First, transfer of molecules to a metastable triplet level by intersystem crossing reduces the total number of molecules in the singlet-state system, and thus reduces the maximum gain possible. Second, if the triplet–triplet absorption overlaps the fluorescence, the associated loss will alter the gain profile and laser properties.

Triplet losses build up toward a steady-state value during the optical pumping and laser action; thus, they set limitations on the rate of pumping required if oscillation is to be achieved, as well as influencing the laser frequencies possible. Snavely, Buettner, Peterson, and co-workers(39,40,43) have investigated triplet effects by using a gain expression similar to Eq. (2). The critical inversion and triplet populations were found by requiring that the following conditions be satisfied:

$$G(\omega_L) - L(\omega_L) = 0 \tag{8a}$$

$$\left.\frac{\partial G}{\partial \omega}\right|_{\omega_L} = 0 \tag{8b}$$

$$N = N_{S_0} + N_{S_1} + N_{T_1} \tag{8c}$$

where for a closed system and in the absence of photochemical processes, the total molecular concentration N is a constant. If the absorption and emission spectral functions required in Eq. (2) and their derivatives are known and the resonator losses $L(\omega)$ established, then Eq. (8) can be solved to determine $N_S{}^c$, the critical population density required for threshold, and N_T, the triplet-state population density,

*If the gain curves in Fig. 4 are expressed in terms of a molar gain coefficient (cm^{-1}/ mole/liter), they become a universal set appropriate to all concentrations.

consistent with stimulated emission at frequency ω_L. Such calculations were performed for rhodamine 6G where the triplet-state absorption cross section was known. The results are shown in Fig. 6. The effect of the triplet accumulation is evident in the larger excited-singlet population required for threshold and the shift of the minimum critical inversion to shorter wavelength. If the wavelength of the laser emission is measured, then by locating the minimum of such plots at that wavelength, the allowable value of N_S^c and N_T can be determined. Values derived from this type of analysis for a $5 \times 10^{-5}M$ ethanol solution rhodamine 6G in a resonator of 7 cm length and 0.99 mirror reflectivity are plotted in Fig. 7.

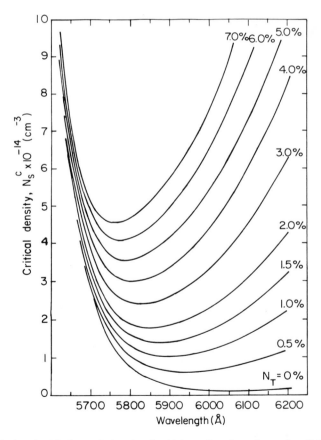

Fig. 6. Calculated critical population density N_S^c as a function of wavelength for various triplet-state populations $N_T(43)$.

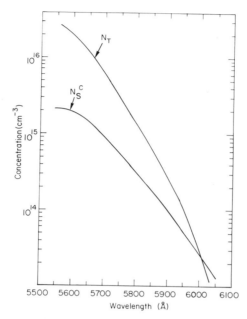

Fig. 7. Dependence of the minimum values of critical population density N_S^c and triplet-state population N_T on the wavelength of laser emission(3).

Calculations of the gain require information about the spectral functions of the dyes and the laser resonator properties. From such data, the dependence of the laser frequency on resonator losses and on the length and concentration of the active medium can be derived. From the critical gain expressions and measured laser frequency, the associated level populations can be determined. Since dye lasers have been operated thus far only on a pulsed basis, it is of interest to examine the time evolution of the population distributions during optical pumping and the dynamic changes in the gain profile resulting from the time-varying losses.

B. Time Dependence

The level populations in the gain expression Eq. (2) at any time after the initiation of pumping are obtained from solutions of the rate equations governing the excitation and relaxation processes. Consider first the buildup of the gain toward the threshold for oscillation. Since the photon field in the laser cavity will be small during this period, stimulated emission processes are omitted. Excitation is assumed to proceed via optical pumping into S_1 and/or higher excited singlet states followed by very rapid decay to S_1. Pumping wavelengths suitable for inducing emission from S_1 to S_0 or exciting molecules from S_1 or T_1 to higher excited states are assumed to be absent. The metastable T_1 level is considered to be located far enough below S_1 that reverse intersystem transitions $T_1 \to S_1$ are negligible. Thus we treat a simple system consisting of three sets of energy levels: the lowest excited singlet (S) and triplet (T) states and the ground state (0).

The rate equations for this effective three-level system are

$$\frac{dN_S}{dt} = -\frac{1}{\tau_S}N_S + P(t)N_0 \tag{9a}$$

$$\frac{dN_T}{dt} = -\frac{1}{\tau_T}N_T + k_{ST}N_S \tag{9b}$$

$$\frac{dN_0}{dt} = -P(t)N_0 + \left(\frac{1}{\tau_S} - k_{ST}\right)N_S + \frac{1}{\tau_T}N_T \tag{9c}$$

where τ_S and τ_T are the singlet and triplet state lifetimes, $P(t)$ is the optical pumping rate, and k_{ST} is the intersystem crossing rate. Radiation trapping arising from the absorption and reemission of a photon is not considered in the present analysis. Computer solutions of the above coupled rate equations can be obtained for an arbitrary pumping pulse. The inputs are the various lifetimes or rate constants and $P(t)$. While the time dependence of an actual pumping pulse can be readily found, the magnitude of the pumping rate at any instant of time is difficult to determine accurately. The effective pumping rate is determined by integration of the spectral distribution of the source over the absorption band(s) of the dye, and is affected by the geometry and scattering in the particular laser cavity used. However, given estimated rate constants and a pumping pulse applied at $t = 0$, the time development of the level populations N_i from their thermal equilibrium values can be calculated. If these populations are inserted into the gain expression Eq. (2) together with the measured spectral functions at a specific temperature, a final computer readout composed of the gain G as a function of frequency ω at selected times during the pumping can be obtained. Such calculations have been carried out by the authors for several representative dyes using actual laser and flashlamp pumping functions (33,41).

If, for a given dye, the intersystem crossing rate is small or the triplet lifetime is short enough with respect to the pumping rate so that no significant triplet population and associated absorptive losses develop (i.e., $k_{ST}\tau_T \ll 1$), one need know only $N_S(t)$ in Eq. (7) to determine the gain and the possible lasing frequency. To achieve oscillation for this medium using an excitation pulse having a known pumping rate, it is necessary only to find the maximum value of $N_S(t)$ by solving the rate equations, evaluating the peak gain, and then selecting an optical resonator having a smaller loss coefficient.

Because the populations change during pumping, the peak of the gain curve $G(\omega)$ will also change. As we have seen, the lasing frequency is dependent upon the losses or the Q of the laser cavity. For cavities and dyes having time-independent losses, the lasing frequency, if threshold is reached, will be independent of the properties of the optical pumping pulse; only the time required to achieve threshold will vary. The faster the pumping, the sooner will threshold be reached. The predicted dependence of the time of threshold on cavity Q and pumping rate have been verified qualitatively for flashlamp-pumped rhodamine 6G in ethanol(33); the results are summarized in Fig. 8.

The time evolution of the gain curves $G(\omega)$ can reveal dramatic changes if time-varying losses such as those arising from excited triplet-state absorption are present. In the above simple case, knowledge of only the $S_0 - S_1$ absorption or emission cross sections, fluorescence lifetime, and quantum yield was required; now, in addition,

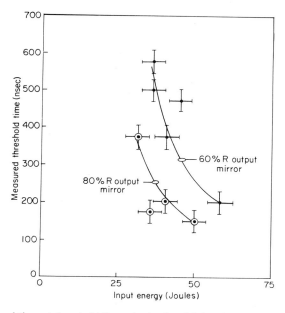

Fig. 8. Observed time at threshold for a rhodamine 6G in ethanol laser as a function of input energy to the flashlamp. A 100% reflectivity mirror and the indicated output mirror formed the resonator(33).

knowledge of the $T_1 - T_i$ absorption cross section, triplet lifetime, and rate of intersystem crossing is required. Unfortunately, such data are usually not available. Calculations have been made for anthracene, however. They show that triplet–triplet transitions can greatly distort the $G(\omega)$ curves(33) and, if k_{ST} is sufficiently large, can make lasing impossible(38).

The transfer of molecules to a metastable triplet level is determined by k_{ST}. If the triplet lifetime is long (that is, $k_{ST}\tau_T \gg 1$), the triplet population at time t after pumping begins will be, from Eq. (9b),

$$N_T(t) \approx k_{ST} \int_0^t N_S(t)\, dt \qquad (10)$$

This is an integrating process and for long pumping pulses the number of molecules in level T can become a sizable fraction of the total population. There are two approaches to minimize losses associated with the triplet system so as to achieve laser action. One is to reach threshold before an appreciable number of molecules accumulate in T by using a very short (pulse width less than k_{ST}^{-1}), intense laser pulse for pumping. A second approach is to decrease the instantaneous triplet population by decreasing the triplet lifetime τ_T. If $\tau_T^{-1} \sim k_{ST}$, Eq. (10) is no longer an appropriate solution of Eq. (9b). As discussed elsewhere, τ_T can be reduced by oxygenating the dye solutions since this increases the rate of intersystem crossing $T \rightarrow S_0$.

To illustrate the effect of triplet losses on the time development of the gain profile, consider again the spectra of rhodamine B in methanol in Fig. 1. Note that a triplet absorption overlaps the region of the fluorescence curve where laser action may be expected. Calculations (33) were performed using several k_{ST} values to vary the rate of build-up. The singlet lifetime of rhodamine B in methanol is 2.0 nsec; the triplet lifetime was assumed to be very much longer than τ_S or the pump pulse width. In Fig. 9, the time-dependent excited singlet- and triplet-state populations are shown together with the pumping pulse $P(t)$ used in solving Eqs. (9). Observe that when k_{ST} is large, the maximum of $N_S(t)$ actually occurs before the peak of $P(t)$. Therefore, for long pumping pulses, if a large fraction of the population is transferred to a metastable triplet system, the maximum gain is not necessarily coincident with that of the pumping intensity.

Gain curves $G(\omega)$ for $5 \times 10^{-5}\,M$ rhodamine B in methanol were computed using the spectra in Fig. 1, the flashlamp curve in Fig. 9, and a peak pumping intensity $(P\tau_S)_{\max} = 10^{-2}$. For $k_{ST}\tau_S \gtrsim 0.4$, the triplet–triplet losses are so dominant that, for practical purposes, the gain is always negative. Gain curves for smaller values of k_{ST} are shown in

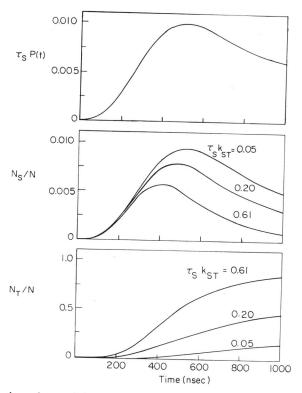

Fig. 9. Time dependence of the normalized pump function $\tau_S P(t)$ and calculated normalized excited singlet- and triplet-state populations N_S and N_T for different choices of intersystem crossing rate, k_{ST}. $\tau_S = 2.0$ nsec, $\tau_T = \infty$ (33).

Fig. 10. Since the triplet-state absorption makes a larger relative contribution to gain reduction in the low-frequency, low-gain portion of the fluorescence spectrum, changes owing to the triplet population appear initially in this region. These figures reveal the sensitivity of the shape of the over-all gain curve and the region of possible laser action to the rate of intersystem crossing.

Measurements of the time and frequency at threshold can be combined with calculated gain curves to obtain estimates of the rate of intersystem crossing. In experiments (33) using a flashlamp-pumped solution of rhodamine B, lasing was recorded at $\sim 16,250$ cm^{-1} approximately 200 nsec after initiation of pumping. The cavity loss coefficient was $\sim 3.5 \times 10^{-3}$ cm^{-1} and the estimated pumping rate corresponded to $(P\tau_S)_{max} \sim 10^{-2}$. Inspection of Fig. 10 indicates that

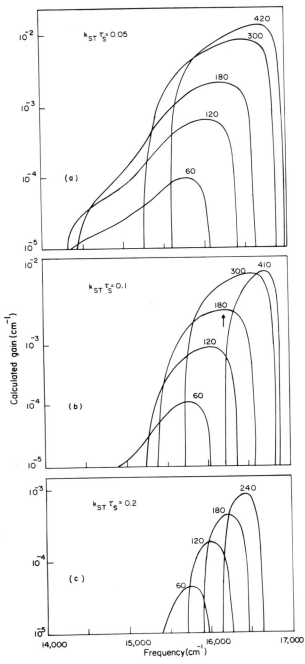

Fig. 10. Calculated gain vs. frequency curves for rhodamine B in methanol for different intersystem crossing rates k_{ST}. The parameter is the time t (in nsec) after application of the pumping pulses $\tau_S P(t)$ shown in Fig. 9 (*33*).

for $k_{ST}\tau_S = 0.1$ the peak in the gain curve has the required value for oscillation and the frequency and time at which threshold gain is reached are approximately correct (see arrow). For $\tau_S = 2$ nsec, this corresponds to a k_{ST} value of 5×10^7 sec^{-1}. The accuracy achievable from such measurements is dependent, of course, upon how well the triplet–triplet absorption cross section, the triplet lifetime, and other parameters are known. Variation in the magnitude and shape of the gain curves may arise also from changes in the value of the triplet lifetime.

The preceding analysis and rate equations are no longer adequate when the photon field in the cavity builds up to the point where the rates of stimulated and spontaneous emission become comparable and laser action begins. To account for stimulated emission processes, a term $qB\Delta n$ can be added to the rate equations, where $q(t)$ is the number of photons in the resonator, B is the stimulated emission coefficient, and Δn is the population difference of the upper and lower laser levels. The time dependence of q is governed by

$$\frac{dq}{dt} = qB\Delta n + \frac{A\Delta n}{p} - \frac{q}{t_c}, \tag{11}$$

where p is the number of modes coupled to the fluorescence line and A is the spontaneous emission probability. The parameter t_c is the decay time constant of the laser cavity(46); for a broadband resonator of length l and mirror reflectivity R, we have $t_c = nl/c(1 - R) = Q/\omega$. The resulting set of coupled nonlinear rate equations have been solved using analog(36) and digital(2,35) computer techniques. Thus far, however, the explicit frequency dependences of the spontaneous and stimulated emission coefficients and the photon field have not been included in the calculations.

In a pair of papers, Sorokin et al.(2,35) obtained computer solutions of equations similar to those in Eqs. (9) and (11). In the first paper(2), short-pulse laser pumping was considered and triplet effects were neglected; in the latter paper on flashlamp-pumped dye lasers, triplet effects were included. The problem was effectively reduced to consideration of a single pair of vibrational–rotational levels in S_0 and S_1; the terminal level was assumed to be a high-lying level which was essentially unpopulated, hence forming a four-level laser scheme. Frequency effects were not considered. Thus lasing at a single frequency ω', as would occur if the resonator Q were of the form $Q(\omega) = Q_0\delta(\omega - \omega')$, was treated. A Gaussian-shaped pump pulse with a half-width of 150 nsec was used. Other parameters were $t_c = 3 \times 10^{-9}$ sec, $\tau_r = 6 \times 10^{-9}$ sec, $\tau_{nr} = \infty$, $k_{ST} = 2.22 \times 10^7$ sec^{-1}, and $\tau_T = \infty$. An example of the resulting time-dependent behaviors of q, N_S, and N_T is shown in Fig. 11. The abrupt buildup of the photon field at threshold, the change in excited-state population, and the continued buildup of triplet population are all well illustrated by the curves. Solutions were obtained for a wide range of parameters. Careful study of the results in the two papers can be very instructive. They show very clearly the transient behavior of the population inversion and photon field throughout the pumping pulse and their dependence on such factors as the

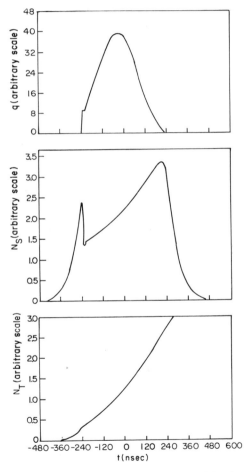

Fig. 11. Computer solutions, obtained by Sorokin et al. (2), of the time dependence of the photon field q, excited singlet population N_S, and triplet population N_T assuming a Gaussian excitation pulse having a width of 300 nsec.

pumping rate, cavity losses, quantum efficiencies, and intersystem crossing rate. For example, to generate fast-rising pulses, the use of a low-Q cavity is indicated.

Calculations of the time dependences can be used also to investigate the laser efficiency. For a pump applied at $t = 0$, Sorokin et al.(2) have defined an efficiency

$$E = \frac{t_c^{-1} \int_0^\infty q(t)\, dt}{N \int_0^\infty P(t)\, dt} \tag{12}$$

This quantity was evaluated as a function of quantum yield ϕ. As expected, the efficiency suffers because of population accumulation in the metastable triplet when k_{ST} is large and τ_T is long compared to the pump pulse. For constant integrated pumping energy, (1) the efficiency increases as the pump width is decreased, thereby reducing the effects of competing decay processes, and (2) for long pulses, the gain may be sufficient for lasing only over a small portion of the pump pulse near its peak intensity, hence lowering the over-all efficiency.

C. OPTICAL PUMPING

If the shape of the pumping pulse, cavity losses, and spectroscopic properties of a dye of interest are known, the rate equations can be solved for various pumping rates $P(t)$ and the gain characteristics calculated as described above. The pumping rate required to achieve oscillation, the time and frequency at threshold, and the minimum pump energy required for lasing can thus all be found simultaneously. In lieu of such calculations, however, there are some simple and useful criteria for determining the optical pumping requirements.

One such criterion is the critical inversion required for oscillation. For an optical resonator ~ 10 cm long having mirror reflectivity $R \sim 0.95$, this number for a typical dye is relatively small: $\sim 10^{14}$ molecules/cm³. To achieve this population in S_1, the pumping rate must be sufficient to compete with the large spontaneous decay rate from S_1. The short, energetic light pulses obtainable from Q-switched lasers are well suited for providing the required high pumping rate.

If the pump radiation is applied in a time that is short compared to the singlet lifetime τ_S, an absorbed pump energy of ~ 0.1 mJ/cm³ should be sufficient to achieve oscillation. However, to maintain the inversion for a dye having a singlet lifetime of 5×10^{-9} sec, a minimum pumping power of ~ 20 kW/cm³ is required. More generally, it is the integrated pumping intensity within some characteristic time that is important for achieving laser action. This time is dependent upon properties of both the dye and the pump pulse (33).

For dyes in which triplet-state absorption is important, the pumping rate must also be sufficient to achieve lasing threshold before this loss predominates. While this is usually not a significant consideration for Q-switched laser pumping, it can be for flashlamp pumping and has been discussed extensively in the literature. Sorokin et al.(2,35) have shown that, for dyes having long triplet lifetimes, the critical inversion must be reached before the triplet loss becomes equal to the singlet gain. The critical time, assuming $\tau_T = \infty$, is given by

$$T_\ell = \frac{2\sigma_S}{k_{ST}\sigma_T}. \tag{13}$$

The condition that a flashlamp attain the maximum intensity in a time less than T_l has been applied (2) to a number of different dyes to predict the rise-time requirements of the pump pulse. This requirement is relevant to efficient pumping or for a constant energy flashlamp. The limitations of this rise-time criterion and the dependence of T_l on the desired lasing frequency have been discussed (33). To achieve laser action, it is the integrated pumping intensity within some characteristic time that is important. In general, the gain increases to a maximum value at a time which is dependent upon properties of both the dye medium and the pump pulse. For oscillation, the pumping rate must be sufficient to reach threshold gain before this time.

If triplet–triplet losses occur in the spectral region of interest, an accumulation of molecules in the metastable triplet may prevent long-pulse or CW laser action. Consider the steady-state solution of Eq. (9b),

$$N_T(t) = k_{ST}\tau_T N_S(t). \tag{14}$$

From this result and measured values of $\sigma_S(\omega)$ and $\sigma_T(\omega)$, it should be possible to predict from Eqs. (2) and (8) whether CW oscillation is possible. From such considerations, Snavely (3) has derived an expression for the upper limit on the triplet-state lifetime consistent with CW oscillation. It is desirable, of course, to keep $N_T(t)$ to a minimum and thereby reduce the triplet-state losses. One approach considered by Snavely and Schäfer (51) is to quench the lowest triplet-state T_1 by oxygenating the solution. It should be noted that both k_{ST} and τ_T in Eq. (14) involve intersystem crossing which arises from an admixing of singlet and triplet states. Thus a simultaneous increase in k_{ST} which would negate the decrease of τ_T should be avoided.

IV. Pumping Techniques

Early work indicated that the pumping rate required for dye laser action could be achieved either by a Q-switched ruby laser or by a flashlamp capable of dissipating several tens of joules in ~ 1 μsec. In 1966, Sorokin and Lankard(*1*) obtained laser action from several dyes using a Q-switched ruby laser as a pumping source. Subsequently, a number of other Q-switched solid-state laser systems, some of which included frequency multipliers(*29,52–59*), and even the pulsed N_2 gas laser(*60*) also were used for dye laser excitation.

Shortly after obtaining laser action with a Q-switched ruby laser pump, Sorokin and Lankard(*61*) obtained dye laser action using a specially constructed short pulse (~ 0.5 μsec FWHP) flashlamp. Later, a number of workers found that laser action could be obtained using linear flashlamps in circuits which produced larger pulses and lower pump rates. Eventually pump pulses over 0.5 msec long were used to obtain laser action(*51*). In short, pumping techniques have advanced to the point where at least some organic dyes can be lased using relatively simple modifications of existing cavity–flashlamp combinations, and the possibility of obtaining CW laser action has been suggested(*51*).

A. LASER PUMPING

Two different pumping configurations have been used with laser excitation: longitudinal or end pumping, where the dye laser cavity is collinear with the pump laser cavity, and transverse pumping, where the axis of the dye cavity is perpendicular to the axis of the pump laser cavity(*1*). Figure 12 shows the two configurations schematically. In transverse pumping, lasing is achieved more easily by focusing the pump light into a line along the dye laser cavity axis with a cylindrical lens(*62*); Fig. 13 shows such an arrangement. Laser-pumped dye lasers often can have gain coefficients in excess of two inverse centimeters. Because of this high gain, an end-pumped dye laser sometimes will oscillate with only the uncoated end windows of the dye cuvette acting as cavity mirrors. If the wavelength of the laser pump light is well matched to the dye absorption band, the optical conversion can be quite efficient; values as high as 25% have been reported(*35*). Quasi-CW dye laser operation was achieved by Soffer and Evtuhov(*63*) who used as a pump a repetitively Q-switched ruby laser operating at a rate of 200 pulses/sec.

The pulsed nitrogen laser, which operates at 3371 Å and has a pulse duration of a few nsec, has recently been used to excite dye lasers(*60*).

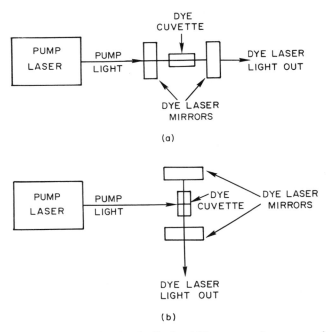

Fig. 12. Schematic diagram of (a) longitudinal and (b) transverse laser-pumped dye lasers.

The rectangular output beam of the nitrogen laser is well suited for focused transverse pumping.

Since laser-pumped dye lasers are quite simple to construct, provided a pump laser is already available, they provide an easy entry into dye laser work. More compact and practical systems require direct excitation using the flashlamp pump sources described below.

B. Flashlamp Pumping

Sorokin and Lankard(*61*) produced laser action in a number of visible emitting dyes using a specially constructed coaxial flashlamp. The development of such a flashlamp was aided by the body of knowledge accumulated by chemists working in the area of flash photolysis (*64*); the review by Porter(*65*) describes some of the earlier work. To achieve high-energy flashes of short duration, the use of low-inductance circuits having low capacity—high-voltage condensers with low internal inductance (<20 nH)—was necessary. Sorokin's lamp, shown in Fig. 14, consists of two coaxial quartz tubes passing through a hole in the center of a 0.5-μF, 20-kV, disk capacitor. Copper sleeves

Cylindrical lens   Pump laser

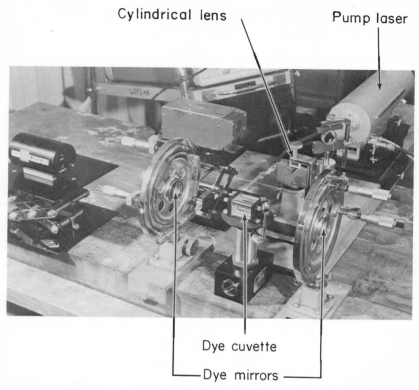

Dye cuvette

Dye mirrors

Fig. 13. Photograph of a transverse-pumped dye laser (courtesy of Y. Miyazoe and M. Maeda).

around the outer quartz tube connect the capacitor to annular stainless steel electrodes at the ends of the tubes; the discharge takes place in the 1-mm-thick annular space between the two quartz tubes. The dye solution is placed within the inner quartz tube. Sorokin's lamp was used for both single-shot and repetitive operation. In single-shot operation the capacitor was charged; then the gas discharge medium (air) was pumped out of the lamp until breakdown occurred. For continuous operation at rates up to one flash per second, a valve at the gas inlet port was used to adjust the flow rate and the pressure so that the lamp would fire at the desired voltage (*35*). This lamp delivered up to 100 J in 0.55 μsec (FWHP), corresponding to an average electrical input power of up to 90 MW, and produced laser action in rhodamine 6G, acridine red, and fluorescein with thresholds as low as 12 J (*61*).

Coaxial flashlamps of this construction are somewhat fragile.

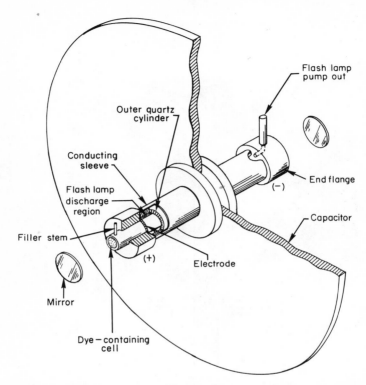

Fig. 14. Diagrammatic sketch of the low-inductance flashlamp design of Sorokin and Lankard used to produce stimulated emission from organic dyes in solution (2).

Sorokin et al. (35) found an outer wall thickness of at least 3 mm was needed to avoid breakage under their input conditions. Alumina outer tubes have been used to take advantage of their greater strength and high visible diffuse reflectivity. An alumina outer tube reflects outgoing light back into the dye; this increases the utilization of the pump light. Unfortunately, after numerous shots the alumina tube blackens and must be removed for cleaning.

Coaxial flashlamps have been made demountable by using O-ring seals throughout. The coaxial disk capacitor of the original design could be replaced by a Plexiglas insulator and other, more readily obtainable, low-inductance capacitors connected by short straps to the condenser mounting flanges. The commercial availability of low-inductance tubular capacitors with a hole running down the center makes it possible to mount the entire flashlamp within the capacitor, as illustrated in Fig. 15. The window at the high-voltage end of the

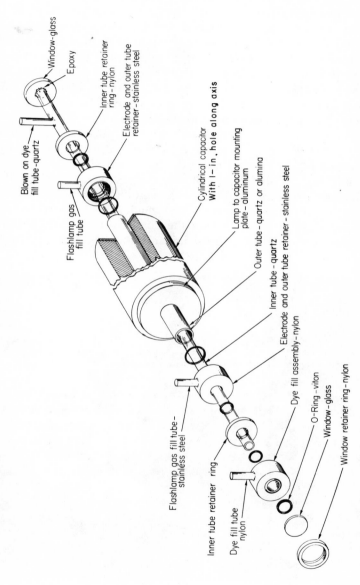

Window-glass

Epoxy

Inner tube retainer
ring - nylon

Blown on dye
fill tube - quartz

Electrode and outer tube
retainer - stainless steel

Cylindrical capacitor
With 1 - in. hole along axis

Flashlamp gas
fill tube

Lamp to capacitor mounting
plate - aluminum

Outer tube - quartz or alumina

Inner tube - quartz

Electrode and outer tube retainer - stainless steel

Flashlamp gas fill tube -
stainless steel

Dye fill assembly - nylon

Inner tube retainer ring

O - Ring - viton

Dye fill tube
nylon

Window - glass

Window retainer ring - nylon

Fig. 15. Exploded view of annular flashlamp using O-ring seals mounted in hole along the axis of a low-inductance tubular capacitor.

inner tube is sealed on with epoxy to minimize the possibility of short-circuiting the flashlamp through the dye solution.

Furumoto and Ceccon(66) have made an extensive study of the parameters affecting the operation of annular flashlamps. When these lamps were self-fired, filaments were found to be formed in the annular discharge region causing several undesirable effects. This filamentation increased the discharge inductance and caused the pulse duration to lengthen. Local high-pressure regions caused by the high-energy filaments led to lamp breakage. Finally, the filaments led to nonuniform pumping of the dye and nonreproducible output.

To eliminate these problems, Furumoto developed the lamp shown in Fig. 16. The lamp was energized by a cylindrical $1.3\text{-}\mu\text{F}$, 25-kV capacitor and fired by a low-inductance pressurized spark-gap switch as shown in Fig. 17; this switch allowed the gas fill and the firing voltage to be controlled independently. Filamentation was eliminated by a proper choice of the gas-fill pressure for the desired discharge voltage. The measured total inductance of the components was 27 nH. The construction of the lamp itself was similar to that described by Sorokin

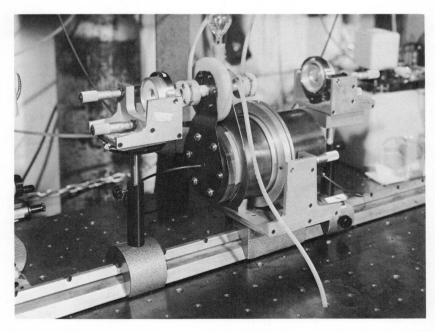

Fig. 16. Assembled view of Furumoto's spark-gap-triggered lamp including dye laser mirrors(66).

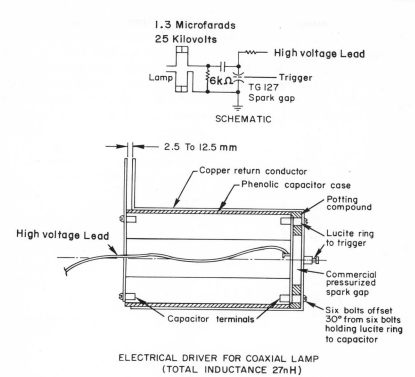

ELECTRICAL DRIVER FOR COAXIAL LAMP
(TOTAL INDUCTANCE 27nH)

Fig. 17. High voltage assembly and circuit diagram for Furumoto's spark-gap-triggered annular lamp(66).

et al.(35), except that copper-impregnated tungsten was used for the electrode material. Quartz was used for both the inner and outer walls of the lamp. Furumoto emphasized that care was taken to keep the discharge volume clean; the quartz tubes were dipped in nitric acid and washed with distilled water before assembly.

Studies were made of a number of lamps with arc-channel thickness varying between 0.25 and 1.0 mm, and with lengths of 6 and 12 cm. Some of the results are shown in Fig. 18; best performance was obtained using xenon as a fill gas. An annulus dimension of about 0.3 mm was optimum for 6-cm-long flashtubes. The input–output curve shown in Fig. 19 indicates an over-all efficiency of somewhat better than 0.4% for a $5 \times 10^{-5} M$ solution of rhodamine 6G pumped with this lamp.

Furumoto's work gave an indication of the ultimate efficiency to be expected from a dye laser operating in the visible. Measurements of the absolute intensity of the radiation from the flashlamp, shown in Fig. 20, indicated that in the visible portion of the spectrum the flash-lamp resembled a 21,000°K blackbody radiator. Since only 2% of the radiation of a 21,000°K blackbody lies within the pump band of rho-damine 6G, this represents an upper limit on the over-all efficiency of such a system. The small portion of the lamp output normally absorbed suggests that the efficiency of flashlamp-pumped dye lasers could be

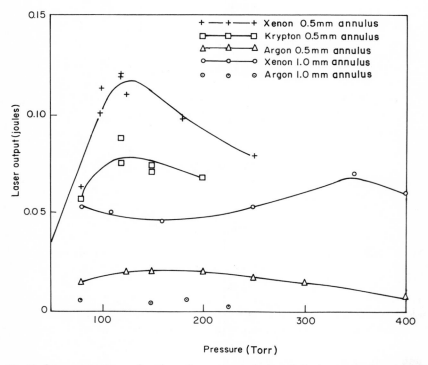

Fig. 18. Laser output as a function of pressure and size of discharge annulus at 50-J input for a 6-cm lamp (66).

improved by using mixtures of dyes which can transfer energy to each other to increase the width of the total absorption band and provide a better match with the spectral distribution of the pump source.

It is interesting to estimate the actual pump rate achievable with a flashlamp. The 6-cm-long flashlamp with 0.3-mm annulus thickness gave rise times to peak of as little as 0.12 μsec and a FWHP of about 0.22 μsec; if discharged uniformly, the lamp could survive inputs

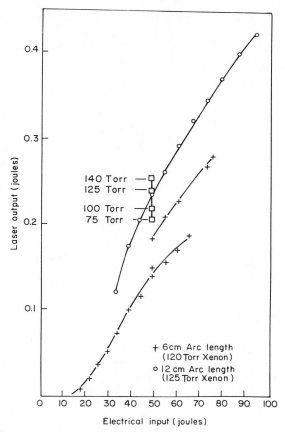

Fig. 19. Dye laser output energy as a function of input energy. Vertical bar (□—□) indicates the effect of a change in gas fill pressure at a fixed pump input of 48 J for the 12-cm lamp. Efficiency at optimum for this lamp is 0.51%(66). The dye was rhodamine 6G.

of up to 100 J, corresponding to an average electrical input of 225 MW.* Assuming a 50% conversion of electrical energy to optical energy and that 2% of the optical output lies in the pump band of rhodamine 6G, an average optical power of ~ 2 MW is available to pump the dye. Thus a very carefully designed flashlamp can produce pumping rates comparable to those of many laser pump sources.

Furumoto(67) subsequently developed an even faster flashlamp for

*This value, obtained by dividing the input energy by the base width, differs from Furumoto's input powers, obtained by multiplying measured current and voltage. It is used to compare the various flashlamps discussed in a uniform, if arbitrary, manner.

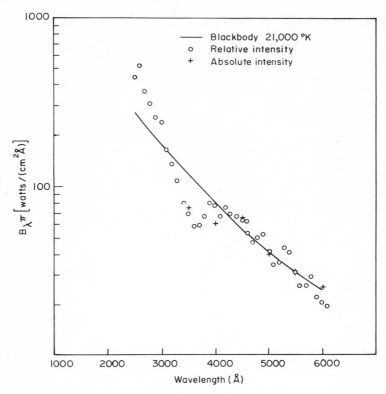

Fig. 20. Absolute and relative spectral brightness (B_λ) of the flashlamp in a 50-J discharge(66).

use in survey work which produced lasing in many blue-emitting dyes. This lamp dissipated 60 J in 70 nsec (FWHP) corresponding to an average electrical input of 430 MW; the output pulse shape is shown in Fig. 21.

Special linear flashlamps, either sealed commercially available units designed for short-pulse work, or specially built laboratory units, have been used to pump dye lasers. Bradley and O'Neill(68), using an air-filled linear flashlamp in the elliptical cavity shown in Fig. 22, obtained the pump and laser emission shown in Fig. 23.

A number of workers have used conventional linear flashlamps in low-inductance circuits as pumps. Schmidt and Schäfer(36), using a low-inductance capacitor to drive a commercial linear xenon flashlamp, obtained a 30-J pulse with duration of 1 μsec (FWHP), corresponding to an average lamp input of ~15 MW. By insertion of a

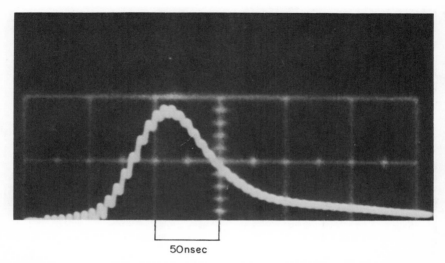

50nsec

Fig. 21. Light output from the dye laser survey flashlamp. (Courtesy of H. Furumoto.)

Fig. 22. Linear-flashlamp-pumped dye laser system (*68*).

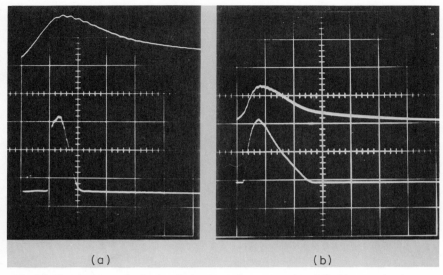

(a)   (b)

Fig. 23. Simultaneous rhodamine B dye laser (lower trace) and flashlamp pulse (upper trace)(*68*). (a) Laser operating near threshold; time scale: $0.5 \, \mu\text{sec/cm}$. (b) Stronger pumping; time scale: $1 \, \mu\text{sec/cm}$.

spark gap in series with the lamp, operation at voltages well above the lamp breakdown potential was obtained. We have modified a commercial elliptical cavity, designed for use with Nd:YAG lasers, by connecting a low-inductance tubular capacitor in series with a spark gap to the 3-in.-long xenon flash tube. The pump pulse had a duration of $3 \, \mu\text{sec}$ (FWHP). With this configuration, thresholds as low as 2–3 J have been observed for rhodamine dyes in liquid and acrylic plastic hosts. Repetition rates of up to 50 pps were achieved by flowing the dye and cooling the flashlamp. Linear flashlamps have been used in other pump configurations; as an example, Fig. 24 shows a linear flashlamp used in a spherical cavity.

Longer pumping pulses also have been used to pump dyes. Aristov and Maslyukov(*69*) obtained laser action from solutions of rhodamine B-extra, cooled to $190°\text{K}$ using pumping pulses $150 \, \mu\text{sec}$ long. Two linear xenon flashlamps, 12 cm long and 2 cm in diameter, in an elliptical cavity, were used for pumping. Snavely and Schäfer(*51*) used a 3-in. spiral flashlamp to obtain laser action lasting $140 \, \mu\text{sec}$ from an oxygenated solution of rhodamine 6G in methanol. The lamp produced a trapezoidal light pulse with a $550 \, \mu\text{sec}$ (FWHP) and an energy of 400 J (~ 3.5 MW average input). Figure 25 shows the apparatus used; the dye cuvette is within the spiral. The helical flashlamp increased

Fig. 24. Dye laser pumped by a commercially available linear flashlamp in a spherical pump cavity. (Courtesy of G. I. Farmer and B. G. Huth.)

the uniformity of pumping and reduced the effect of pump-induced thermal inhomogeneities.

Ferrar(*70*) has reported parameter studies of the improved linear flashlamps shown in Fig. 26. Flexible end pieces were used to absorb the very large mechanical shock usually borne by the lamp envelope during a high-energy short-duration discharge. The explosion limit of this linear lamp is much higher than that of conventional lamps for

Fig. 25. Dye laser pumped by a commercially available spiral flashlamp. (Courtesy of F. P. Schäfer.)

pulses of less than 10 μsec in duration. For example, a 3-mm bore by a 70-mm-long lamp made of 1-mm-wall quartz had a measured explosion limit of 110 J for a 1.7-μsec pulse. This compares with 30 J for a conventional linear lamp of the same dimensions. By using five 3-mm-i.d. by 200-mm-long lamps in parallel, Ferrar obtained over 1 J of laser light from ethanolic solutions of rhodamine 6G and 7-diethylamino-4-methylcoumarin. The total stored electrical energy was 700 J and the pump pulse duration was ~ 5 μsec (FWHP), corresponding to a total lamp average power of ~ 70 MW.

High repetition rates may be obtained with flashlamp pumping if

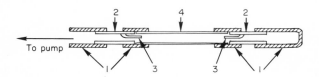

Fig. 26. Construction of an improved linear flashlamp with shock-absorbing ends. Components: (1) semiflexible plastic tubing, (2) metal sleeve, (3) wire electrode, (4) quartz tubing(70).

adequate cooling of the dye is provided. This is usually done by pumping the dye solution axially down the tube and through a heat exchanger. However, axial pumping is not always convenient for high repetition rates, since high flow velocities are required and since the dye cell configuration makes it difficult to remove the liquid between the inlet of the cell and the window. Boiteux and de Witte(71) constructed a flashlamp-pumped dye laser in which the dye flows transverse to the optical cavity. Their system was used repetitively to pump rhodamine 6G at 50 pps. The frequency was limited by the spark gap used to drive the tubes; the liquid itself was completely removed 200 times/sec.

V. Properties of Dye Lasers

A. Compilation of Reported Dye Lasers and Wavelengths

Table 4 summarizes the dyes in which laser action has been observed through October 1969 and gives a representative wavelength for each. Since the exact wavelength obtained in a particular experiment depends upon the losses present, where a number of wavelengths has been reported or a tuning range given, a center wavelength is used. It is clear that far fewer dyes have been lased with flashlamp pumping than with laser pumping, which is to be expected since the high pumping rate with laser pumping produces large inversions and minimizes the effect of triplet–triplet absorption.

B. Laser Spectral Properties

1. *Bandwidth*

An unusual aspect of organic dye lasers is that the oscillating bandwidth in a typical broadband cavity is of the order of 10–100 Å. The narrower bandwidths generally occur with the blue- and UV-emitting dyes. A number of authors have noted the presence of channeled spectra in the output which are attributed to subsidiary Fabry–Perot resonators formed by different surfaces within the cavity(19,49). Furumoto and Ceccon(73) have attributed the "hole" found in the output spectrum of rhodamine B(72) to the presence of a small (< 1%) variation in the reflectivity of the nominally 100% reflector(73).

TABLE 4

Reported Dye Lasers and Wavelengths

Organic compound	Reference	Typical solvent	Laser-pumped output wavelength Å	Flashlamp-pumped output wavelength Å
Para-terphenyl	59,67	Cyclohexane	3410	3410
2,5-diphenyl-1,3,4-oxadiazole	67	Dioxane		3480
Isopropyl-2-phenyl-5(4-biphenylyl)-1,3,4-oxadiazole	67	Ethanol		3698
2,5-Diphenylfuran	67	Dioxane		3710
p-Quaterphenyl	67	DMF		3740
2,5-Diphenyloxazole	67	Dioxane		3810
Diphenylbutadiene	59	Toluene	3830	
2-Biphenyl-5-styryl-1,3,4-oxadiazole	59	Toluene	3905	
αNND or 2,5-di(α-naphthyl)-1,3,4-oxadiazole	59	Toluene	3910	
αNPO or 2-phenyl-5-α-naphthyl-1,4-oxazole	29	Toluene	3995	4000
1,2-Di-4-biphenyl-ethylene	57	Toluene	4080	
BBO or 2,5-dibiphenylyloxazole	29,59	Benzene	4085	
p,p'-Diphenylstilbene	29,60,67	Benzene	4085	4090
1,4-Distyrylbenzene	57,59	Toluene	4150	
1,4-Di[2-(5-phenyloxazolyl)]-benzene	57	Toluene	4170	
Bis-MSB or p-bis(o-methylstyryl)-benzene	29	Ethanol	4190	
4,4'-Dichlo-1,4-distyryl-benzene	57	Toluene	4200	
Aquafluor[a]	99	Ethanol		4200
POPOP or p-bis[2-(5-phenyloxazolyl)]benzene	29,60,99	Ethanol, toluene	4210	4190
Liquiflor[a]	99	Ethanol		4220
Dimethyl POPOP or 1,4-bis-2-(4-methyl-5-phenyloxazolyl) benzene	29,57	Cyclohexane, ethanol	4230,4310	

Compound	References	Solvent		
Bis-MSB or p-bis-(0-methylstyryl)-benzene	29,99	Toluene, benzene	4250	4240
2'-Methoxy-1,4 distyryl benzene	38	Toluene	4250	
1,2,-Di-(α-naphthyl)-ethylene	57	Toluene	4260	
2,2' Dimethoxy-1,4-distyryl benzene	57	Toluene	4300	
1-Styryl-4[ω-vinyl-(n-biphenylyl)]benzene	57	Toluene	4320	
9,10 Diphenylanthracene	60,62	Cyclohexane	4325	
Acridone	62,99	Ethanol	4370	4350
2.5 Bis [5-tert-butylbenzoxazolyl(2)]thiophene	68	Benzene		4370
3-Ethylaminopyrene-5,8,10-trisulfonic acid	100	Water	4410	
9-Aminoacridine hydrochloride	102	Ethanol		4585
4-Methyl umbelliferone or 4-methyl-7-hydroxycoumarin	2,5,60,101	Water (pH ~ 9)	4500	~ 4600
7-Diethylamino-4-methyl coumarin	5,60,83	Ethanol	4500	4600
Esculin	5,101	Water (pH ~ 9)		4600
7-Hydroxycoumarin or umbelliferone	33,39	Water (pH ~ 9)		~ 4600
Benzyl β-methyl umbelliferone	102	Ethanol		4600
2,4,6-Triphenyl-pyrilium fluoroborate	100	Methanol	4850	
3-Aminophthalimide	4	Isoamyl alcohol	5000	
Trypaflavin	4	Ethanol	5050	
Acriflavin hydrochloride	67,103	Ethanol	5100	5175
Fluorescein	57,60,100	Aqueous alkaline	5180	
Na-fluorescein	2,61	Water, ethanol	5270	~ 5450
Eosin	61	Ethanol	5400	
2,7-Dichloro-fluorescein	2	Ethanol	5400	~5450
Lachs	4	Glycerol	5400	

TABLE 4 (Continued)

Organic compound	Reference	Typical solvent	Laser-pumped output wavelength Å	Flashlamp-pumped output wavelength Å
Rhodamine 6G	2,4,36,39,49, 51,57,60, 61,67,90, 103	Ethanol, polymethyl-methacrylate	5550	5950
Monobromofluorescein	4	Glycerin	5600	
Uranine	4,35	Ethanol	5600	
Pina (orthol)	4	Ethanol	5650	
Dibromofluorescein	4	Glycerin	5680	
Rhodamine B	2,4,33,36,39, 41,61,72, 90,100, 101,103	Ethanol	5770	6200
Rhodamine C	49,57	Glycerin	5700	6200
Pyronin B	36,57,102	Acetone, polymethyl-methacrylate	5760	yellow
Acridine red	2,61	Ethanol	5800	6150
Rhodamine G	102,103	Ethanol	5850	
Rhodamine S	102	Ethanol		5860
Pyronin G	4,49	Isoamyl alcohol	5900	6000
Lucegenin	4	Water and sulfuric acid	6000	

Dye		Solvent	
Eosin	4	Ethanol	6000
Saphranine-T	57	Ethanol	6100
Isoquinoline red	4	Water	6200
Rapid-filter gelt	4	Isoamyl alcohol	6200
Violet rot	4	Isoamyl alcohol	6200
Rhodamine 3 B	4,49	Isoamyl alcohol	6200, 6150
3,3′-Diethyloxadicarbocyanine iodide	100	Methanol	6580
3,3′-Diethyl-2,2′-thiadicarbocyanine iodide	54	Acetone	7110
3,3′-Diethyl-10-chloro-2,2′-(5,6,5′,6′-dibenzo)thiadicarbocyanine iodide	58	Acetone	7140
3,3′-Diethyl-2,2′-(5,5′-dimethyl)thiazolinotricarbocyanine iodide	58	Glycerin	7170
3,3′-Diethyloxytricarbocyanine iodide	56	Ethanol	7285
3,3′-Diethyl-5,5′-dimethoxy-6,6′-bis(methylmercapto)-10-methyl, thiadicarbocyanine bromide	56	Ethanol	7330
3,3′-Diethyl-2,2′-oxatricarbocyanine iodide	58	Acetone	7440
1,1′-Dimethyl-11-bromo-2,2′-quinodicarbocyanine iodide	58	Glycerin	7450
1,1′-Diethyl-4,4′-quinocarbocyanine iodide or cryptocyanine	49,53,58	Glycerin	7450
1,1′-Dimethyl-4,4′-quinocarbocyanine iodide	58	Glycerin	7490
1,1′-Diethyl-2,2′-dicarbocyanine iodide	58	Glycerin	7500
Echtblau B	4	Glycerin	7530
1,1′-Diethyl-4,4′-quinocarbocyanine bromide	58	Glycerin	7540
Dicyanine	4	Glycerin	7560
Napthalene green	4	Glycerin	7560
Rhoduline blue 6G	4	Glycerin	7580
Magnesium phthalocyanine	4	Quinoline	7590
Brilliant green or malachite green G	49	Isoamyl alcohol	7600
1,1′-Diethyl-γ-cyano-2,2′-dicarbocyanine-tetrafluoroborate or DTCDCT	75	Pyridine	7600

TABLE 4 (Continued)

Organic compound	Reference	Typical solvent	Laser-pumped output wavelength Å	Flashlamp-pumped output wavelength Å
Chloroaluminum phthalocyanine	35	Dimethyl sulfoxide	7615	
3-Ethyl-3'-methylthiathiazolinotricarbocyanine iodide	56	Ethanol	7700	
3,3'-Diethyl-10-chloro-2,2'-(4,5,4',5'-dibenzo)thiadicarbocyanine iodide	58	Acetone	7740	
3,3'-Diethyl-11-methoxythiatricarbocyanine iodide	56	Ethanol	7855	
1,3,3,1',3',3'-Hexamethylindotricarbocyanine iodide	56	Ethanol	7935	
Rapid-filter grün	4	Glycerin	7950	
1,1'-Diethyl-γ-acetoxy-2,2'-dicarbocyanine tetrafluoroborate	54	Methanol	7970	
1,1'-Diethyl-γ-nitro-4,4'-dicarbocyanine-tetrafluoroborate or DTNDCT	75	Pyridine	8000	
Blatt grün	4	Sulfuric acid	8000	
3,3'-Diethylthiatricarbocyanine iodide	56	Ethanol	8075	
Victoria blue	4	Glycerin	8090	
Victoria blue R	4	Glycerin	8140	
1,1'-Diethyl-11-bromo-2,2'-quinodicarbocyanine iodide	58	Glycerin	8150	
1,3,3,1',3',3'-Hexamethyl-2,2'-indotricarbocyanine iodide	58,67	Acetone	8190	8000
3,3'-Diethylthiatricarbocyanine iodide or DTTC	35	Dimethyl sulfoxide	8220	
Methylene green	4	Sulfuric acid	8230	
1,3,3,1',3',3'-Hexamethyl-4,5,4',5'-dibenzoindoctricarbocyanine perchlorate	56	Ethanol	8245	
3,3'-Diethyl-2,2'-selenatricarbocyanine iodide	58	Acetone	8260	
Methylene blue	4	Sulfuric acid	8290	
1,1'-Diethyl-11-bromo-4,4'-quinodicarbocyanine iodide	58	Methanol	8300	

3,3'-Diethylthiatricarbocyanine bromide	*54*	Methanol	8350
Methylene blue	*4*	Sulfuric acid	8350
3,3'-Diethyl-6,7,6',7'-dibenzothiatricarbocyanine	*56*	Ethanol	8385
Toluidine blue	*4*	Sulfuric acid	8480
Thionin	*4*	Sulfuric acid	8500
3,3'-Diethyl-2,2'-(5,6,5',6'-tetramethoxy) thiatricarbocyanine	*58*	Acetone	8530
3,3'-Diethyl-6,7,6',7' dibenzo-11-methylthiatricarbocyanine iodide	*56*	Ethanol	8560
3,3'-Diethyl-2,2'-(e.r,4',5'-dibenzo)thiatricarbocyanine	*58*	Acetone	8600
Phthalocyanine	*4*	Sulfuric acid	8630
1,1'-Diethyl-2,2'-quinotricarbocyanine iodide	*56,58*	Acetone	8980
1,1'-Diethyl-4,4'-quinotricarbocyanine iodide	*58*	Acetone	10,000
Pentacarbocyanine analogs	*104*	Nitrobenzene	10,950–11.750

2. Tuning

One of the most important and useful properties of organic dye lasers is that their output frequency can be tuned over a range of the order of 400 Å and that the normally broad output spectrum can be narrowed by a number of techniques without appreciable loss of output energy.

a. *Tuning and Narrowing with a Frequency Selector.* One set of techniques involves the insertion of a frequency-selective element into the cavity. The cavity losses $L(\omega)$ thus can be adjusted so that laser action occurs only at the desired frequency. One simple way to accomplish this is to insert a glass filter into the cavity(41). This produces frequency selection without narrowing the output substantially. Yamanaka(74) obtained both frequency tuning and line narrowing using an intracavity interference filter. Murakawa et al.(50) tuned a laser-pumped dye laser over a range of 400 Å; the spectral width of the output was reduced to about 20 Å by use of a Littrow-mounted prism as the frequency-selector element.

Frequency narrowing has been reported by Soffer and McFarland (20) who used a diffraction grating in a Littrow arrangement as one reflector in the cavity of a laser-pumped rhodamine 6G laser. The tuning obtained is illustrated in Fig. 27. The spectral width of the laser output was reduced from 60 to 0.6 Å while the output energy was about 60% of the value obtained using broadband dielectric mirrors. A tunable range of about 400 Å was observed. Sorokin et al.(2) used a diffraction grating to tune a flashlamp-pumped laser and obtained similar narrowing.

Further narrowing was obtained by Bradley et al.(75) using the laser pumping configuration shown in Fig. 28. By introducing a Fabry–Perot etalon into a cavity formed by an Echelle grating and a dielectric mirror, they obtained output in a single longitudinal mode with a linewidth of less than 0.01 Å (see Fig. 29). The output power was reduced by a factor of about 2 when one mirror was replaced with the Echelle grating, and by a factor of 4 when the Fabry–Perot transmission filter was introduced. Thus spectral narrowing by a factor of $\sim 10,000$ was accompanied by an output power reduction of a factor of only 8.

Sorokin et al.(76) locked the frequency of a flashlamp-pumped rhodamine 6G laser to the D lines of sodium vapor by means of the large Faraday rotation that appears in the vicinity of an absorption line.

b. *Loss Tuning.* A number of other parameters such as dye concentration and temperature, number of unexcited molecules in the cavity,

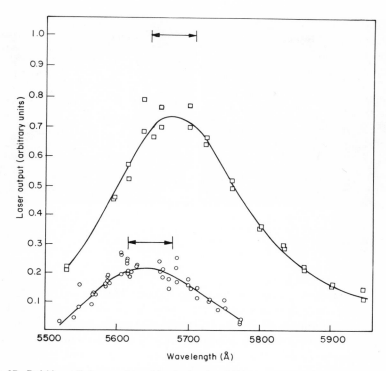

Fig. 27. Relative efficiency of tunable rhodamine 6G laser vs. wavelength(20). Grating in cavity dielectric reflectors. Optical density: for upper curve, 1.3 (347 mμ); for lower curve, 0.35 (347 mμ). Arrows indicate spectral bandwidth for each case.

cavity Q, and the time of opening of an internal electrooptic Q switch has been used to vary the output frequency of dye lasers. Before discussing these effects it should be pointed out that they are all related to the same physical phenomenon. Because the long-wavelength tail of a singlet–singlet absorption band usually overlaps the fluorescence

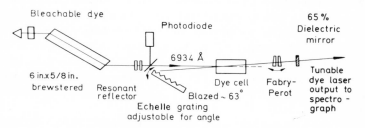

Fig. 28. "Single mode" laser-pumped dye laser(75).

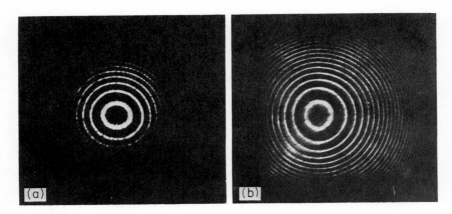

Fig. 29. Interferograms of (a) He–Ne laser (6328 Å) and (b) dye laser with 3-mm Fabry–Perot filter in cavity(75). Recording interferometer free spectral range 4.5 GHz. Dye laser linewidth < 500 MHz.

band, it introduces a loss in the lasing region. This loss is time dependent, since it is determined by the population of the ground-state singlet, which varies throughout the pumping pulse. Schäfer et al.(54) observed that the emission wavelengths of a dye laser increased with increasing cavity Q. This behavior is predicted by the gain-vs.-frequency curves obtained in Section III and shown in Fig. 4.

Because lasing occurs later in the pump pulse in a low-Q cavity than in a high-Q cavity, the loss due to the long-wavelength tail of the singlet–singlet absorption is less than the gain extends to shorter wavelengths. Figure 30 shows a plot of the wavelength for maximum gain vs. time. This was derived(41) from a family of gain-vs.-wavelength curves with time as a parameter calculated for flashlamp-pumped rhodamine B assuming $k_{ST} = 0$. This curve suggested that by inserting an electrooptic Q switch into the dye laser cavity, as shown in Fig. 31, and opening it at successively later times after the initiation of the pumping pulse, one could vary the oscillating wavelength. Figure 32 shows the results of this experiment(33). As the delay between the initiation of the pumping pulse and the opening of the electrooptic Q switch is increased, the oscillating wavelength first decreases and then increases. The effect is caused by the time-varying singlet–singlet absorption losses which decrease up to the peak of the pumping pulse and then increase again.

Unless the Stokes shift of the dye is so great that the absorption and

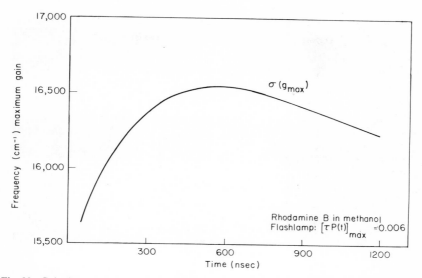

Fig. 30. Calculated frequency at which the gain is a maximum for $5 \times 10^{-5}\,M$ rhodamine B as a function of time for flashlamp pumping(*41*).

fluorescent bands do not overlap appreciably, any unexcited dye in the cavity will introduce a loss at the short-wavelength side of the fluorescence curve. The number of unexcited dye molecules in the cavity may be changed by variation of the number of dye molecules not exposed to pump light or by variation of the number of dye molecules in the active region. As the concentration of dye molecules in the active

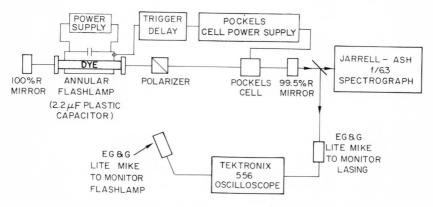

Fig. 31. Experimental arrangement used for *Q*-switched laser studies(*33*).

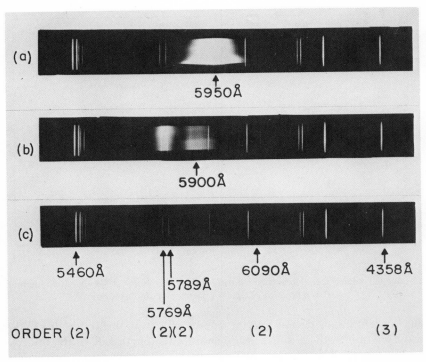

Fig. 32. Output spectrum of a *Q*-switched rhodamine 6G in ethanol laser using the experimental arrangement in Fig. 31 *(33)*. (a) No voltage on Pockels cell; (b) Pockels cell opened at 1.04 μsec after initiation of pumping; (c) Pockels cell opened at 1.13 μsec. Since the Pockels cell does not always act as a perfect shutter, prepulsing can occur; the emission at $\approx$ 5780 Å in (b) is owing to oscillation in the lower *Q* cavity formed when the *Q* switch is still mainly closed.

region is increased while the pump intensity is kept constant, the number of unexcited (and hence absorbing) dye molecules increases. Farmer et al.*(42)* built a laser-pumped DTTC laser whose output wavelength could be tuned over 600 Å by varying the optical path length from 1 to 20 cm. Their apparatus is shown in Fig. 33. A number of workers have found also that the wavelength of a dye laser increases with increasing dye concentration*(35,49,54)*. Figure 34 illustrates an example of such concentration tuning.

Schappert et al.*(48)* have described the temperature tuning of a dye laser. They were able to decrease the wavelength of a 6×10^{-5} *M* solution of DTTC in ethanol about 200 Å by lowering the solution temperature from 78° to −117°C. As discussed in Section III,A, the effect can be understood in terms of the reduced overlap of the absorption

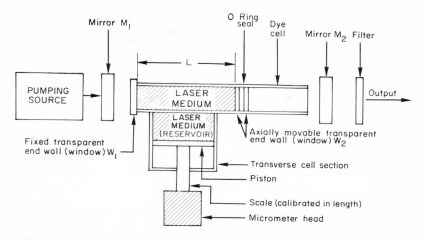

Fig. 33. Tunable dye laser. The output wavelength of this device is continuously adjustable over a 700-Å range by changing of the optical path length, L, in the active medium. This length is adjusted by positioning of the piston with the micrometer screw. The filter is of use in some cases to eliminate the residual pumping beam(*42*).

and fluorescence bands with decreasing temperature. Observe also, as shown in Fig. 5, that the stimulated emission spectrum is narrower at lower temperatures.

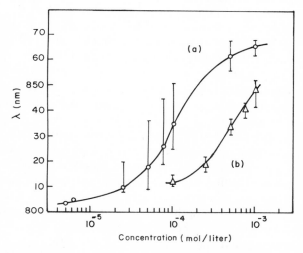

Fig. 34. Lasing wavelength of a DTTC-bromide laser-pumped dye laser(*54*); (a) high Q, (b) low Q.

If the Stokes shift is so large that there is no significant overlap of the absorption and the fluorescent bands, no loss-tuning effects should be observed. This is indeed shown by 7-hydroxycoumarin, which has a Stokes shift of $\sim 5000 \; cm^{-1}$ and exhibits neither concentration nor cavity-Q tuning(33).

3. FREQUENCY SWEEPING

Bass and Steinfeld(77) observed the time development of the dye laser intensity at several wavelengths and noted that the most intense output wavelength of DTTC-iodide and cryptocyanine laser-pumped dye lasers swept from short to long wavelengths during lasing. To explain this phenomenon they proposed that if relaxation of these molecules in the ground electronic state was slow compared to the 30-nsec duration of the lasing, then molecules which had lased would absorb at the laser wavelength. The laser gain would then become greater at a longer wavelength and the center wavelength would shift in time. Further measurements on the same system by Farmer et al. (105), who used a streak camera to obtain time-resolved spectra, indicated that at relatively high dye concentrations ($\sim 5 \times 10^{-4} \, M$) "hooking" occurred. That is, the output first swept from short to long wavelengths, as observed by Bass and Steinfeld, but then swept back to shorter wavelengths. Similar results were observed by Gibbs and Kellock(78) in the cryptocyanine and chloroaluminum phythalocyanine dye systems using laser pumping.

The hooking phenomenon can be qualitatively understood by extending the arguments of Bass and Steinfeld. As the lasing sweeps to long wavelengths, the gain eventually decreases because of both the decrease of $f(\omega)$ and the transient increase in absorption mentioned above. The dye laser then has greater gain at a shorter wavelength where $f(\omega)$ is larger and so the time-varying spectrum hooks back to shorter wavelengths.

Furumoto and Ceccon(73) made extensive studies of frequency sweeping in flashlamp-pumped dye lasers. They observed the output of solutions of rhodamine 6G, rhodamine B, acridine red, fluorescein, and 4-methyl umbelliferone with a streaking camera and found a variety of behaviors. For example , the rhodamines showed sweeping from long to short wavelengths, whereas 4-methyl umbelliferone first swept to short wavelengths and then hooked back.

Other physical phenomena may contribute to frequency sweeping. If the triplet–triplet absorption overlaps the lasing region, growth of the triplet population with time can cause frequency sweeping from long to short wavelengths. Such behavior has been observed by Ferrar(*28*) in a flashlamp-pumped sodium fluorescein dye laser. Other evidence, such as the measured triplet absorption spectrum and the termination of lasing while at pump levels well above threshold, support the conclusion that this wavelength shift is due to the growth of triplet absorption.

It should be noted that thus far the published analyses of dye laser gain properties have not been extended to treat the question of frequency sweeping. This requires consideration of both the temporal and frequency characteristics of the photon field $q(\omega, t)$ after the onset of stimulated emission. The computer solutions of the rate equations by Sorokin et al. referred to in Section III show that, depending upon the pumping rates and resonator losses, the population inversion may either decrease continuously after laser action starts or first exhibit a decrease and then increase with the pumping pulse. The effect of the time-dependent and frequency-dependent contributions of singlet absorption and emission to the gain will determine the direction of sweeping. Since a possible cause of frequency sweeping is slow thermalization of the vibrational level populations in S_0, the rate equations require further modification to include this effect in the analysis. A more complete treatment of sweeping must include also possible effects of time-varying triplet losses.

In the case of pumping by long-pulse flashlamps, heating of the dye solution can produce time-dependent refractive index changes and thermal gradients which can alter the cavity Q and the lasing wavelength(*51*).

C. TEMPORAL BEHAVIOR

In the case of laser pumping, threshold is usually achieved early in the pumping pulse because of the high pump rate, and the pump pulse is usually over before an appreciable triplet concentration has accumulated. Therefore, the output of a laser-pumped dye laser will generally follow the pumping pulse rather closely and have a time duration approximately equal to that of Q-switched laser pulse(*35*). In the case of flashlamp excitation the pump rate is usually lower than for laser pumping and hence the time to threshold becomes longer. As the flashlamp pumping continues, triplet-state absorption may increase or thermal inhomogeneities may develop which quench the lasing. Thus, as shown in Fig. 23, laser action with flashlamp pumping is generally somewhat shorter in duration than the pumping pulse. Sorokin et al.(*2*) demonstrated the effect of the triplet system absorption in quenching laser action by using a flashlamp-pumped rhodamine 6G laser to optically pump a DTTC laser. The fact that the DTTC laser output did not follow the pumping pulse was attributed to triplet losses. Snavely and Schäfer(51) showed that the effect of the triplet may be reduced by oxygenating the solutions, thus decreasing the triplet lifetime. They also suggested that thermally induced optical inhomogeneities may be responsible for some of the laser quenching effects that have been observed. They noted, further, that the transmission of a gas laser beam through a dye cell, excited by a 3-in. linear flashlamp in a close-coupled cavity, was destroyed by thermal effects in approximately 30 μsec(51).

Dye lasers have been mode locked by many of the same techniques with other lasers. One of the simplest techniques is to use a mode-locked pump laser source, as several workers have shown(75,79,80). Glenn et al.(79) used this approach to mode lock a mixture of rhodamine B and rhodamine 6G; they found it necessary to have the optical length of the exciting laser cavity an integer multiple of that of the dye laser cavity. This requirement can be avoided by operating the dye laser in a superradiant traveling wave mode. Mack(81) used the wedged dye cell shown in Fig. 35a to avoid cavity effects; since a high concentration was needed to obtain superradiant operation the optical density in the 2-cm test cell was 6.0. The temporal and spatial output from this laser is shown in Fig. 35b–d. Schmidt and Schäfer(82) and Bradley and O'Neill(68) have reported using an organic dye as a saturable absorber in the cavity to mode lock the output from a flashlamp-pumped rhodamine 6G laser. An acoustic modulator within the cavity also has been used to mode lock a flashlamp-pumped coumarin dye laser(83). A two-photon fluorescence technique was used by Soffer and Linn(84) to detect the presence of picosecond pulses in the output of a tunable mode-locked rhodamine 6G laser. A frequency-doubled mode-locked neodymium glass laser was used as the pump and the dye laser output was tuned with an intracavity grating. The mode-locked pulses from the dye were approximately 10^{-11} sec wide, even when the lasing bandwidth was narrowed from 150 to 6 cm^{-1}.

D. Spatial Properties

1. Polarization

Sorokin et al.(35) performed an analysis which indicated that for transverse pumping the relationship between the polarization of the

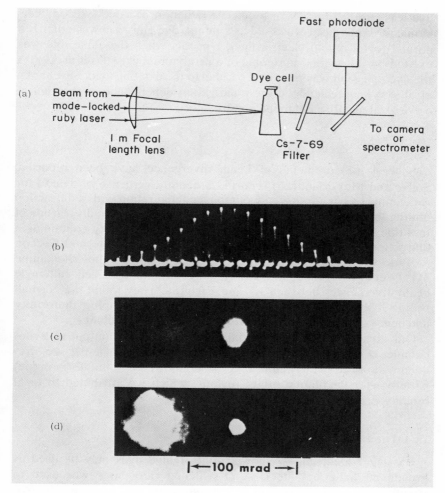

Fig. 35. Traveling-wave dye laser(*81*). (a) Experimental arrangement; (b) Temporal output; (c) Spatial output—far field—3-mm-thick entrance window; (d) Spatial output—far field—1–5-mm-thick entrance window.

dye laser beam and that of the pump laser depended upon the symmetry of the absorbing and emitting transition. For DTTC-iodide in a transverse pumping configuration, the polarization of the laser output was parallel to that of the ruby pump laser, as the theory had predicted. McFarland(*55*) found that for transversely pumped rhodamine 6G in ethanol, the polarization of the output light depended upon whether the dye was pumped by UV or by visible radiation. When the second harmonic of a ruby laser was used as a pump, dye emission was polarized normal to the pump electric field, whereas when the same dye was excited by the second harmonic of a neodymium glass laser at 5300 Å the dye emission was polarized parallel to the exciting field. Sevchenko et al.(85) have calculated the polarization dependence of the gain of a dye excited with a polarized pump.

2. Beam Divergence

Several measurements of beam divergence have been reported. Soffer and McFarland(20) reported a beam divergence of 5 mrad for rhodamine 6G in a transverse pumping configuration with a diffraction grating in the cavity. Bradley et al.(75) obtained a beam divergence of less than 0.5 mrad using a longitudinally pumped cavity containing a diffraction grating and a Fabry–Perot etalon. Furumoto and Ceccon (66) measured the beam divergence of a flashlamp-pumped rhodamine 6G laser with broadband cavity mirrors, and obtained full-angle beam divergences between 1.7 and 2.6 mrad. They found that a small beam divergence was strongly dependent on pumping uniformity and hence on eliminating filaments in the flashlamp discharge.

Our observations of the spatial distribution of the output of a rhodamine 6G laser longitudinally pumped by a 1-pulse-per-sec, frequency-doubled neodymium YAG laser showed a nonuniform output pattern with fluctuating bright regions, which we attributed to local bright spots on the source laser.

E. OTHER PROPERTIES

Bass and Deutsch(19) showed that organic dyes may be used as broadband light amplifiers. A Q-switched ruby laser was used to stimulate Raman emission in an organic liquid and to excite the dye amplifier, simultaneously. The experimental setup is shown in Fig. 36. For a solution of cryptocyanine in glycerol, gain was observed at 330 Å from the center of the dye laser oscillating range (Fig. 37), demonstrating the broadband nature of the light amplifier. Derkacheva and Sokolovskaya(86) investigated the amplifying properties of DTTC using the Raman emission from calcite.

Huth et al.(87) have generated UV light from a rhodamine 6G flashlamp-pumped dye laser by frequency doubling with a KDP crystal. The second harmonic had a peak power of about 40 W which was tunable from 2900 to 3000 Å. Derkacheva et al.(88) observed that

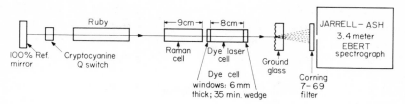

Fig. 36. Experimental apparatus used to observe the properties of dye solutions acting as light amplifiers(*19*).

when a polymethine dye was pumped with a Nd–glass laser, a narrow line, whose frequency depended on the angle between the pump laser and the dye cell, appeared for orientations beyond a certain critical angle. This line was present in addition to the normal broad dye emission, and its wavelength varied between 1.07 and 1.09 μ. The effect was interpreted as due to a four-photon resonant parametric interaction.

Mixtures of dyes may be used to obtain more efficient flashlamp-pumped dye lasers. If a dye with a shorter wavelength absorption band can transfer energy efficiently to the lasing dye, then the effective absorption band of the combined system will be broadened and a larger portion of the spectral output of the flashlamp will be utilized. Peterson and Snavely(*104*) studied the increase in laser intensity obtained by using combinations of dyes; their results are summarized in Table 5. Snavely(*3*) has pointed out that it is important to choose dyes which are chemically compatible. In case of dyes whose fluorescent intensity is pH-dependent, the combinations are limited to dyes fluorescing at the same value of pH. Glenn et al.(*79*) have varied the output of a laser over a range of about 300 Å by changing the composition of a mixture of organic dyes. Mixtures of dyes have been used also to extend the spectral range over which lasing occurs(*58*).

Most dye powders obtained commercially are adequate for good, reproducible laser action. This does not mean that no improvement in operation can be obtained by further purification. Farmer and Huth (*89*) report that the output of their rhodamine 6G solutions increased by a factor of nearly three when the dye was purified by recrystallization. Figure 38 shows an apparatus for purifying a dye by filtration and recrystallization or by vapor crystallization. The results of Farmer and Huth indicate the need for some caution in comparing laser performance obtained in different laboratories because of the possibility of varying dye purity.

Care must be taken not to contaminate the dye solution with unwanted impurities. The O rings, pumps, tubing, and any other material

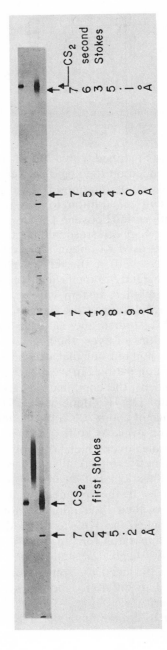

Fig. 37. Spectrographic data on the amplification of the first and second Stokes-shifted lines from CS$_2$ by a 2×10^{-6} M solution of cryptocyanine in glycerol. The laser was fired 8 times to obtain good exposures for each line. The reference lines are neon lines except for the one at 7635.1 Å, which is argon. The first line shows the Stokes emission, the second line the dye laser spectrum, and the third line the amplified Stokes emission *(19)*.

TABLE 5

Dye Combinations with Experimental Concentrations (*104*)

Lasing dye concentration ($\times 10^{-5} M$)		Pumping dye concentration ($\times 10^{-5} M$)		Increase in laser intensity (I_{comb}/I_{pure})
Rhodamine B	10	Rhodamine 6G	10	4
Acridine red	10	Rhodamine 6G	10	9
Dichlorofluorescein	3	Fluorescein	2	1.4
Fluorescein	5	7 Hydroxycoumarin	40	5
Dichlorofluorescein	4	⎰ Fluorescein	3	9
		⎱ 7 Hydroxycoumarin	40	

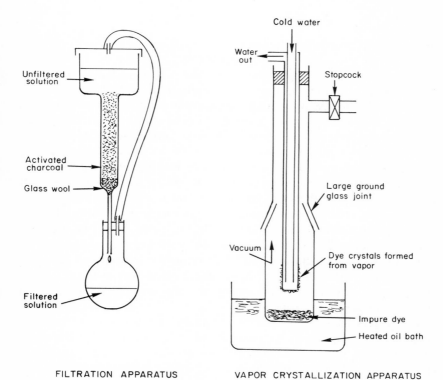

FILTRATION APPARATUS   VAPOR CRYSTALLIZATION APPARATUS

(NOT TO SCALE)

Fig. 38. Apparatus for the purification of dyes: (a) filtration and recrystallization; (b) vapor crystallization.

which comes in contact with the solution should be kept as clean as possible. It has been found that some types of plastic tubing, for example Tygon, react with the dye molecules and make lasing impossible. Thus the solution-handling apparatus must be made of nonreactive materials; nylon, teflon, polypropylene, and polyethylene are acceptable pump and tubing materials for dye solutions.

Dyes can be embedded in solid as well as liquid hosts and made to lase. A number of workers have found that commercial fluorescent plastics can be lased with either laser or flashlamp pumping(20,90). The rhodamine dyes have been mixed in polymethylmethacrylate and epoxy hosts to make laser rods. However, the optical quality of plastic or epoxy dye lasers is not yet high enough to give performance comparable to liquid solutions.

VI. Applications of Dye Lasers

Dye lasers are distinguished from other coherent light sources by the tunability of their output wavelength. This property makes dye lasers ideal sources of light for applications requiring specific optical wavelengths. For example, Yamaguchi et al.(91) have demonstrated resonant scattering of dye laser light from atomic sodium. Since the resonant light-scattering cross sections of an atomic or molecular species is in general about ten orders of magnitude as large as that for nonresonant processes such as Rayleigh scattering, resonant scattering is one of the most sensitive means of detecting that species; Fig. 39 shows the results of using a dye laser to measure this extremely wavelength-selective effect. Yamaguchi et al. pointed out that resonant scattering of a tunable dye laser light can be used to detect atmospheric constituents and suggested this technique for measuring pollutant levels in air-pollution control applications. Bowman et al.(92) have, in fact, used a dye laser(93) in an optical radar system to detect sodium in the atmosphere.

Dye lasers have been used also to study self-induced transparency in potassium vapor(94), to measure the wavelength dependence of laser-induced gas breakdown(95), to provide a pump source for excited-state spectroscopy(96), and to produce laser action in Cs, Rb, and Sr vapor(97,98). Further uses of dye lasers may include illumination for displays, underwater illumination, and optical frequency amplification.

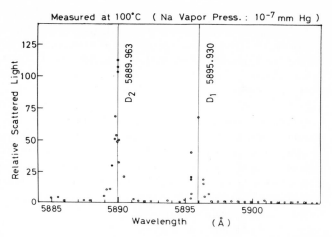

Fig. 39. Resonant scattering of tuned dye laser light from atomic sodium(*91*).

VII. Recent Developments

Since the body of this review was written two significant developments have been reported: (1) CW operation of dye lasers and (2) lasing from excited state complexes which results in extended tuning ranges.

Peterson et al.(*106*) obtained CW operation from a solution of rhodamine 6G in water using the 5145-Å line from an argon laser for pumping. A flowing dye solution was used in a cell designed to minimize heating-induced optical inhomogeneities. With longitudinal pumping the observed lasing threshold was 200 mW and an output of 30 mW at 5970 Å was obtained for a 960-mW pump power. Similar results were subsequently obtained by Banse et al.(*107*) also using rhodamine 6G and an argon laser. They overcame heating effects by using only a small portion of the total volume of a rotating dye cell.

The excited state complex (exciplex) laser(*108*) is based on the observation that 4-methylumbelliferone can exist in basic, neutral, and acidic forms, each one having a different emission band and each one capable of lasing throughout a different range of wavelengths. It was proposed that in the acidic form dye molecules in an excited state accept protons to form an excited state complex or exciplex. By careful adjustment of the solution's pH, it is possible to find a hydrogen ion concentration that allows both neutral and acidic forms to exist simultaneously and to lase. With transverse pumping by a pulsed N_2 laser, a tuning range of 1760 Å was obtained (from 3910 to 5670 Å). Further experiments(109) have supported the exciplex hypothesis and shown similar effects with 7-hydroxycoumarin.

LIST OF SYMBOLS

A	probability for spontaneous emission
a	absorption coefficient per unit length
B	Einstein induced transition coefficient
c	velocity of light in vacuum
E	laser efficiency
$e(\omega)$	stimulated emission coefficient per unit length
f	normalized fluorescence function
FWHP	full width at half maximum power
G	gain per unit length
k	Boltzmann constant
k_{ST}	rate of intersystem crossing
L	resonator loss coefficient
l	length of optical resonator
N	total population density
N_i	population density of level i
$N_S{}^c$	critical population density required for threshold
n	index of refraction
Δn	population inversion
P	optical pumping rate
p	number of resonator modes coupled to fluorescence line
Q	quality factor
q	number of optical quanta in resonator
R	mirror reflectivity
S_i	singlet electronic state i
t	time
t_c	optical cavity decay time
T	absolute temperature
T_i	triplet electronic state i
T_l	characteristic time when triplet loss equals singlet gain
δ	delta function
ϵ	molar extinction coefficient
ϵ_S, ϵ_T	molar extinction coefficient for singlet and triplet states, respectively
ϕ	quantum yield

σ^a	absorption cross section per molecule
σ^e	stimulated emission cross section per molecule
τ_i	lifetime of state i
τ_r	radiative lifetime
τ_{nr}	nonradiative lifetime
ω	angular frequency of photon

ACKNOWLEDGMENTS

We wish to thank D. Bua, R. Lotti, P. Meyer, T. Varitimos, and D. Woodward for their contributions to our experimental efforts. We are indebted to W. Doherty for her aid in the mathematical calculations and to J. Barker and R. Sewall for editorial assistance. In addition, we thank M. Bennett and J. Grenda for their help both in scanning the current dye literature and in locating obscure references. Our gratitude is extended to M. O'Leary for her careful preparation of the illustrations, and to E. Wareing and A. Immonen for their intrepid efforts in typing the manuscript. Finally, very useful discussions with H. Furomoto, S. Protopapa, J. Steinfeld, B. Snavely, and O. G. Peterson are acknowledged.

Many dye laser researchers generously provided us with descriptions of their latest work, and thus allowed this review to be as up-to-date as possible. So that questions concerning their unpublished work or personal communications cited in the references may be posed directly to those researchers, their affiliations are given below.

D. J. Bradley, Physics Department, Queen's University, Belfast, United Kingdom

L. D. Derkacheva, P. N. Lebedev Institute, Moscow, U.S.S.R.

G. I. Farmer and Dr. B. G. Huth, IBM Federal Systems Division, Federal Systems Center, Gaithersburg, Maryland

C. M. Ferrar and Dr. M. Mack, United Aircraft Research Laboratories, East Hartford, Connecticut

H. Furumoto, DOT-SRC, Cambridge, Massachusetts

W. E. Gibbs, Australian Defense Scientific Service, Department of Supply, Defense Standards Laboratory, Melbourne, Australia

D. Gregg, Lawrence Radiation Laboratory, Livermore, California

J. A. Meyer and R. D. Sharma, Avco Everett, Research Laboratory, Everett, Massachusetts

Yasushi Miyazoe and Mitsuo Maeda, Department of Electrical Engineering, Kyushu University, Hakozaki, Fukuoka, Japan

Shin-ichi Murakawa, Gentaro Yamaguchi, and Chiyoe Yamanaka, Department of Electrical Engineering, Faculty of Engineering, Osaka University, Miyakojima, Osaka, Japan

F. P. Schäfer, Physikalisch-Chemisches Institut, Universität Marburg, Marburg, Germany

B. B. Snavely and Dr. O. G. Peterson, Research Laboratories, Eastman Kodak Company, Rochester, New York

B. H. Soffer, Hughes Research Laboratory, Malibu, California

P. P. Sorokin, IBM Watson Laboratories, Yorktown Heights, New York

O. deWitte, Ecole Polytechnique, Paris, France

REFERENCES

1. P. P. Sorokin and J. R. Lankard, *IBM J. Res. Develop.*, **10**, 162 (1966).

2. P. P. Sorokin, J. R. Lankard, V. L. Moruzzi, and E. C. Hammond, *J. Chem. Phys.*, **48**, 4726 (1968).

3. B. B. Snavely, *Proc. IEEE*, **57**, 1374 (1969).

4. B. I. Stepanov and A. N. Rubinov, *Soviet Phys. – Usp.*, **11**, 304 (1968).

5. P. P. Sorokin, *Sci. Am.*, **220** (2), 30 (1969).

6. O. N. Witt, *J. Chem. Soc.*, (London) (1876).

7. I. B. Berlman, *Handbook of Fluorescence Spectra of Aromatic Molecules*, Academic Press, New York, 1965.

8. *Organic Electronic Spectral Data*, Vol. 1, 1946–52 (M. J. Kamlet, ed.), Wiley, New York, *ibid.*, Vol. 2, 1953–55 (H. E. Ungnade. ed.); *ibid.*, Vol. 3, 1956–57 (L. A. Kaplan and O. H. Wheeler, ed.); *ibid.*, Vol. 4, 1958–59 (J. P. Phillips and F. C. Nachod, eds.).

9. A. Schmillen and R. Legler, "Luminescence of Organic Substances," Vol. 3 of *Landolt–Börnstein Numerical Data and Functional Relationships in Science and Technology (New Series); Group II: Atomic and Molecular Physics* (K.-H. Hellwege, ed.), Springer-Verlag, Berlin, 1967.

10. *Color Index*, 2nd ed., British Soc. Dyers and Colourers and American Assoc. Textile Chemists and Colorers, 1956–8.

11. H. M. Hershenson, *Ultraviolet and Visible Absorption Spectra*, Index for 1930–54, Academic Press, New York, 1956; *ibid.*, Index for 1955–59, 1961; *ibid.*, Index for 1958–62, 1964.

12. J. N. Murrell, *The Theory of the Electronic Spectra of Organic Molecules*, Methuen and Co., London, 1963.

13. H. Suzuki, *Electronic Absorption Spectra and Geometry of Organic Molecules*, Academic Press, New York, 1967.

14. A. Jablonski, *Z. Physik*, **94**, 38 (1935).

15. J. Franck, *Trans. Faraday Soc.*, **21**, 536 (1926).

16. E. U. Condon, *Phys. Rev.*, **32**, 858 (1928).

17. P. M. Rentzepis, *Chem. Phys. Letters*, **2**, 117 (1968).

18. M. E. Mack, *J. Appl. Phys.*, **39**, 2483 (1968).

19. M. Bass and T. F. Deutsch, *Appl. Phys. Letters*, **11**, 89 (1967).

20. B. H. Soffer and B. B. McFarland, *Appl. Phys. Letters*, **10**, 266 (1967).
21. D. Eastwood, L. Edwards, M. Gouterman, and J. Steinfeld, *J. Mol. Spectry.*, **20**, 381 (1966).
22. J. R. Novak and M. W. Windsor, *Proc. Roy. Soc. (London)*, **A308**, 95 (1968).
23. B. H. Soffer and B. B. McFarland, *Appl. Phys. Letters*, **8**, 166 (1966).
24. C. R. Giuliano and L. D. Hess, *Appl. Phys. Letters*, **9**, 196 (1966).
25. M. Hercher, W. Chu, and D. L. Stockman, *IEEE J. Quantum Electron.*, **4**, 954 (1968).
26. L. G. S. Brooker, *Rev. Mod. Phys.*, **14**, 275 (1942).
27. Y. Miyazoe and M. Maeda, private communications.
28. C. M. Ferrar, *IEEE J. Quantum Electron.*, **5**, 621 (1969).
29. T. F. Deutsch and M. Bass, *IEEE J. Quantum Electron.*, **5**, 260 (1969).
30. A. S. Cherkasov, *Bull. Acad. Sci. USSR, Phys. Ser.*, **24**, 597 (1960).
31. N. G. Bakhsheiv, *Bull. Acad. Sci. USSR, Phys. Ser.*, **24**, 593 (1960).
32. P. Pringsheim, *Fluorescence and Phosphorescence*, Wiley, New York, 1949.
33. M. J. Weber and M. Bass, *IEEE J. Quantum Electron.*, **5**, 175 (1969).
34. L. Lindqvist, *Arkiv Kemi*, **16**, 79 (1960).
35. P. P. Sorokin, J. R. Lankard, E. C. Hammond, and V. L. Moruzzi, *IBM J. Res. Develop.*, **11**, 130 (1967).
36. W. Schmidt and F. P. Schäfer, *Z. Naturforsch.*, **22a**, 1563 (1967).
37. B. I. Stepanov and A. N. Rubinov, *Bull. Acad. Sci. USSR, Phys. Ser.*, **32**, 1189 (1968) and references therein.
38. V. D. Kotsubanov, L. Ya. Malkes, Yu. V. Naboikin, L. A. Ogurtsova, A. P. Podgornyi, and F. S. Pokrovskaya, *Bull. Acad. Sci. USSR, Phys. Ser.*, **32**, 1357 (1968).
39. B. B. Snavely and O. G. Peterson, *IEEE J. Quantum Electron.*, **4**, p. 540 (1968).
40. A. V. Buettner, B. B. Snavely, and O. G. Peterson, *Proceedings of the International Conference on Molecular Luminescence*, (E. C. Lim, ed.), Benjamin, New York, 1969, p. 403.
41. M. Bass, T. F. Deutsch, and M. J. Weber, *Appl. Phys. Letters*, **13**, 120 (1968).
42. G. I. Farmer, B. G. Huth, L. M. Taylor, and M. R. Kagan, *Appl. Opt.*, **8**, 363 (1969).
43. O. G. Peterson, W. C. McColgin, and J. H. Eberly, *Phys. Letters*, **29A**, 399 (1969).
44. S. A. Ahmed and C. Yu, *Proc. IEEE*, **57**, 1686 (1969).
45. D. E. McCumber, *Phys. Rev.*, **134**, A299 (1964).
46. A. Yariv and J. P. Gordon, *Proc. IEEE*, **51**, 4 (1963).
47. D. E. McCumber, *Phys. Rev.*, **136**, A954 (1964).
48. G. T. Schappert, K. W. Billman, and D. C. Burnham, *Appl. Phys. Letters*, **13**, 124 (1968).
49. A. N. Rubinov and V. A. Mostovnikov, *Bull. Acad. Sci. USSR, Phys. Ser. (English Transl.)*, **32**, 1348 (1968).
50. S. Murakawa, G. Yamuguchi, and C. Yamanaka, *Japan. J. Appl. Phys.*, **7**, 681 (1968).
51. B. B. Snavely and F. P. Schäfer, *Phys. Letters*, **28A**, 728 (1969).
52. P. P. Sorokin, W. H. Culver, E. C. Hammond, and J. R. Lankard, *IBM J. Res. Develop.*, **10**, 401 (1966).
53. M. L. Spaeth and D. P. Bortfeld, *Appl. Phys. Letters*, **9**, 179 (1966).
54. F. P. Schäfer, W. Schmidt, and J. Volze, *Appl. Phys. Letters*, **9**, 306 (1966).
55. B. B. McFarland, *Appl. Phys. Letters*, **10**, 208 (1967).

56. L. D. Derkacheva, A. I. Krymova, A. F. Vompe, and I. I. Levkoev, *Opt. Spectry.* (*USSR*), **25**, 404 (1968).

57. V. D. Kotzubanov, Yu. V. Naboikin, L. A. Ogurtsova, A. P. Podgornyi, and F. S. Pokrovskaya, *Opt. Spectry.* (*USSR*), **25**, 406 (1968).

58. Y. Miyazoe and M. Maeda, *Appl. Phys. Letters*, **12**, 206 (1969).

59. G. A. Abakumov, A. P. Simonov, V. V. Fadeev, L. A. Kharitonov, and R. V. Khokhlov, *Soviet Phys.—JETP Letters*, **9**, 9 (1969).

60. J. A. Myer, C. L. Johnson, E. Kierstead, R. D. Sharma, and I. Itzkan, *Appl. Phys. Letters*, **16**, 3 (1970).

61. P. P. Sorokin and J. R. Lankard, *IBM J. Res. Develop.*, **11**, 148 (1967).

62. B. G. Huth and G. I. Farmer, *IEEE J. Quantum Electron.*, **4**, 427 (1968).

63. B. H. Soffer and V. Evtuhov, *IEEE J. Quantum Electron.*, **5**, 386 (1969).

64. S. Claesson and L. Lindqvist, *Arkiv Kemi*, **12**, 1 (1958).

65. G. Porter, in *Technique of Organic Chemistry* (S. L. Friess, E. S. Lewis, and A. Weissberger, eds.), Wiley, New York, 1963, p. 8.

66. H. W. Furumoto and H. L. Ceccon, *Appl. Opt.*, **8**, 1613 (1969).

67. H. W. Furumoto, private communication, 1969.

68. D. J. Bradley and F. O'Neill, *Opto-Electron.*, **1**, 69 (1969).

69. A. V. Aristov and Yu. S. Maslyukov, *Opt. Spectry.* (*USSR*), **24**, 450 (1968).

70. C. M. Ferrar, *Rev. Sci. Instr.*, **40**, 1436 (1969).

71. M. Boiteux and O. de Witte, *Appl. Opt.*, **9**, 514 (1970).

72. T. F. Deutsch, M. Bass, P. Meyer, and S. Protopapa, *Appl. Phys. Letters*, **11**, 379 (1967).

73. H. Furumoto and H. Ceccon, *Appl. Phys. Letters*, **13**, 335 (1968).

74. C. Yamanaka, private communication, 1969.

75. D. J. Bradley, A. J. F. Durrant, G. M. Gale, M. Moore, and P. D. Smith, *IEEE J. Quantum Electron.*, **4**, 707 (1968).

76. P. P. Sorokin, J. R. Lankard, V. L. Moruzzi, and A. Lurio, *Appl. Phys. Letters*, **15**, 179 (1969).

77. M. Bass and J. I. Steinfeld, *IEEE J. Quantum Electron.*, **4**, 53 (1968).

78. W. E. K. Gibbs and H. A. Kellock, *IEEE J. Quantum Electron.*, **4**, 293 (1968).

79. W. H. Glenn, M. J. Brienza, and A. J. DeMaria, *Appl. Phys. Letters*, **12**, 54 (1968).

80. L. D. Derkacheva, A. I. Krymova, V. I. Malyshev, and A. S. Markin, *Opt. Spectry.* (*USSR*), **26**, 1051 (1969).

81. M. E. Mack, *Appl. Phys. Letters*, **15**, 166 (1969).

82. W. Schmidt and F. P. Schäfer, *Phys. Letters*, **26A**, 558 (1968).

83. C. M. Ferrar, "Mode-Locked Flashlamp Pumped Coumarin Dye Laser at 4600 Å," *IEEE J. Quantum Electron*, **QE-5**, 550 (1969).

84. B. H. Soffer and J. W. Linn, *J. Appl. Phys.*, **39**, 5859 (1968).

85. A. N. Sevchenko, A. A. Kovalev, V. A. Pilipovich, and Yu. V. Razvin, *Soviet Phys.—Doklady*, **13**, 226 (1968).

86. L. D. Derkacheva and A. I. Sokolovskaya, *Opt. Spectry.* (*USSR*), **25**, 244 (1968).

87. B. G. Huth, G. I. Farmer, L. M. Taylor, and M. R. Kagan, IBM Center for Exploratory Studies TM 48.68.012.

88. L. D. Derkacheva and A. I. Krymova, *Soviet Phys.—JETP Letters*, **9**, 343 (1969).

89. G. I. Farmer and B. G. Huth, private communication, 1969.

90. O. G. Peterson and B. B. Snavely, *Appl. Phys. Letters*, **12**, 238 (1968).

91. G. Yamaguchi, S. Murakawa, H. Tanaka, and C. Yamanaka, *Japan. J. Appl. Phys.*, **8**, 1265 (1969).

92. M. R. Bowman, A. J. Gibson, and M. C. W. Sandford, *Nature*, **221**, 456 (1969).

93. A. J. Gibson, *J. Sci. Instr. (J. Phys. E)*[2] **2**, 802 (1969).

94. D. J. Bradley, paper presented at *Belfast Conference on Dye Lasers and Non-linear Optics, 1969*.

95. A. J. Alcock, C. DeMichelis, and M. C. Richardson, *Appl. Phys. Letters*, **15**, 72 (1969).

96. T. J. McIlrath, *Appl. Phys. Letters*, **15**, 41 (1969).

97. P. P. Sorokin and J. R. Lankard, *J. Chem. Phys.*, **51**, 2929 (1969).

98. P. P. Sorokin and J. R. Lankard, "Infrared Laser Action in Sr Vapor Resulting from Optical Pumping and Inelastic Collisions," *Phys. Rev.*, **186**, 342 (1969).

99. H. Furomoto and H. Ceccon, *J. Appl. Phys.*, **40**, 4204 (1969).

100. F. P. Schäfer, W. Schmidt, and K. Marth, *Phys. Letters*, **24A**, 280 (1967).

101. B. B. Snavely, O. G. Peterson, and R. F. Reithel, *Appl. Phys. Letters*, **11**, 275 (1967).

102. D. W. Gregg and S. J. Thomas, *IEEE J. Quantum Electron.*, **5**, 302 (1969).

103. H. Samelson, *Electronics*, **41**, 142 (1968).

104. O. G. Peterson and B. B. Snavely, *Bull. Am. Phys. Soc.*, **13**, 397, (1968).

105. G. I. Farmer, B. G. Huth, L. M. Taylor, and M. R. Kagan, *Appl. Phys. Letters*, **12**, 136 (1968).

106. O. G. Peterson, S. A. Tuccio, and B. B. Snavely, *Appl. Phys. Letters*, **17**, 245 (1970).

107. K. Banse, G. Bret, J. Furxer, H. Gassman, and W. Seelig, Post. Deadline paper 22.4P, *6th Intern. Quantum Electronics Conf., Kyoto, Japan, September 1970*.

108. C. V. Shank, A. Dienes, A. M. Trozzolo, and J. A. Myer, *Appl. Phys. Letters*, **16**, 405 (1970).

109. A. Dienes, C. V. Shank, and A. M. Trozzolo, *Appl. Phys. Letters*, **17**, 189 (1970).

Author Index

Numbers in parentheses are reference numbers and indicate that an author's work is referred to although his name is not cited in the text. Numbers in italics give the page on which the complete reference is listed.

A

Abakumov, G. A., 303 (59), 318 (59), 328 (59), *344*

Abrahams, M. S., 54 (110), 55 (110), 58 (115, 117, 121–123), 61–63 (115), 64 (122), 83 (117), 83 (1646), *105, 106, 107*

Abrams, R. L., 113 (26, 27), 164, 166 (26, 27), 196 (26, 27), 204 (81), 224 (199), 233 (26, 27), 235 (26, 27), *262, 266, 267*

Adams, A., Jr., 94 (233),

Adams, M. J., 29 (280), 70 (33), *103, 110*

Ahern, W. E., 21 (131), 23 (131), 69 (131), 99, (267), *106, 109*

Ahmed, S. A., 286 (44), 291 (44), *343*

Alcock, A. J., 338 (95), *335*

Alexander, F. B., 89 (183), *107*

Alferov, Zh. I., 14, 19 (59), 74 (128b), 75, 76 (128b), *103, 104, 106*

Amick, J. A., 16 (47), *104*

Anderson, R. L., 56 (113), *105*

Anderson, W. W., 25, 26 (78), 29 (78), *104*

Andreev, M. V., 14, 74 (128b), 75, 76 (128b), *103, 106*

Ansimova, I. D., 89 (187), *107*

Antonov, N. V., 50, *105*

Aristov, A. V., 314, *344*

Armstrong, J. A., 80 (155), 234 (208), 235 (208), *106, 267*

Ashley, K. L., 12 (38), 77 (38), *103*

Austin, J. W., 113 (41), 183 (41), 187 (41), *263*

B

Bachert, H., 99, *109*

Bailey, L. G., 89, *107*

Bakhsheiv, N. G., 285, *343*

Banse, K., 339, *345*

Barchewitz, P., 112 (5, 6), 113 (37), 128 (5, 6), 130, 138, 139, *262, 265*

Barger, R. L., 219, *266*

Barker, E. F., 117 (117, 118), *265*

Barnes, C. E., 59, (284), *110*

Barr, T. A., 113 (63), 211, 212 (63), *263*

Baryshev, N. S., 89, *107*

Basov, N. G., 79, 83 (168), 90 (193, 194), 94 (220, 226), 95, 98, 99, *106–109*, 113 (78), 238, 239 (78), 250, *263*

Bass, M., 270 (19), 276 (19), 285 (33), 286 (41), 288 (41), 289 (33), 294 (41), 295–298 (33), 302 (33), 303 (29), 317 (72), 318 (29), 319 (33), 320 (33), 320 (41), 324 (41), 326 (33, 41), 327 (33, 41), 328 (33), 330 (33), 334, 335 (19), 336 (19), *342, 343, 344*

Beaulieu, J. A., 222, *267*

Becke, W., 21, *104*

Beneking, H., 18, *104*

Bennett, W. R., 145, *265*

Berkeyheiser, J. E., 45, *105*

Berkovskii, F. M., 94, *108*

Berlman, I. B., 272 (7), 273 (7), 276 (7), *342*

Bernard, M. G. A., 6, *103*

Besson, J., 89 (188), *107*

Bevacqua, S. F., 83, *107*

347

Subject Index

A

Aluminum gallium arsenide
  band structure, 4,5
  internal quantum efficiency, 87
  laser diodes
    double heterojunction, 89
    single heterojunction, 83–89
  lattice constant, 83
  material preparation, 16–19
  spontaneous diode emission, 87
  vapor phase epitaxy, 18, 88
Amplifier gain in CO_2 lasers, 168–183
  current and gas temperature dependence, 170–171
  effects of gas flow, 173–175
  gain coefficient, 169
  gain distribution, 177–178
  numerical analysis of, 179–183
  pressure and bore dependence, 171–174
Amplifiers, coupled laser diodes, 99–100
Annular flashlamps, construction, design and operation of, 304–312
Applications of dye lasers, 338
Auger effect, 93
Avalanche-pumped semiconductor lasers
  gallium arsenide, 98–99
  Gunn-domain formation, 98–99
  impact ionization, 98
  indium antimonide, 98
  indium phosphide, 99
Average power of laser diodes
  CW, 69
  pulsed, 69–70

B

Band-edge tailing, 7–11
Band-to-band transitions
  direct, 3, 61
  indirect, 3–4
Bottleneck 01^10 level, 151–155

  collisional relaxation in various gas mixtures, 152–155
  deactivation, probability of, 151–153
  relaxation rate constants, 154

C

Cadmium mercury telluride lasers, 93–94
Cadmium phosphide lasers, 93–94
Cadmium selenide lasers, 93–95
Cadmium sulfide lasers, 91–93
Cadmium telluride lasers, 93–94
Cadmium tin phosphide lasers, 93–94
Cathodoluminescence
  equipment, 95
  experimental results, 91–98
  pumping schemes, 91–92
Chemical kinetics in CO_2 lasers, 145–167
  energy transfer processes, 146–149
  resonance transfer, *see* Excitation processes
  rotational level relaxation, 163–167
  vibrational level relaxation, *see* Relaxation processes
Classification of laser diode types, 12
  double heterojunction, 14
  heterojunction lasers with multiple wavelengths, 15
  homojunction laser, 13
  large optical cavity laser (LOC), 15–16
  single heterojunction (CC), 14
Close-confinement laser diode, *see* Single heterojunction (CC) laser diode
CO_2 molecular structure, 116–126
  energy level diagram, 124, 127
  normal modes, 116–120
  rotational energy levels, 120–121
  rotational wave functions, 121
  selection rules for rotational-vibrational transitions, 126
  vibrational wave functions, 122

£10-75

This book is to be returned on or before
the last date stamped below.

27 FEB 1991

DUE

07 MAY 1992

1 2 MAY 2007

- 6 MAY 1993

3 0 JUN 1994

- 5 JUL 1994

2 4 JUN 1996

0 2 JUN 2004

LIBREX —